WITHDRAWN

AF443554

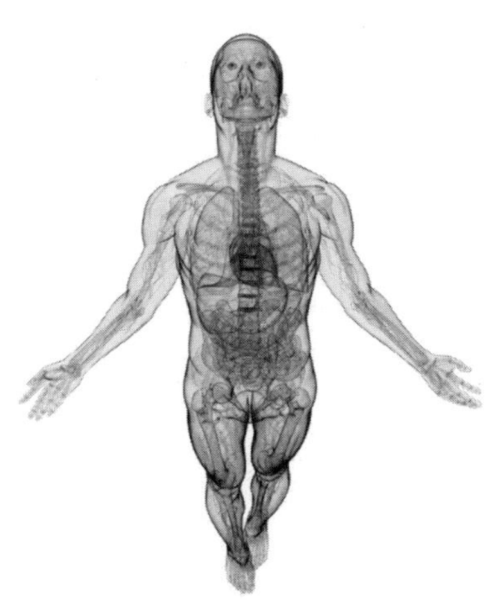

# NEW MATERIALS AND TECHNOLOGIES FOR HEALTHCARE

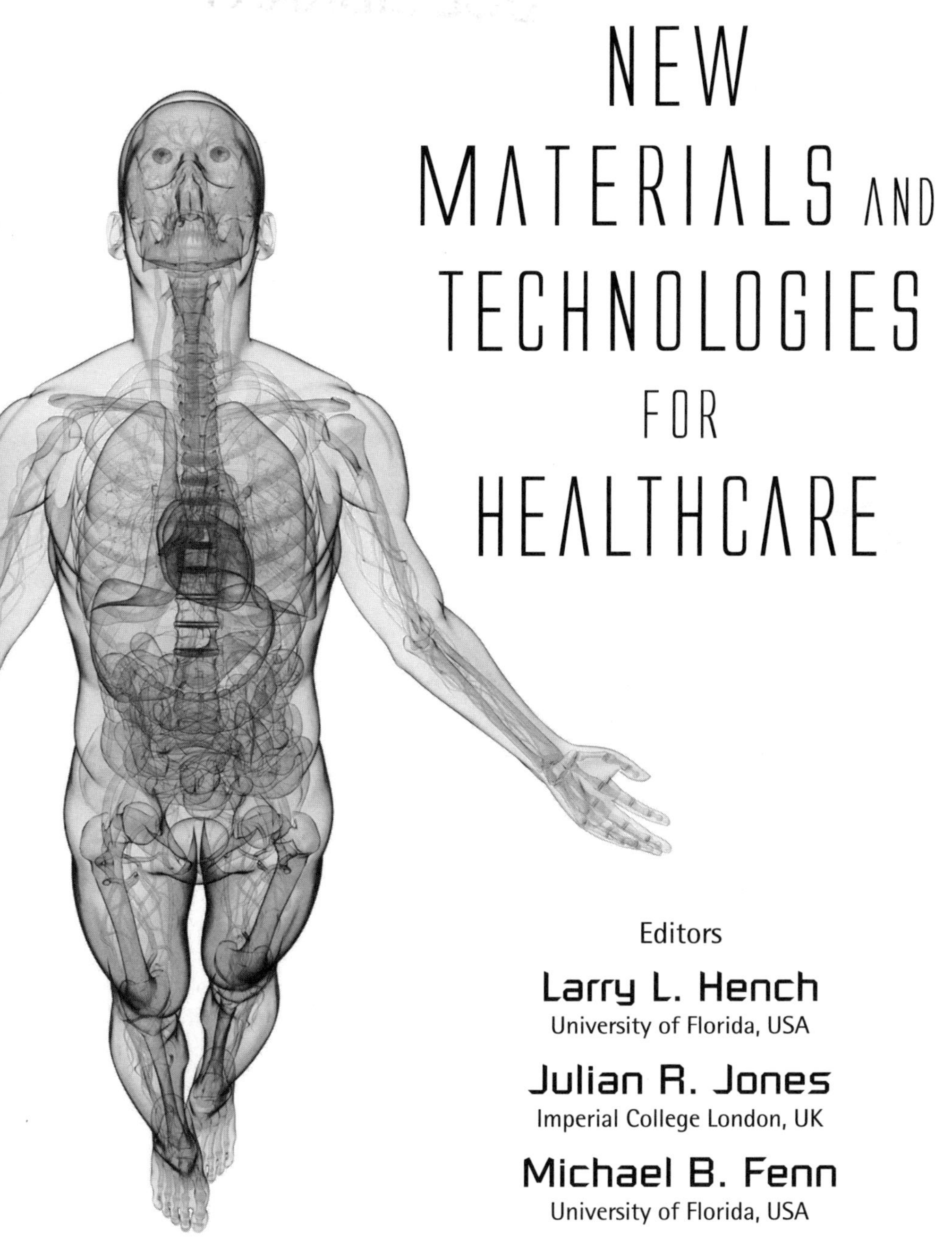

# NEW MATERIALS AND TECHNOLOGIES FOR HEALTHCARE

Editors

**Larry L. Hench**
University of Florida, USA

**Julian R. Jones**
Imperial College London, UK

**Michael B. Fenn**
University of Florida, USA

Imperial College Press

*Published by*

Imperial College Press
57 Shelton Street
Covent Garden
London WC2H 9HE

*Distributed by*

World Scientific Publishing Co. Pte. Ltd.
5 Toh Tuck Link, Singapore 596224
*USA office:* 27 Warren Street, Suite 401-402, Hackensack, NJ 07601
*UK office:* 57 Shelton Street, Covent Garden, London WC2H 9HE

**British Library Cataloguing-in-Publication Data**
A catalogue record for this book is available from the British Library.



ISBN-13 978-1-84816-558-8
ISBN-10 1-84816-558-7

Typeset by Stallion Press
Email: enquiries@stallionpress.com

Printed in Singapore by World Scientific Printers.

# Contents

# 21st Century Challenges for Biomaterials



## CHAPTER 1

Larry L. Hench

## 1.1. Introduction

During the 20th century a revolution in public health took place. Clean drinking water, disposal of sewerage, and immunization resulted in a greatly increased life expectancy. The large increase in average lifetime has led to a new challenge for the 21st century: how to maintain a high quality of life in an aging population. Now, fewer children die in birth or during their early years. Antibiotics prevent death by many infections. As a consequence, in the 21st century tens of millions of people aged between 60 and 100 are alive. Life expectancy increases every year in the developed world. All people, regardless of age, desire a high quality of life. However, achieving this desire is difficult and expensive.

The problem facing an aging population is that all tissues and organs begin a progressive path of deterioration from the age of 30 onwards. Replacement of aged, diseased or damaged tissues has become routine. This is a result of the development of reliable and affordable biomaterials, and the perfection of surgical procedures for implantation of prostheses along with subsequent reliable methods for rehabilitation of patients. Nevertheless the biomaterials in use today are a compromise compared with the healthy normal tissues they replace. All biomaterials have limitations: mismatches in elastic moduli (stiffness), deterioration of the tissue–material interface, fatigue, wear, loss of blood supply, and other factors lead to 15–50% failure of many medical devices over a 10–30 year lifetime. The data, compiled from a large number of published peer-reviewed papers,[1] show that few prostheses need to be replaced in the first ten years. During the next ten years many factors lead to a progressive loss of devices.[1] Today, more and more patients outlive their replacement parts. Thirty years of research and development of new biomaterials, new device designs, rigorous quality control, and governmental regulations has had only limited success in extending the lifetime of most prostheses. In general, the extension of patient lifetimes has increased faster than improvement in the lifetime of many medical devices.

Thus, an alternative approach to the replacement of tissues is needed. One new approach to this challenge is regeneration of tissues instead of replacement of tissues, a concept called regenerative medicine. This attempts to stimulate use of the body's own repair mechanisms to regenerate new tissues, and is a great challenge for cross-disciplinary research. Tissue engineering of replacement parts that are composed of living cells grown in temporary man-made templates (scaffolds) is one aspect of regenerative medicine.[2] Numerous chapters in this book address the progress made in using tissue engineering to repair or replace parts of the body. Stem cell technology is also a key component in regenerative medicine. Several chapters address the potential advantages of stem cell therapy, together with the barriers and difficulties in making the technology clinically viable. Chapter 2 discusses some of the controversial aspects of stem cell technology.

There are many factors to address in meeting the challenge towards achieving an affordable long-term quality of life in an aging population, and the 34 chapters in this book discuss many of these. The primary objective of this introductory chapter is to present some of the broad issues involved in modern healthcare technology. Details are presented in subsequent chapters.

## 1.2. Impact of the Media

In a previous book, *Science, Faith and Ethics*,[3] I point out that a major issue in present-day technology-driven healthcare is a growing perception that modern materials and devices provide "miracle cures." All too often patients crave replacement parts but do not understand the risks or long-term compromises in life-style that are involved after they receive the implants. Economics and the drive for increased market share and profits by health-care companies frequently lead to so-called "advances" with little long-term evidence to support the claims. This causes uncertainty for the surgeon and the patient who both desire the best outcome. Litigation often results when failure occurs, leading to even greater costs to patient and society.

Articles in the popular press that emphasize these uncertainties have become commonplace but offer few, if any, solutions. I cite a few examples to illustrate the issue of media impact on public perception of health-care technology. The 27 November 2006 issue of *Forbes* magazine, a leading business journal, featured on its cover a surgeon (or actor?) in full operating theatre scrubs and mask, holding up a heart stent implant in his gloved hand.[4] The text on the cover was: "STENTS, DEFIBRILLATORS, SPINAL DISCS, ARTIFICIAL KNEES: ARE THESE AS SAFE AS YOU THINK?"

A headline that is even more frightening to the reader — and to past or potential patients — appears with the article on page 94, written by Matthew Herper and Robert Langreth. In 36-point type, it reads: "DANGEROUS DEVICES. *Twenty million Americans walk around with high tech medical gear grinding away inside them. Are they safe?*"

The article opens with data on the wide-spread use of drug-coated stents that are engineered to prevent reclogging (restenosis) of arteries. The authors report that since 2003, drug-coated stents have been inserted into more than 4 million patients, generating US$5 billion/year in sales for Johnson & Johnson and Boston Scientific, the companies that manufacture and sell the devices to the hospitals. The controversy highlighted by the article is that some critics claim there is increased risk of heart attacks from drug-coated stents compared with bare metal stents. Also, the article suggests that there is growing concern that the lucrative stent implant business leads to many more people receiving such implants than actually benefit from the devices. These issues are discussed in Chapter 18.

The *Forbes* article cites a similar concern of alleged overuse for the quarter of a million patients implanted with defibrillators. This device is designed to prevent sudden cardiac death by sensing abnormal heart rhythms and shocking the heart back into action with a jolt of electricity. The article presents alarming stories of several patients, and describes other medical devices that are claimed to have been rushed into overuse.

A major point emphasized in the article is the fact that 79 devices have been removed from the medical device market by the Food and Drug Administration (FDA) in the past five years because of potentially fatal side effects. Lesser complications have also led to the recall of 2,300 devices. The FDA (510K) route to gaining regulatory status for a new medical device is to show that the device to be marketed is substantially equivalent to a device marketed prior to enactment of the 1976 Medical Device Amendment. A device approved in this manner is often referred to as a "grandfathered device." The authors also maintain that device approval is less complicated and the length of trials is shorter than that required for drugs. Consequently, the reader is led to believe that thousands of devices are in the US$85 billion/year medical device market-place with minimal long-term data guaranteeing safety or efficacy. What is the reality? Regulatory procedures are continually becoming more and more rigorous, which has the effect of reducing the number of new devices entering the clinical arena, as companies are less willing to risk the investment of putting new devices through the expensive regulatory pathways. Chapter 32 explores these issues.

Tracking detailed survivability of the myriad medical devices in use today is a near-impossibility, especially since there are few common criteria for evaluation and comparison of performance. A previous book edited by Hench and Wilson, *Clinical Performance of Skeletal Prostheses*, compared published survivability data for some of the most commonly used devices, such as total hip and knee joints, dental implants, and middle ear prostheses.[1] The published results showed large variability in performance even of similar prostheses, depending upon the source of reporting. Surgical teams supported by implant manufacturers often reported superior results. The article in *Forbes* discusses the complicated issue of surgeon–manufacturer relationships and the potential effect on survivability and risk/reward for the patient. One purpose of this book is to present published evidence for the successful use of many of these new medical materials and technologies; the authors of the chapters have no economic ties to the manufacturer or the clinician.

Most medical device manufacturers recognize the importance of long-term follow-up despite the implications in the *Forbes* article. For example, Medtronic, a leading medical device manufacturer, has recently introduced stainless steel prostheses for damaged discs in the neck. The implant is designed to mitigate one of the most common human ailments, pain in the neck and other parts of the spine. More than 12 million Americans suffer from chronic or acute problems related to deterioration or damage to the spine, and aging increases the incidence of vertebral deterioration and pain. More than 200,000 cervical procedures are performed annually in the USA, according to the American Association of Neurological Surgeons. Spinal fusion by use of bone grafts is the common method of repair but this procedure can lead to enhanced stress on neighboring discs and also often results in restricted range of motion. Chapter 16 discusses these topics.

The new Medtronic device is a two-piece ball and trough design, anchored by screws, that preserves much of the normal range of motion between the vertebrae. According to the manufacturer, overall success rates are 80% after 24 months, compared to 68% by spinal fusion. Neurological success for the implant *versus* fusion is purported to be even better. Important in the context of this introductory chapter is that Medtronic reports it will continue to conduct a seven-year study to evaluate the long-term safety and effectiveness of the implant, as well as a five-year enhanced surveillance study. This is essential because there have been numerous alternative vertebral replacement devices put into the market, with high rates of failure.

It is always necessary to ask, "What percentage of success is sufficient for a new device? Is 80% success good enough?" For eight of ten patients the answer is obviously "Yes!" However, two of ten patients will continue to suffer. Unfortunately the same situation exists for most medical devices. Lack of success — that is, failures — is a natural consequence of the use of unnatural means to repair or replace diseased or damaged living tissues. Surgeons, patients, and families must accept that risk and should not expect "miracle cures" and zero rates of failure.

## 1.3.  Organ Replacement

Failure of organs is, as yet, unresolved by the use of man-made replacement parts. Modest progress has been made to augment or supplant the function of the heart, lungs, liver or kidney, but for the present the only viable alternative for the terminal patient is a transplant, a living replacement from a donor. This book does not cover the current status of organ transplants. For the interested reader, a previous book by the author, *Science, Faith and Ethics*,[3] discusses the socio-economics, ethics, supply, personal costs, and risk/reward issues involved in organ transplants.[3] Books cited in the reference list of that book provide insight into the technical and socio-economic issues of transplants. Availability of viable, healthy organs suitable for transplant is the dominant issue that restricts this approach to saving life. A new name is added to the organ transplant waiting list every 18 minutes, and tens of thousands of patients die every year worldwide while on an organ transplant waiting list.

There is a growing interest in addressing the organ supply problem. Some governments have already acted to require organ donation without patient or family consent. Others, such as that of the UK, are considering enacting such regulations. This is, in part, because "Four out of ten organs that are perfectly suitable for use are not donated because the family does not consent," as reported in a *Telegraph Magazine* article, 26 January 2008, titled "The Donor Dilemma."[5] This continues the debate launched earlier when a headline in the *Sunday Times* of 15 July 2007 proclaimed "Doctors: We Must All Donate Organs."[6] The article was based upon a report that the chief medical officer of the UK, Sir Liam Donaldson, is supporting a change in the British law to create an organ donation system that will presume patients have given consent for their body parts to be available for transplantation. Such a system would require that individuals who want to opt out would have to register in a similar way to those who now carry organ donor cards.

This is a controversial and fundamental change in the bio-ethical position of the UK government because it essentially gives the state new powers over people's bodies. Critics maintain such a law violates the first principle of ethics, the right of autonomy. Proponents of the change, including the British Medical Association (BMA), maintain:

> Each year many people die waiting for an organ transplant. At the same time bodies are buried or cremated complete with organs that could have been used to save lives, not because the deceased objected to organ donation but simply because they never got around to signing up to the NHS (National Health Service) Organ Donor Register or informing their relatives of their wishes.[6]

Other countries in the European Union (EU) have already intro-duced "presumed consent" laws that have led to a large increase in donor organs. Belgium passed such a law in 1986 and Spain in 1989. However, following Sir Liam Donaldson's report to the British Parliament recommending a "presumed consent" law, an editorial appeared in *The Times* on 18 July 2007 arguing against the change.[7] *The Times* maintains that a change to presumed consent will not solve the problem. An important limitation, the editorial asserts, is the safeguards needed to ensure that organs are removed only from patients who have absolutely no hope of life. Two doctors are required to certify that the patient is brain dead. The doctors must also ask permission of next of kin, an essential provision of the so-called "soft opt-out" system. This is a big barrier because families often refuse donation. There is a short time of viability of donor organs, depending on which organ is to be transplanted. Thus, a rapid decision by a family is required for the organ to be useful. A decision at an emotionally distressed time can lead to additional trauma for the family. Refusal is the easier choice. The UK's present system of registered donation avoids much of the uncertainty, whereas a presumed consent system does not.

Another issue in the organ transplant debate is illegal trafficking of organs — organs for sale. Sales of organs in developed countries are illegal. However, sales of organs in so-called third world or developing countries have all too frequently become commonplace on the black or blood market. Kidneys are available for approximately US$2,000, more than a year's income for many donors in such countries. Corneas are sold for equivalent sums. The recipient, however, runs great risk due to potential infection, uncertain disease transmission, and lack of tissue matching. Transplant failure rates are high in the best of circumstances, even in transplant centers in the US and Europe, and highly dangerous to the patient if illegal organs are used.

Recently a major criminal case of theft of body parts came to light in the USA, with the head of the business, Biomedical Tissue Services, found guilty and convicted with a long prison sentence. The cadaver bodies came from funeral homes in New York, Pennsylvania, and New Jersey. The company had harvested bone, skin and tendons from the bodies and shipped them to three publicly traded firms, as well as to two non-profit agencies, for tissue processing. The illegal parts were used in vertebral disc replacements, knee operations, dental implants, bone grafts, and various other surgical procedures performed by unsuspecting surgeons in the USA and Canada. Court records indicate that about 10,000 people received tissues illegally procured by Biomedical Tissue Services. The extent of this type of illegal trade in human tissue is unknown. The consequences to the families of the stolen body parts and the patients receiving the parts are also unknown.

If a safe, ethically sound, affordable alternative source of viable tissues and organs can be created, as is the goal of regenerative medicine, there will be no need for illicit trade in body parts. The next chapter addresses one of the newest approaches to achieving this goal and meeting the challenges of the 21st century, stem cell technology and tissue engineering. A guide to the various experimental approaches to developing engineered tissues and organs is the book *Future Strategies for Tissue and Organ Replacement* edited by Julia M. Polak, Larry L. Hench, and Paul Kemp.[8] This book addresses the following topics:

1. the clinical need for organ replacement;
2. future advances in organ transplantation;
3. revolutionary therapies being developed; and
4. regulatory issues of such living products.

Progress continues to be made in meeting the challenge of new approaches to organ replacement, but routine clinical use is still to be achieved. The references in Chapter 2 are a guide to the potential and the problems of creating a new approach to repair and replacement of human body parts. The remaining chapters in this book provide a starting point for achieving such an understanding.

## References


1. Hench, L.L. and Wilson, J. (1993). *Clinical Performance of Skeletal Prostheses*, Chapman and Hall, London.
2. Hench, L.L. and Jones, J.R. (eds) (2005). *Biomaterials, Artificial Organs and Tissue Engineering*, Woodhead Publishing Ltd, Cambridge, UK.
3. Hench, L.L. (2001). *Science, Faith and Ethics*, Imperial College Press, London.
4. Herper, M. and Langreth, R. (2006). Dangerous devices, *Forbes*, 27 November 2006, p. 94.
5. Lambert, V. (2008). The donor dilemma, *Telegraph Magazine*, 26 January 2008, 14–16.
6. Templeton, S.-K. (2007). Doctors: we must all donate organs, *The Sunday Times*, 15 July 2007, 1–2.
7. Anon. (2007). Legacy of life, *The Times*, 18 July 2007, p. 2.
8. Polak, J.M., Hench, L.L. and Kemp P. (2002). *Future Strategies for Tissue and Organ Replacement*, Imperial College Press, London.

# Stem Cell Technology: Hope or Hype?



## CHAPTER 2

Larry L. Hench and Julian R. Jones

## 2.1. Introduction

As indicated in Chapter 1, there is a potential long-term solution to the survivability of prostheses, the organ donor dilemma, and illegal trafficking in body parts. The proposed solution is to use regenerative medicine to repair organs or tissues *in situ*, or to grow replacement tissues or organs *in vitro*. These approaches towards regenerative medicine are addressed throughout this book. Many of the approaches involve the use of stem cells, which are cells that can become (that is, differentiate into) mature cells that can produce tissue. The needs, technical concerns, and ethical issues regarding the use of stem cells to repair or grow tissues and organs are also addressed in numerous popular books[1–10] and a broad range of scientific articles and books.[11–17] The references listed below[1–17] provide an overview of the breadth and depth of this subject. This chapter gives the reader a starting point for exploring the complexity of the issues involved.

In 1998 the headlines of newspapers throughout the world[18] trumpeted: "Hearts to be Grown in Lab within Ten Years." More than a decade later this great hope for humans has not yet been realized. On 14 January 2008 another newspaper headline[19] proclaimed: "Working heart is grown in lab." Is this later headline hope or hype? The scientific article that was the basis for the news release was published in one of the world's leading scientific journals. Strong peer reviews insure, with a few exceptions to be discussed later, that the published results are accurate and credible. The article describes results of a team at the University of Minnesota that re-grew a rat heart in the lab and started it beating and pumping liquid.

The experiments involved the researchers removing all the cells from eight new-born rat hearts, leaving the extracellular matrix with a heart shape, and containing interconnecting conduits where blood vessels had been. The viable rat heart muscle cells and endothelial cells which form blood vessels were then injected back into the proteinaceous matrix which acted as a template (scaffold) for the cells. Within two days the muscle cells covered the matrix walls and lined up together. The endothelial cells migrated inside to coat the blood vessels. The hearts were then stimulated electrically by the researchers. Dr Doris Taylor, Director of the Center for Cardiovascular Repair at the University of Minnesota, reported to the Associated Press, "By two days, we saw tiny microscopic con-tractions and by seven to eight days there were contractions of the heart large enough to see by the naked eye."[19] The tiny hearts could pump liquid at a rate of about one-fourth the rate of a normal fetal rat's heart. "Obviously, we have a long way to go,"[19] Dr Taylor cautions.

A long way indeed; creating a man-made scaffold that behaves *in vitro* in a manner equivalent to the extracellular matrix of the cell-depleted rat heart is far beyond present research capabilities.

However, within the past decade the heart hype is being trans-formed to hope for the hundreds of thousands who die of heart failure annually. Although progress is being made, tissue-engineered human hearts or other organs are still many years away from being available for routine clinical use. Why? This second introductory chapter summarizes some of the technical barriers to be overcome in order to produce viable, three dimensional (3D), vascularized tis-sues and organs.

The goal of tissue engineering (TE) is the design and *in vitro* construction of living, functional cellular components that can be implanted to maintain, regenerate or replace malfunctioning tissues or organs. A primary focus of TE is the creation of bio-hybrid organs. Success requires a temporary biocompatible 3D matrix (scaffold) and a source of cells capable of multiplying and self-assembly throughout the scaffold to create a 3D construct.[11–17] Ideally, the construct would possess all the functions of a tissue or organ. Achieving a viable combination of multiple cell types organized with anisotropic structures that are characteristic of all living tissues and organs is a barrier yet to be overcome.[13] Without a stable blood supply the tissues or organs will die. Creating and maintaining a viable blood supply — that is, vascularization — within large 3D constructs has also yet to be achieved. Blood vessels will not grow into large porous scaffolds without some kind of stimulation. However, recent work at Harvard has created a vascular network inside small porous polymer scaffolds by growing endothelial cells and bone marrow stem cells together in a collagen gel within the scaffold, which, when implanted into mice, connected up to the existing blood vessel network.[14] This provides clear evidence of the need for TE approaches.

A present barrier to TE success is the absence of control of biomechanical stimuli on cell-scaffold constructs during TE processing.[17] Simultaneous micro-mechanical and micro-biochemical gradients must be supplied to cells. Within a tissue or organ there is an intricate 3D cytoskeletal network that provides continual signaling and feedback between all the cells. The biomechanical stimuli induce biochemical responses by the cells that control the state of health of the tissue. Small fluctuations of the physical distortion of cell shapes throughout a 3D construct thus can lead to cell growth, cell differentiation, de-differentiation, or even cell death depending upon the vector and strength of the stimuli. These cytoskeletal responses to micro-stimuli are under genetic control and are especially difficult to replicate in TE bioreactors. There are many different cell types in the body and many of them send signals to other cells in order for tissues to grow and function. These complex interactions are not yet understood, let alone mastered, in TE applications. *In vitro* research has demonstrated the importance of these physical factors but much work is required to incorporate the understanding in design of scaled-up reactors capable of producing full scale constructs.

## 2.2. Technical Challenges for Regenerative Medicine

The challenges for successful TE spread across many scientific disciplines, from biology (cell type and control), to materials science (scaffold design), to biomechanical (mechanical stimuli), and to chemical engineering (bioreactor design). Once the technical challenges have been surmounted there remain issues of surgical technique, and the (possibly temporary) fixation method. Socio-economic challenges are equally complex: governmental regulations, insurance issues, ethical factors, and cost/risk ratios all must be considered if the field is to become the revolution in healthcare that is often proclaimed. A brief summary of some of these challenges follows.

### 2.2.1. *Cell source*

Perhaps the most important technical challenge for TE is a reliable source of the multiple types of cells needed to grow various tissues and organs. The ideal cell source is something that is discussed at length among scientists and surgeons.

One problem is where to obtain the cells. They can either come from the same patient, from human donors (dead or alive), or from an animal source. The donor source is important as cells from other donors may trigger immuno-rejection in the same way that an organ might be rejected. Ideally, cells would be harvested from the patient, seeded on a scaffold to grow a TE construct and then the construct would be implanted into the same patient. The problem with this strategy is that it may take a several weeks for the cells to populate the scaffold and lay down matrix. This may be viable for patients with organs that are deteriorating slowly, but trauma or cancer patients will need an implant faster. Stem cell scientists believe that embryonic stem cells may hold the key as they may not be rejected, even if they are from a non-matched donor.

Therefore, another issue is what cell type will be the most effective in growing the tissue or organ. The choices are between primary cells and progenitor cells. Primary cells are the mature tissue cells, such as the osteoblasts in bone, which are responsible for bone production. For bone TE applications osteoblasts would need to be harvested from the host bone, for instance from the femoral heads removed in total hip replacement operations. Primary cells have the advantage that they are tissue-specific (bone cells are likely to produce bone), but they are slow to proliferate. Although they are easy to harvest from cadavers available in storage banks (e.g. bone banks) they are difficult to harvest from a patient who needs surgery.

Progenitor cells are stem cells that differentiate into the cell type needed to grow or repair a tissue. A stem cell will, for example, differentiate into an osteoblast under the right conditions, and this will produce bone matrix. Adult stem cells are always present in the body. They are multipotent, which means they can differentiate into selected cells types. Mesenchymal stem cells (MSCs) are found in several sites, such as the blood and bone marrow. They can differentiate into several cell types including osteoblasts (bone), chondrocytes (cartilage), adipocytes (fat), astrocytes (brain), muscle, and blood cells. They are critical to tissue regeneration as they also release protein growth factors and provide signals for mature cells. MSCs are therefore an important cell source for tissue engineering strategies.

Embryonic stem (ES) cells are only present at the early stage of a fetus's development and are totipotent, which means they can differentiate into any cell type, dictated by the signals they receive. They are usually harvested from the blastocyst (the collection of cells that precedes the formation of a fetus), sourced from unwanted pregnancies or from cloning, which is why they are the subject of ethical debate and are illegal in some countries. To avoid having to harvest from the blastocyst, a technique of reprogramming cells has been developed, where the DNA in fibroblasts can be genetically reprogrammed to behave like an ES cell. Great potential advantages of stem cells are:

1. they can be multiplied indefinitely without losing their potency;
2. non-autologous cells may not be rejected by the body; and
3. they are totipotent.

A practical problem is that ES cells can easily lose their stem cell potency in cell culture. In order to keep them as stem cells they have to be grown on a cell layer of mouse fibroblasts, known as a feeder layer. There is a risk that some of the genetic material from the mouse cells could be transferred to a patient, which would cause complications. A further concern is whether complete control over the destiny of ES cells can be achieved. There is risk of tumor formation from rapid proliferation of undifferentiated ES cells remaining in a TE construct. More research is needed to eliminate these concerns, and at the time of writing, direct implantation of undifferentiated ES cells is not permitted.

Other types of stem cells include fetal cells, such as hematopoietic stem cells that can be harvested from umbilical cord blood, which are thought to be strongly multipotent. These are the cells that parents may choose to harvest and freeze when their baby is born. All these cell sources are the subject of many research projects.

### 2.2.2. *Stable 3D constructs*

All tissues and organs have a complex interdependence of cell types with an interconnected 3D architecture. At the present stage of technology most TE constructs only involve one, or at most two, cell phenotypes grown primarily in a 2D configuration. This compromise in structure and absence of multiple cell interactions limits the differentiation, function, and clinical viability of the constructs. Chapter 9 discusses the concepts of design of 3D constructs from bioactive, resorbable biomaterials, and shows current progress towards achieving an ideal scaffold material for TE of bone.

### 2.2.3. *Vascularization*

All tissues and organs must have a continual supply of nutrients and a pathway for removal of waste products of cellular metabolism. Generally, these metabolic requirements are achieved by an interpenetrating network of blood vessels connected to the circulatory system. Cartilage is a notable exception. Aside from the work in mice at Harvard,[14] TE constructs lack a vascular network when they are transplanted to a host site. Surgeons would prefer an off-the-shelf device that can be implanted to regenerate tissue, but in order to survive, the host tissues must quickly infiltrate a TE graft with a blood supply. The trouble is that blood vessels will take the easier path, usually around an implant. They will also not grow into a scaffold and stay there unless there is cell function inside the scaffold, creating a "chicken and egg" situation regarding what should happen first. The process of forming a vascular network is called angiogenesis. A major challenge of TE technology is to achieve vascularization rapidly after implantation, and to maintain a viable nutrient supply and elimination of waste products as the construct becomes integrated. If the 3D scaffold used to grow the construct is bioresorbable (degrades safely in the body) then the pathway for elimination of the degradation products becomes crucial. Growth of 3D constructs containing angiogenic stimuli and an interconnecting network of channels for rapid infiltration of blood vessels after implantation offers a possible solution to this challenge. Because of its importance, Chapter 4 (Biomaterials to Control Angiogenesis) is devoted to this topic.

### 2.2.4. *Cell-construct interfacial stability*

Problems of stability at the host tissue–TE construct interface are a common barrier that has yet to be overcome. Shrinkage, infiltration by new tissue, de-differentiation of cells towards a scar tissue lineage (fibroblastic), and micro-mechanical breakdown of extracellular matrix at the interface are all limitations to long-term clinical viability at present. Creating a viable blood supply across the interface (see above) is also difficult.

### 2.2.5. *Growth media*

Specialized types of media are used for the culture of most cell types, to give them the best chance of growing well. The media can include supplements that stimulate mature cells to lay down matrix or maintain their phenotype, or to stimulate stem cells to differentiate towards a specific cell lineage. The supplements can be growth factors or hormones. However, the supplements can be expensive, and the challenge of whether or not the signals continue to be present comes when cells are transferred into the body. An ideal scenario would be that the scaffold on which the cells are growing provides the signals that the cells require, while the media provides other nutrients. A further challenge arises when co-culture strategies are used, for example when MSCs are cultured onto a scaffold to produce bone tissue and endothelial cells are added to create blood vessels which will eventually feed the new bone construct. In this case the MSCs need a particular culture media and the endothelial cells need a different media. Consequently, success requires the creation of a unique hybrid media suitable for the co-cultures.

### 2.2.6. *Bioreactor design*

When a 3D construct is being grown on a porous scaffold *in vitro*, the cells must penetrate the entire scaffold, nutrients must reach those cells, and waste products must be taken away. *In vivo*, this role is carried out by the blood vessel network and most *in vivo* sites are dynamic flow environments. *In vitro*, mechanisms are needed that can carry out this role artificially before blood vessels form. Cells can initially be encouraged into scaffolds by applying a negative pressure, using a vacuum pump that will remove air from within the scaffolds and encourage media containing cells to penetrate porous scaffolds. The simplest method of encouraging fluid transport is to use an orbital shaker plate to agitate the media; however, much work is being done to develop bioreactors that provide controlled fluid flow and continual refreshment of the media and its nutrients, at 37°C. A key design feature when using porous scaffolds is that the fluid must flow through the scaffold, which will only happen if it is not allowed to flow around the scaffold.

### 2.2.7. *Sterilization*

Maintaining sterility of a TE construct that contains living cells is a complex challenge for manufacturing, handling, storage, transport, implantation, and regulation. Most methods used for sterilization of non-living implants and devices, such as gamma ray irradiation and autoclaving, kill cells as well as pathogens. Sterility must be achieved during processing and maintained until implantation is complete. Procedures for TE constructs that are equivalent to those used for organ transplant must be created and implemented on a large scale. This is a daunting challenge and the costs of meeting it are hard to calculate.

### 2.2.8. *Cost*

All of the above factors add to manufacturing costs and presently limit many TE applications to exploratory patients. A big concern is the difficulty for insurance companies and healthcare providers, such as Medicare, to approve costly new technologies when the outcomes regarding effects on mortality and follow-on complications are unknown. Cost/risk ratios are nearly impossible to calculate at the present time for most TE clinical applications.

### 2.2.9. *Survivability*

Long-term survivability of TE constructs is uncertain and is likely to be so for many years to come. Consequently, in many cases use is restricted by ethical and legal considerations to applications where no other procedure is available. These "worst-case" surgical scenarios make it difficult to assess the viability and success of the new procedures. The limited number of cases and the low probability of success due to poor general health of the patients make it difficult to determine what went wrong, when failure occurs.

### 2.2.10. *Regulatory factors*

Tissue-engineered products will be subjected to the same regulatory procedures as non-living biomaterials and devices, with perhaps even more stringent requirements. Applications for regulatory approval of most TE devices have not yet been submitted. At present, few TE products have been produced with GMP (good manufacturing process) or ISO (International Standards Organization) standards and approved for routine clinical use. Costs and risk/benefit factors are usually hard to predict because of this uncertainty of regulatory approval.

## 2.3. *In Vivo* Stem Cell Therapies

An alternative approach in regenerative medicine to the growth of TE constructs *in vitro* is to transplant stem cells directly into the diseased or damaged tissue or organ; that is, *in vivo* stem cell therapy. This concept is especially prone to "hype and hope" publicity. *Time*,[20] an influential weekly news magazine, had the following heading on its color cover on 7 August 2006: "THE TRUTH ABOUT STEM CELLS. The hope, the hype and what it means for you."

The author, Nancy Gibbs, begins the lead story, "The debate is so politically loaded that it's tough to tell who's being straight about the real areas of progress and how breakthroughs can be achieved."

The article provides a service to the public by summarizing in a few pages the steps required to achieve stem cell therapies; the technical difficulties; and the political, religious, and ethical controversies. Several companies claim in the article to be ready to apply for FDA approval for clinical trials; however, the article correctly summarizes: "But so far, no HESCs (human embryonic stem cells) have been differentiated reliably enough that they could be safely transplanted into people."

Umbilical cord blood stem cells have become a viable alternative for bone marrow transplants for blood disorders, especially when a bone marrow match cannot be found. Adult stem cells obtained from a variety of sites from a patient such as <u>fatty tissues</u>, as well as from bone marrow, also offer promise; but most of the technical difficulties outlined above for TE constructs still must be overcome for the stem cell transplant technology to be used clinically on a routine basis.

Proof of the level of business interest in the outcome of the scientific debate is another cover story, again from *Forbes*. In its 16 June 2008 issue,[21] Dr Ed Baetge, Novocell Chief Scientist, is on the cover with the headline: "STEM CELLS GET REAL. The Maverick Scientists Who Are Cashing In And Targeting the Biggest Diseases."

The article written by Robert Langreth and Matthew Herper is thought-provoking. It begins: "Stem cells are the fountain of youth — Or the tools of Satan. It depends on whom you ask. Somewhere in between are pioneering companies making incredible strides in medicine."

Dr James Thompson, Professor at the University of Wisconsin, and colleagues, published the seminal paper on human ES cells more than a decade ago.[22] He has decided that the time has come to go commercial, the *Forbes* article announces. Thompson has founded Cellular Dynamics International, a biotech company that uses ES cells to generate human heart cells. The cells provide a fresh *in vitro* model for testing the potential cardiac toxicity of cancer drugs in early stages of development. This application of stem cell technology offers promise of bypassing some of the expensive and often uncertain animal testing of drugs. Will stem cells be the basis for new therapies? — new drugs? — new billionaires?

It is far too early to tell. Many of the chapters in this book provide an overview of what may be possible. The reading list that follows provides an additional overview of the technology, current status of understanding, status of clinical developments, and some of the ethical issues of this new and exciting field.


## 2.4. New Developments

Discussions of the applications and ethical issues of stem cells are of widespread interest, as noted in the numerous popular books listed in the References. The books generally provide a useful overview of the subject. However, it is a problem to separate scientific reality from personal perspective in many of these books; the problem is the rapid pace of science in the field. That is perhaps best illustrated by referring the reader to the article "Breaking ground on translational stem cell research" by Hall *et al.*, published in the prestigious Annals of the New York Academy of Sciences in 2010. The article is a summary of a two-day conference sponsored by the New York Stem Cell Foundation, with attendees representing 15 countries. It is pointed out in the paper that within a 12-month period in 2009 there was considerable progress on numerous scientific fronts regarding stem cell research, including:

1.  development of disease-specific cell lines;
2.  creation of pluripotent stem cells (iPS) cell lines with technologies that are compatible with eventual clinical use; and
3.  FDA approval of the first clinical trial using cells derived (differentiated) from human embryonic stem cells (HESC) for spinal cord surgery.

Conference sessions and sections of the paper are devoted to diabetes, cancer, and blood diseases; heart and muscle disease; neurodegeneration; and spinal cord injury. Discussions of recent changes in public policy are also reviewed. The article provides 54 reference citations which give the reader the opportunity to follow scientific progress in the field during the last few years.

For the beginner, we recommend the book *Stem Cells for Dummies*, which appears in the References.[7] In spite of the "dumb down" title the book is highly informative and discusses most of the important scientific issues, social and ethical concerns, clinical potential, and problems still to be surmounted. The senior author, Lawrence S.B. Goldstein, is a distinguished researcher in the field and has presented the topics in a style that is easy to follow. The second book we recommend for an exploration of the scientific and technical topics raised in this chapter in more detail is *Advances in Tissue Engineering*, edited by Professor Dame Julia Polak and colleagues.[17] The debates presented in the volume *The Stem Cell Controversy* (second edition) edited by Ruse and Pynes are illuminating, and provide an overview of the diversity of opinion on this important subject.[9]

## References


1. Anon. (1999). *Organ Procurement and Transplantation*, The National Academies Press, Washington, DC.
2. Caplan, A.L. (1998). *The Ethics of Organ Transplants: The Current Debate*, Prometheus Books, Amherst, NY.
3. Larue, G.A. and Bayly, R. (1992). *Long Term Care in an Aging Society*, Prometheus Books, Amherst, NY.
4. Kirkwood, T. (1999). *Time of Our Lives*, Weidenfeld and Nicolson, London.
5. Herold, E. (2006). *Stem Cell Wars: Inside Stories from the Front Lines*, Palgrave Macmillan, New York, NY.
6. Scott, C.T. (2006). *Stem Cell Now*, Plume, Penguin Group, New York, NY.
7. Goldman, L.S.B. and Schneider, M. (2010). *Stem Cells for Dummies*, Wiley Publishing Inc., Hoboken, NJ.
8. Peters, T. (2007). *The Stem Cell Debate*, Fortress Press, Minneapolis, MN.
9. Ruse, M. and Pynes, C.A. (eds) (2006). *The Stem Cell Controversy: Debating the Issues*, 2nd ed., Prometheus Books, Amherst, NY.
10. Pano, J. (2006). *Stem Cell Research: Medical Applications and Ethical Controversy*, Checkmark Books, New York, NY.
11. Fisher, O.Z., Khademhosseini, A., Langer, R., *et al.* (2010). Bio-inspired materials for controlling stem cell fate, *Acc. Chem. Res.*, **43**, 419–428.
12. Badylak, S.F., Freytes, D.F. and Gilbert, T.W. (2009). Extracellular matrix as a biological scaffold material: structure and function, *Acta Biomater.*, **5**, 1–13.
13. Polak, J.L. and Hench, L.L. (2005). Gene therapy progress and prospects, *Gene Ther.*, **12**, 1725–1733.
14. Tsigkou, O., Pomerantseva, I., Spencer, J.A., *et al.* (2010). Engineered vascularized bone grafts, *Proc. Natl. Acad. Sci. USA*, **107**, 3311–3316.

15. Place, E.S., Evans, N.D. and Stevens, M.M. (2009). Complexity in biomaterials and other challenges in the translation of tissue engineering, *Nat. Mater.*, **8**, 457–470.
16. Hall, Z.W., Kahler, D., Manganiello, M., *et al.* (2010). Breaking ground on translational stem cell research, *Ann. N. Y. Acad. Sci.*, **1189**, E1–E15.
17. Polak, J.M., Mantalaris, S. and Harding, S.E. (2008). *Advances in Tissue Engineering*, Imperial College Press, London.
18. Anon. (1998). Hearts to be grown in labs in ten years, *The Sunday Times*, 1 April 1998.
19. *The News Press* (2008). Working heart is grown in lab, *The News Press*, 14 January 2008, p. 1.
20. Gibbs, N. (2006). Stem cells: The hope and the hype, *Time*, 7 August 2006, 27–35.
21. Langreth, R. and Herper, M. (2008). Stem cells get real, *Forbes*, 16 June 2008, p. 86.
22. Thompson, J.A., Itskovitz-Eldor, J., Shapiro, S.S., *et al.* (1998). Embryonic stem cell lines derived from human blastocysts, *Science*, **282**, 1145–1147.

# Bioactive Materials for Gene Control

CHAPTER **3**

Larry L. Hench

## 3.1. Introduction

The function of biomaterials is to replace diseased, damaged, and aged tissues. First-generation biomaterials, including metals such as stainless steel and titanium; biopolymers, such as poly(methyl methacrylate), ultrahigh molecular weight poly(ethylene), and silicone rubbers; and bioceramics, such as alumina and zirconia, were selected to be as inert as possible in order to minimize the thickness of interfacial scar tissue. These bioinert materials are still the most commonly used biomaterials. Millions of patients have received bioinert implants and devices with excellent success for the first ten years of use. Longer-term suvivabilities of bioinert implants (from 10–20 years) decrease, as the length of service increases. Details of their biological behavior and clinical applications are described in reviews and biomaterials textbooks.[1–3] Survivability data for many types of implants used for skeletal repair have been compiled.[4]

Bioactive glasses, termed second-generation bioactive materials, have provided an alternative type of material and biomaterial–tissue interface from the 1970s onwards. Bioactive materials are capable of bonding implants to tissues without formation of interfacial scar tissue. Examples of bioactive materials include bioactive glasses, such as 45S5 Bioglass®, synthetic hydroxyapatite (HA), and bioactive glass-ceramics (e.g. Cerabone). Resorbable materials, such as poly (lactic acid) and poly (glycolic acid) and tri-calcium phosphate (TCP) were developed in the 1970s and 1980s and

25

are another category of second-generation biomaterials. Biodegradable polymers, for example, are used as resorbable sutures. Details of the characteristics of bioactive materials and resorbable materials are also presented in numerous biomaterials textbooks and handbooks.[1–5]

## 3.2. Third-Generation Biomaterials

During the last decade, the concepts of bioactive materials and resorbable biomaterials have converged into a new, third-generation of biomaterials: bioactive materials are being made resorbable, and resorbable polymers are being made bioactive.[6] Molecular modifications of resorbable polymers and bioactive composite systems elicit specific interactions with cell integrins and thereby direct cell proliferation, differentiation, and extracellular matrix production and organization. Third-generation bioactive glasses, composites, hybrid materials, and macroporous foams are being designed to activate genes that stimulate regeneration of living tissues.

This chapter reviews the discovery that controlled release of biologically active Ca and Si species released from bioactive glasses leads to the up-regulation and activation of seven families of genes in osteoprogenitor cells that give rise to rapid bone regeneration. Because of this, silicon has also been incorporated into hydroxyapatite bone graft to perform the same role.[7] This finding offers the possibility of creating a new generation of gene-activating biomaterials designed specially for tissue engineering (TE) and *in situ* regeneration of tissues.

Two alternative routes of repair are now available with the use of third-generation, molecularly tailored biomaterials.

### 3.2.1. *Tissue engineering*

The aim of this strategy is to seed progenitor cells onto biologically active resorbable scaffolds, such as those described in Chapter 9. The cells are grown outside the body, become differentiated, and produce extracellular matrix, forming a TE construct, which is then implanted into patients to replace diseased or damaged tissues. With time the scaffolds are resorbed and replaced by host tissues that include a viable blood supply and nerves. The living tissue-engineered constructs adapt to the physiological environment and should provide long-lasting repair. Clinical applications including repair of articular cartilage, skin, and other tissues are described in many chapters of this book and are discussed in a recent comprehensive review article.[8]

### 3.2.2. *In situ tissue regeneration*

This approach involves the use of biomaterials in the form of powders, solutions, doped microparticles, porous granules or scaffolds to stimulate local tissue repair directly, inside the body. Bioactive materials release chemicals in the form of ionic dissolution products, or growth factors such as bone morphogenic protein (BMP), at controlled rates by diffusion or network breakdown, and these activate the cells in contact with the stimuli. The cells produce additional growth factors that in turn stimulate multiple generations of growing cells to self-assemble into the tissues *in situ* along the biochemical and biomechanical gradients that are present. NovaBone®, Perioglas®, and NovaMin® products are all third-generation bioactive glass products based upon 45S5 Bioglass®.

### 3.3. Bioactive Glasses: 45S5 Bioglass®

The grandfather composition of all bioactive glasses — 45S5 Bioglass® — was discovered in 1969.[1,9] The glass composition of 45% $SiO_2$, 24.5% $Na_2O$, 24.5% CaO, and 6% $P_2O_5$ (in percentage weight) was selected to provide a large amount of CaO with $P_2O_5$ in a $Na_2O\text{-}SiO_2$ amorphous network. The glass was first tested in a rat femoral implant model and was shown to bond strongly to bone within six weeks. This finding was the basis for the first paper published in 1971 in the *Journal of Biomedical Materials Research*, summarizing the *in vivo* results and the *in vitro* tests that provided an explanation for the interfacial bonding of the implant to bone.[9]

The *in vitro* tests showed that the 45S5 Bioglass® composition developed a hydroxycarbonate apatite (HCA) layer in test solutions that did not contain calcium or phosphate ions. The surface phase of HCA that was formed *in vitro* was equivalent to the interfacial HCA crystals observed *in vivo* by Dr T.K. Greenlee's transmission electron micrographs of the bonded interface.[1,9] The HCA crystals were bonded to layers of collagen fibrils produced at the interface by osteoblasts. The chemical bonding of the HCA layer to collagen created the strongly bonded interface between glass and bone.[1,9,10]

A key review article that summarizes the mechanisms behind bone–Bioglass® bonding was published in 1982. It is "Adhesion to Bone" by L.L. Hench and A.E. Clark,[10] and documents in Part A the time sequence of bonding of Bioglass® in rat femur and tibia. In Part B, the bonding of Bioglass® implants to the femur in canine and monkey bones is summarized. Part C reviews the data of bonding of mandibular and maxillar bone of primates and swine to Bioglass® implants. All species exhibited stable bone bonded implants. Bonding occurred rapidly, regardless of species, within weeks of implantation.

## 3.4. Bioglass®–Bone Bond Strength

One of the most difficult topics studied in the first decade of Bioglass® experiments was determining the strength of the bond to bone. Eight different biomechanical test models were developed. A quantitative evaluation of interfacial shear strength in rat and monkey models showed that the strength of the interfacial bond between Bioglass® and cortical bone was equal to or greater than the strength of the host bone.[11] Weinstein *et al.* published a key paper describing the biomechanics of the bonded interface between bone and Bioglass.[12]

## 3.5. Bioglass® Surface Reactions

Bone bonding occurs as a result of a rapid sequence of chemical reactions on the surface of the Bioglass® implant when inserted into living tissues. New analytical techniques, such as cryogenic Auger electron spectroscopy, analytical scanning transmission electron microscopy, cryogenic electron microprobe analysis, and the application of Fourier transform infrared relection spectroscopy (FTIR), were developed and made it possible to determine the kinetics of the surface reactions with great precision.

There are 11 stages in the process of bonding of bioactive glass with bone:

1. Rapid exchange of $Na^+$ and/ or $Ca^{2+}$ with $H^+$ or $H_3O^+$ from solution, causing hydrolysis of the silica groups, which creates silanols:

$$Si\text{-}O\text{-}Na^+ + H^+ + OH^- \rightarrow Si\text{-}OH^+ + Na^+(aq) + OH^-$$

   The pH of the solution increases as a result of $H^+$ ions in the solution being replaced by cations.

2. The cation exchange increases the hydoxyl concentration of the solution, which leads to attack of the silica glass network. Soluble silica is lost to the solution in the form of $Si(OH)_4$, resulting from the breaking of Si-O-Si bonds and the continued formation of Si-OH (silanols) at the glass solution interface:

$$Si\text{-}O\text{-}Si + H_2O \rightarrow Si\text{-}OH + OH\text{-}Si$$

   Stages 1 and 2 occur within 1 or 2 hours of immersion into body fluid.

3. Condensation and repolymerisation of a $SiO_2$-rich gel layer on the surface, depleted in alkalis and alkali-earth cations.

4. Migration of $Ca^{2+}$ and $PO_4^{3-}$ groups to the surface through the $SiO_2$-rich layer, forming an amorphous $CaO\text{-}P_2O_5$-rich film on top of the $SiO_2$-rich layer. The layer grows by incorporating ions from solution.

5. Crystallization of the $CaO\text{-}P_2O_5$ film by incorporation of $OH^-$ and $CO_3^{2-}$ anions from solution to form a mixed HCA layer.

   Stages 3–5 occur within 8–24 hours for 45S5 Bioglass®.

6. Adsorption and desorption of biological growth factors in the HCA layer, to activate differentiation of stem cells.
7. Action of macrophages to remove debris from the site, allowing cells to occupy the space.
8. Attachment of stem cells on the bioactive surface.
9. Differentiation of stem cells to form bone growing cells, such as osteoblasts.
10. Generation of extra-cellular matrix by the osteoblasts to form bone.
11. Crystallization of inorganic calcium phosphate matrix to enclose bone cells in a living, composite structure.

A series of papers describe the kinetics studies.[13–16] Slow resorption of the glass surface releases soluble ionic species and creates a high surface area hydrated silica gel due to a condensation reaction between neighboring Si-OH groups. Nucleation of a crystalline HCA layer follows. Controlled release of soluble Si and Ca species is especially important. Recent molecular biology studies, discussed below, show that the Si species and Ca ions at specific concentrations are responsible for the genetic response to the bioactive glasses, and lead to rapid growth of new bone. The mechanism for the HCA layer bonding to bone involves protein adsorption and cell attachment. The result is growth of new bone with an architecture and biomechanical properties equivalent to normal bone.

Effects of composition of bioactive glasses and glass-ceramics on cell-surface reactions were established by a series of studies done in laboratories throughout the world.[1,14–16]

An important modification of bioactive glasses was the development of A/W (apatite/wollastonite) bioactive glass-ceramic (Cerabone) by Professors T. Yamamuro and T. Kokubo and colleagues at Kyoto University, Kyoto, Japan.[17] A unique hot-press processing method produced a very fine-grained glass-ceramic composed of very small apatite (A) and wollastonite ($W = CaSiO_3$) crystals bonded by a bioactive glass interface. Mechanical strength, toughness, and stability of AW glass-ceramics (AW-GC) in physiological environments are excellent. The team in Kyoto showed that bone bonds to A/W-GC implants with high interfacial strength. Numerous animal tests led to approval for use of the AW-GC material in orthopaedic applications in Japan, with particular success in vertebral replacement and spinal repair, the special interests of Professor Yamamuro. He reports clinical success in more than 3,000 cases of vertebral prostheses, 12,000 cases of laminoplasty, and 20,000 cases of iliac crest prostheses using AW-GC. Details and historical references are reviewed by Professor Yamamuro in Chapters 8 and 14 of this book.

A third group that confirmed the bonding and clinical effectiveness of bioactive glasses was led by Dr Orjan Andersson and Professors Kai Karlsson and Antti Yli-Urpo at Abo Academy and the University of Turku, Finland. Glasses modified from the 45S5 compositional range were designed by Karlsson and Andersson in the 1980s and implanted in animal models.[18] Compositions within boundaries similar to 45S5 Bioglass® bonded to bone; glasses outside the bioactive boundary did not bond. Clinical use in head and neck surgical repair has been successful for many years.

## 3.6. Classes of Bioactivity

Bioactive materials used for either tissue replacement or tissue regeneration must possess controlled chemical release kinetics that synchronize with the sequence of cellular changes occurring in natural wound repair.[1,14–16] If dissolution rates are too rapid, the ionic concentrations are too high to be effective. If the rates are too slow, the concentrations are too low to stimulate cellular proliferation and differentiation. Large differences in rates of *in vivo* bone regeneration and extent of bone repair, documented in papers by Oonishi *et al.*[19] and Wheeler *et al.*[20] indicate that there are two classes of bioactive materials. Class A bioactivity leads to both osteoconduction and osteostimulation as a consequence of rapid reactions on the bioactive glass surface. The surface reactions involve ionic dissolution of critical concentrations of soluble Si and P species and Ca and Na ions that give rise to both intracellular and extracellular responses at the interface of the glass with its physiological environment. Class B bioactivity occurs when only osteoconduction is present; that is, when bone migrates along an interface. This limited tissue response is due to slower surface reactions and minimal ionic release. Only extracellular responses occur at the interface of Class B bioactive tissue interfaces.[1,16] Differences between Class A and B bioactive materials are summarized in several sources.[14–16] The consequences of these differences clinically are slower bone growth and less bone in a graft site, as discussed later in the section on clinical applications.

## 3.7.  First Bioactive Glass Clinical Products: Tissue Replacement

The first Bioglass® device cleared for clinical use in the United States was one used to treat conductive hearing loss by replacing the bones of the middle ear. The device was called the "Bioglass® Ossicular Reconstruction Prosthesis," and trademarked "MEP®," and was cleared via the 510K process in January 1985. It was a solid, cast Bioglass® structure that acted to conduct sound from the tympanic membrane to the cochlea. The advantage of the MEP® over other devices in use at the time was its ability to bond with soft tissue (the tympanic membrane) as well as to bone tissue, such as the stapes footplate. Clinical studies showed that the MEP® outperformed other bioceramic, biopolymer or metal prostheses.[1,21] A modification of the MEP design was made to improve handling in the surgery and it is used clinically with the trademark name of the Douek MED, after Mr Ellis Douek, Professor of Ear, Nose and Throat (ENT) surgery at Guy's Hospital, London, who pioneered the design and tested the improved device. Other uses in head and neck surgery of bioactive glasses are described in more than 20 citations in an historical review.[1]

The second Bioglass® device to be placed into the market was the Endosseous Ridge Maintainence Implant (ERMI®), which was cleared via the 510(k) process, in November 1988. The device was designed by Professor Harold Stanley and dental faculty colleagues at the University of Florida, to support labial and lingual plates in natural tooth roots and to provide a more stable ridge for denture construction following tooth extraction. The devices were simple cones of 45S5 Bioglass® that were placed into fresh tooth extraction sites. They bonded to the bone tissue and proved to be extremely stable, with much lower failure rates than other materials that had been used for the same purpose. Numerous clinical studies have been published.[1]

## 3.8.  Bioactive Control of Genes and the Osteoblast Cell Cycle

For many years it was assumed that formation of a biologically active HCA surface reaction layer was the critical requirement for bioactive behaviour.[1,14–16] However studies from 2000 to 2003 showed formation of a surface HCA layer to be a useful but not the critical stage of reaction for bone regeneration. The controlled rates of release of ionic dissolution products, especially critical concentrations of soluble Si species and calcium ions, is the critical requirement for osteoproduction, now called osteostimulation, as first described by June Wilson and Sam Low.[22]

In order for new bone to form it is essential for osteoprogenitor cells (adult stem cells) to undergo cell division (mitosis). The osteoprogenitor cells that are present must receive the correct chemical stimuli from their local environment that instruct them to enter the active segments of the cell cycle. Numerous papers document the critical importance of local chemical environment on osteoblast cell growth *in vitro* and *in vivo*.[1,23,24] Figure 3.1 shows the osteoblast progenitor cell cycle. Resting cells are in the $G_0$ phase. Every new cell cycle begins after a cell has completed the preceding mitosis. If the local chemical environment is suitable, and following a critical period of growth in the $G_1$ phase, the cell enters the S phase when DNA synthesis begins and leads to duplication of all the chromosomes in the nucleus. During a second growth phase, $G_2$, the cell prepares to undergo division and checks its replication accuracy using DNA repair enzymes.

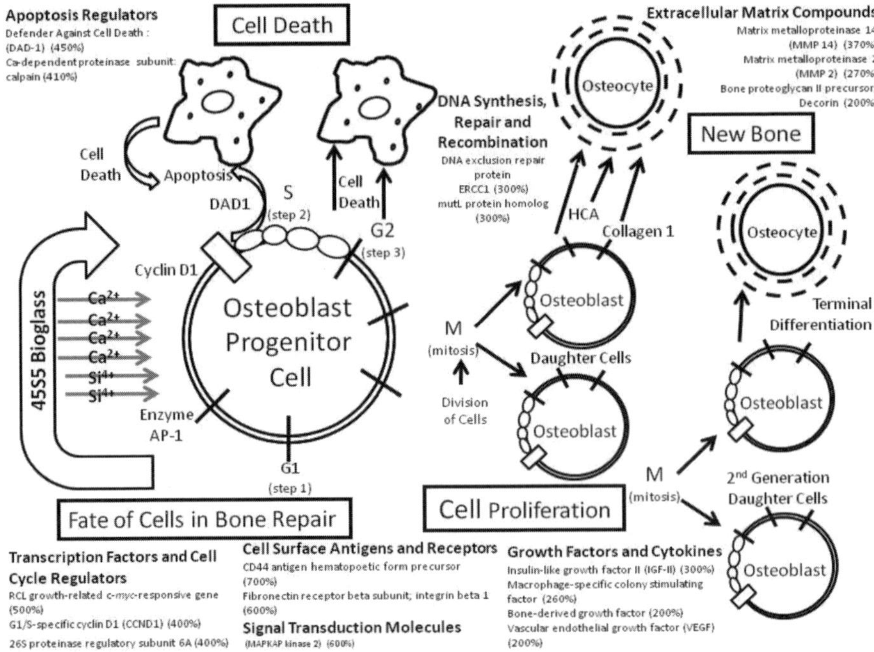


**Figure 3.1.**  Schematic of osteoblast progenitor cell cycle leading to 1) programmed cell death (apoptosis); 2) mitosis and cell proliferation or 3) terminal cell differentiation towards an osteocyte.

Details of the feedback controls and cell cycle checkpoints have been described.[23–26] If the local chemical environment does not lead to completion of the $G_1$ phase or the $G_2$ phase then the cell proceeds to programmed cell death, apoptosis. Bioinert materials or Class B bioactive materials do not produce the local chemical environment to enable the few osteoprogenitor cells present to pass through these cell cycle checkpoints. Only Class A bioactive materials produce rapid new bone formation *in vivo*, osteostimulation.[19,23–26] Approximately 17–20 µgml$^{-1}$ of soluble silica and 88–100 µgml$^{-1}$ of soluble Ca ions are required for the interfacial environment to be osteogenic. The ions are provided by controlled dissolution of a bulk implant or the 45S5 Bioglass® particles in a particulate such as NovaBone® or Perioglas®.

Molecular biology studies by Xynos *et al.* in Professor Dame Julia Polak's group at Imperial College London showed that the bioactive shift of osteoblast cell cycle is under genetic control.[1,26–30] Within a few hours' exposure of human primary osteoblasts to the soluble chemical extracts of 45S5 Bioglass®, several families of genes are activated, including genes encoding nuclear transcription factors and potent growth factors, especially IGF-II, along with IGF-binding proteins and proteases that cleave IGF-II from their binding proteins.[26] There is a 200–500% increase in the expression of these genes over those of the control cultures.[27] Activation of several immediate early response genes and synthesis of growth factors is likely to modulate the cell cycle response of osteoblasts to bioactive glasses. These findings indicate that Class A bioactive glasses enhance new bone formation (osteogenesis) through a direct control over genes that regulate cell cycle induction and progression. Bioactive induction of the transcription of extracellular matrix components, and their secretion and self-organization into a mineralized matrix appears to be responsible for the rapid formation and growth of bone nodules, and differentiation of the mature osteocyte phenotype in the presence of Class A bioactive materials, such as 45S5 Bioglass® and sol-gel-derived bioactive gel glasses of compositions 58S (60 mol% $SiO_2$, 36 mol% CaO, and 4 mol% $P_2O_5$) and 70S30C (70 mol% $SiO_2$, and 30 mol% CaO) tested by Bielby *et al.*[29]

Several studies have confirmed the results of the early findings by Xynos *et al.* and extended the generality to include several types of precursor cells and differing sources of ionic stimuli.[30] Gene-array analyses of five different *in vitro* models using five different sources of inorganic ions provide the experimental evidence for a genetic theory of osteogenic stimulation.[25] All seven experiments showed enhanced proliferation and differentiation of osteoblasts towards a mature, mineralizing phenotype without the presence of any added bone growth proteins, such as dexamethasone or BMP. Shifts in osteoblast cell cycles were observed as early as six hours, with elimination (by apoptosis) of cells incapable of differentiation. The remaining cells exhibited enhanced synthesis and mitosis. The cells quickly committed to generation of extracellular matrix (ECM) proteins and mineralization of the matrix.[31] Gene-array analyses at 48 hours showed early up-regulation or activation of seven families of genes that favored both proliferation and differentiation of the mature osteoblast phenotypes, including: transcription factors and cell cycle regulators (six with increases of 200–500%); apoptosis regulators (three at 160–450% increases); DNA synthesis, repair and recombination (four at 200–300%); growth factors (four at 200–300%) including IGF-I1 and VEG F); cell-surface antigens and receptors (four at 200–700%, especially CD44); signal transduction molecules (three at 200–600%); and ECM compounds (five at 200–370%).

A summary of the seven families of genes activated or up-regulated from the experiments is given in Table 3.1.

## 3.9. Design Concepts for Genetic Control of Bone Regeneration

Two developments make it possible to design a new generation of biomaterials that can control gene expression *in vitro* and *in vivo*. The first is the enhanced understanding of the role of controlled release of ionic dissolution products from bioactive glasses in controlling the molecular biology of osteoprogenitor cells, as reviewed above. The second is the use of sol-gel processing of bioactive glasses to achieve additional control of the rates of ionic release of biologically active stimuli.

Compositions and textures of sol-gel derived glasses can be varied over wide ranges and thereby be used to control the rates and concentrations of soluble Si and Ca species in the physiological solutions. Details of sol-gel processing of bioactive gel-glasses, textural analyses, and bioactivity studies are presented in prior publications.[32,33] Sol-gel processing makes it possible to produce hierarchical microstructures with nanometer-scale pores in the solid webs of 3D scaffolds, while creating an interconnected pore network with greater than 100 µm passages

**Table 3.1.** Families of genes in primary human osteoblasts activated or up-regulated by ionic dissolution products of bioactive glasses.[27]

| Transcription factors and cell cycle regulators | Activation (%) |
| --- | --- |
| RCL growth-related c-*myc*-responsive gene | 500 |
| G1/S-specific cyclin D1 (CCND1) | 400 |
| 26S proteinase regulatory subunit 6A | 400 |
| Cyclin-dependent kinase inhibitor 1 (CDKN1A) | 350 |
| cAMP-dependent transcription factor ATF-4 | 240 |
| Cyclin K | 200 |
| **DNA synthesis, repair and recombination** | **Up-regulation (%)** |
| DNA exclusion repair protein ERCC | 300 |
| mutL protein homolog | 300 |
| High-mobility-group protein (HMG-1) | 230 |
| Replication factor C 38-kDa subunit (RFC38) | 200 |
| **Apoptosis regulators** | **Up-regulation (%)** |
| Defender against cell death 1 (DAD-1) | 450 |
| Ca-dependent proteinase small (regulatory) subunit; calpain | 410 |
| Deoxyribonuclease II (Dnase II) | 160 |
| **Growth factors and cytokines** | **Activation (%)** |
| Insulin-like growth factor II (IGF-II) | 300 |
| Macrophage-specific colony stimulating factor (CSF1; MCSF) | 260 |
| Bone-derived growth factor | 200 |
| Vascular endothelial growth factor precursor (VEGF) | 200 |
| **Cell surface antigens and receptors** | **Activation (%)** |
| CD44 antigen hematopoetic form precursor | 700 |
| Fibronectin receptor beta subunit; integrin beta 1 | 600 |
| N-sam; fibroblast growth factor receptor-1 precursor | 300 |
| Vascular cell adhesion protein-1 precursor (V-CAM1) | 200 |
| **Signal transduction molecules** | **Activation (%)** |
| MAP kinase-activated protein kinase 2 (MAPKAP kinase 2) | 600 |
| Dual specificity nitrogen-activated protein kinase 2 | 200 |
| ADP-ribosylation factor 1 | 200 |
| **Extracellular matrix compounds** | **Activation (%)** |
| Matrix metalloproteinase 14 precursor (MMP 14) | 370 |
| Matrix metalloproteinase 2 (MMP 2) | 270 |
| Metalloproteinase 1 inhibitor precursor (TIMP 1) | 220 |
| TIMP 2 (MI) | 220 |
| Bone proteoglycan II precursor; decorin | 200 |

between macropores of 200–600 μm in diameter. Chapter 9 discusses these materials and hierarchical scaffolds in detail. Gough, Jones, and colleagues demonstrated that such bioactive 3D scaffolds support osteoblast growth and induced differentiation of the cells without use of supplementary organic growth factors.[31,34]

The Beilby *et al.* investigations[29] were especially significant because the cell source was embryonic stem (ES) cells. Soluble Si and Ca species released from 58S sol-gel-derived glasses stimulated gene expression in the murine ES cells characteristic of a mature phenotype in primary osteoblasts. The osteogenic effect of the bioactive gel-glass extracts was dose-dependent. The conclusion was that the bioactive gel-glass material was capable of stimulating differentiation of ES cells toward a lineage with therapeutic potential in TE. Christodoulou *et al.* extended even further the scientific basis of the genetic effect of the dissolution products of 58S gel-glasses on osteogenesis, which were added to cultures of primary osteoblasts derived from human fetal long bone explants cultures (hFOBs).[30] The over 200% up-regulation of gp130 and MAPK3 and down-regulation of IGF-1 were confirmed by real-time RT-PCR analysis. These data suggest that 58S ionic dissolution products possibly mediate the bioactive effect of the gel-glass through components of the IGF system and MAPK signalling pathways. The results from human fetal osteoblasts confirm many of the findings reviewed above, and shown in Table 3.1 using primary human osteoblast cultures derived from excised femoral heads of elderly patients, and thereby demonstrate the generality of the findings of genetic stimulation by the ionic dissolution products of bioactive glasses and gel-glasses.

The implications of the above studies are that it is now feasible to design the dissolution rates and architecture of bioactive, resorbable inorganic scaffolds to achieve specific biological effects *in vivo* that synchronize with the progenitor cell population present *in situ*. For the first time this offers the potential to design biomaterials for specific patients and their clinical needs.

## 3.10. Third-Generation Bioactive Glass Clinical Products

Second-generation Bioglass® implants performed well in replacing diseased or missing hard tissue. The discovery that Bioglass® could stimulate osteoblasts to produce more bone tissue earlier than other synthetic biomaterials led to the innovative concept of *osteostimulation*, the basis for third-generation biomaterials. Development of third-generation Bioglass® products focused on using particles rather than monolithic shapes. The products are manufactured and used under the trademarks PerioGlas® and NovaBone®. Table 3.2 summarizes the time line for development of clinical products based upon 45S5 Bioglass®.

The first NovaBone® particulate material cleared for sale in the United States was PerioGlas®, which was cleared via the 510(k) process in December 1993. In 1995, PerioGlas® obtained a CE mark and marketing of the product began in Europe. The initial indication for the product was to restore bone loss resulting from periodontal disease in infrabony defects. In 1996, additional indications for use were cleared by the FDA, including use in tooth extraction sites and for alveolar ridge augmentation.[1]

During a 17-year clinical history, PerioGlas® has demonstrated excellent clinical results with virtually no adverse reactions to the product. Numerous clinical studies have demonstrated the efficacy of the product in multiple uses. A historical review lists many of these clinical studies.[1]

Building on the successes of PerioGlas® in the market, a Bioglass® particulate for orthopedic bone grafting was introduced into the European market in 1999, under the trade name NovaBone®. Early studies by Dr June Wilson *et al.* in a canine model showed effective bone regeneration with uses of 45S5 Bioglass® particulate.[1] Other animal models followed in various laboratories worldwide.[1] The product was cleared for general orthopaedic bone grafting in non-load bearing sites in February 2000. To date, PerioGlas® and NovaBone® products are sold in

**Table 3.2.**   Time line for development of clinical products based upon 45S5 Bioglass®.

| | |
|---|---|
| 1969 | Discovery of bone bonding to 45S5 Bioglass® |
| 1971 | First peer-reviewed publication[9] |
| 1972 | Bonding of Bioglass® bone segments and coated femoral stems in monkeys |
| 1975 | Bioglass®-coated alumina bone bonding to sheep hip implants (in Germany) |
| 1975 | Bioglass® dental implants bonded in baboons |
| 1977 | Bonding of Bioglass® implant in guinea pig middle ear |
| 1977 | Bioglass® coating patent applications filed |
| 1981 | Discovery of soft connective tissue bonding to 45S5 Bioglass® |
| 1981 | Toxicology and biocompatibility studies (20 *in vitro* and *in vivo*) published to establish safety for FDA clearance of Bioglass® products |
| 1985 | First medical product (Bioglass® Ossicular Reconstruction Prosthesis) (MEP) cleared by FDA via the 510(k) process |
| 1987 | Discovery of osteoproduction (osteostimulation) in use of Bioglass® particulate in repair of periodontal defects |
| 1988 | Bioglass® Endosseous Ridge Maintenance Implant (ERMI) cleared by FDA via the 510 (k) process |
| 1991 | Development of sol-gel processing method for making bioactive gel-glasses |
| 1993 | Bioglass® particulate for use in bone grafting to restore bone loss from periodontal disease in infrabony defects (Perioglas®) cleared by FDA via the 510 (k) process |
| 1995 | Perioglas® obtained CE mark in Europe |
| 1996 | Use of Perioglas® for bone grafts in tooth extraction sites and alveolar ridge augmentation cleared by FDA via the 510 (k) process |
| 1999 | European use of 45S5 particulate for orthopedic bone grafting (NovaBone®) |
| 2000 | FDA clearance for use of NovaBone® in general orthopedic bone grafting in non-load-bearing sites |
| 2004 | FDA clearance of 45S5 particulate for use in dentinal hypersensitivity treatment (NovaMin®) |
| 2005 | Development of variety of dental maintenance products (NovaMin®) |
| 2009 | Anniversary of 1 million doses of NovaBone® bone graft product and 1 million tubes of NovaMin® toothpaste |
| 2010 | NovaMin® products sold to Glaxo-Smith Kline for use in Sensodyne Repair and Protect OTC toothpaste |

over 35 countries, and the manufacturer (NovaBone Products LLC, FL, USA) reports that the product has been used in more than one million surgeries. Table 3.3 lists the various medical and dental products used clinically.

**Table 3.3.**   Clinical medical and dental products based upon 45S5 Bioglass®.

**Orthopedics**

Trauma
> *Long-bone fracture (acute and/or comminuted); alone and with internal fixation*
> *Femoral non-union repair*
> *Tibial plateau fracture*

Arthroplasty
> *Filler around implants (acetabular reconstruction)*
> *Impaction grafting*

General
> *Filling of bone after cyst/tumor removal*

Spine fusion
> *Interbody fusion (cervical, thoracolumbar, or lumbar)*
> *Posterolateral fusion*
> *Adolescent idiopathic scoliosis*

**Cranial–facial**

Cranioplasty
> *Facial reconstruction*

General oral/dental defects
> *Extraction sites*
> *Ridge augmentation*
> *Sinus elevation*
> *Cystectomies*
> *Osteotomies*
> *Periodontal repair*

**Dental–maxillofacial–ENT**

Toothpaste and treatments for dentinal hypersensitivity
Pulp capping
Sinus obliteration
Repair of orbital floor fracture
Endosseous ridge maintenance implants
Middle ear ossicular replacements (Douek MED)

## 3.11. Synthetic Bioactive Glass Bone Graft or Autograft?

A publication in the *Journal of Pediatric Orthopaedics* by Ilharreborde *et al.* in 2008 illustrates the value of NovaBone® as an alternative to the use of auto grafts.[35] At the present time iliac crest autograft is considered to be the gold-standard material for spinal fusion. However, it is well known that there are limitations of use of autografts, such as additional operative time, increased loss of blood, morbidity, length of time of healing of the second operative site, and pain in the donor site. Chapter 7 compares the use of autografts with synthetic graft materials in surgery. The aim of the Ilharreborde *et al.* study was to compare bioactive glass (NovaBone®) with iliac graft autograft as bone substitutes in the treatment of thoracic adolescent idiopathic scoliosis (AIS). Eighty-eight consecutive patients underwent posterior spinal fusion for progressive AIS, with 40 patients implanted with autograft, and 48 patients receiving NovaBone® particulate. A minimum two-year follow-up was required with medical data and radiographs analyzed retrospectively and compared statistically. The results showed that the Bioglass® particulate was as effective as iliac crest graft in achieving fusion and maintenance of correction in the AIS patients. Fewer complications were seen in the Bioglass® group. Thus, the authors concluded, "Bioactive glass can be proposed in the treatment of AIS, avoiding the morbidity of iliac crest harvesting."[35]

## 3.12. Synthetic Bioactive Glass or Calcium Phosphate Bone Graft?

Another example of use of NovaBone® bioactive glass particulate as a synthetic bone graft is in lumbar spinal fusions. A spinal surgical group performed 88 cases of lumbar spinal fusion with all cases consisting of intervertebral body arthrodesis, the majority of which were also combined with postererolatral fusion.[36] All cases used pedicale screws and rods for stabilization. Direct comparison with Vitoss®, tri-calcium phosphate (TCP) bio-matieral, were made in 26 of the postreolateral fusion cases, with NovaBone® implanted on the left side and Vitoss® on the right side. The results, based upon CT myelograms at six months, showed 100% interbody fusion for patients grafted with NovaBone® particulate. In addition to the interbody fusion there was also postereolateral fusion for 50% of the NovaBone® grafts within six months. In contrast, only 11% of the patients achieved postlateral fusion in the Vitoss® (TCP) sites by six months. Direct review of grafted sites during re-entry supported the observation that there was greater bone formation at the NovaBone® grafted sites than for Vitoss® sites.[36]

## 3.13. Preventive Medicine and Dentistry

More recently, Bioglass® particulate has been used for the pre-venion of dentinal hypersensitivity.[37] This application of third-generation biomaterials heralds the onset of a new revolution in health-care preventive medicine and dentistry. Tooth hypersensi-tivity is a problem that affects an estimated 15–20% of the population of the United States, and similar numbers in Europe, occuring when the root portion of the tooth, which is dentin, becomes exposed around the gum line. The dentin has small openings, or tubules, that communicate with the pulp chamber. If the dentinal tubules become exposed, hot or cold or pressure can transmit the sensations to the nerves in the pulp, causing pain. The Bioglass® material used in this application is a very fine particulate that is incorporated into toothpaste, or used with an aqueous vehi-cle and applied to the tooth surface around exposed root dentin. When Bioglass® particles are put in contact with dentin, they adhere to the surface, rapidly form a HCA layer and occlude the tubules, thereby relieving the pain. Studies have shown that the Bioglass® particulate performs better than current therapies when it is used in relatively low concentrations. In 2004 the FDA cleared two products for sale through the 510(k) process, and product sales began in mid 2004. In Chapter 32 Dr David Greenspan discusses details of the development and application of 45S5 Bioglass® in these new dental clinical applications.

## 3.14. Other Applications

Other dental and maxillofacial applications include pulp capping, sinus obliteration, and repair of orbital floor fracture.[1] Use of 45S5 Bioglass® particulate as an injectable for treatment of urinary incontinence has also been tested *in vivo*.[1]

Recent studies have shown that controlled release of specific concentrations of ionic dissolution products from 45S5 Bioglass® particles can also be used to stimulate the onset of vascularization; that is, angiogenesis.[38] These findings indicate that it may be possible to design third-generation biomaterials for cardiovascular repair as well as musculo-skeletal repair. Chapter 6 reviews the current understanding of this important subject.

## 3.15. Implications for the Future

A cellular and molecular basis for development of third-generation biomaterials provides the scientific foundation for molecular design of scaffolds for tissue engineering and for *in situ* tissue regeneration and repair, with minimally invasive surgery. The economic advantages of these new approaches may aid in solving the problems of caring for an aging population. It should be feasible to design a new generation of gene-activating biomaterials tailored for specific patients and disease states. Tissue-engineered constructs based on a patient's own cells may be produced that can be used to select an optimal pharmaceutical treatment. The results suggest that bioactive stimuli may be used to activate genes in a preventive treatment to maintain the health of tissues as they age. Only a few years ago this concept would have seemed unimaginable, but we need to remember that only 40 years ago the concept of a material that would not be rejected by living tissues also seemed unimaginable. Thus, Bioglass® provides a starting point for prevention as well as regeneration.

## 3.16. Summary

After more than 40 years of basic research, extensive pre-clinical *in vitro* and *in vivo* studies in many mammalian species, and approval by regulatory agencies followed by commercial contracts, sales, and marketing relationships, Bioglass® medical and dental products are sold world-wide in more than 35 countries. More than a million patients have benefited from the clinical use of Bioglass® devices and particulate with almost no complications reported to the manufacturers. A wide variety of other bioactive materials are also now in use world-wide. Problems encountered during the technology transfer process cited in the historical review[1] have been overcome. The socio-economic benefits of a new approach to regenerative repair of the body are just begining to be realized.

## References


1. Hench, L.L., Wilson, J. and Greenspan, D.C. (2004). Bioglass: a short history and bibliography, *J. Aust. Ceram. Soc.*, **40**, 1–42.
2. Ratner, B.D., Hoffman, A.S., Schoen, F.J., *et al.* (2004). *Biomaterials Science: An Introduction to Materials in Medicine*, 2nd ed., Academic Press, New York, NY.
3. Park, J.B. and Lake R.S. (1990). *Biomaterials: An Introduction*, Plenum, New York, NY.
4. Hench, L.L. and Wilson J. (eds) (1996). *Clinical Performance of Skeletal Prostheses*, Chapman and Hall, London.
5. Hench, L.L. (1991). Bioceramics: from concept to clinic, *J. Am. Ceram. Soc.*, **74**, 1487–1510.
6. Hench, L.L. and Polak, J.M. (2002). Third-generation biomedical materials, *Science*, **295**, 1014–1017.
7. Thian, E.S., Huang, J., Best, S.M., *et al.* (2006). The response of osteoblasts to nanocrystalline silicon-substituted hydroxyapatite thin films, *Biomaterials*, **27**, 2692–2698.
8. Place, E.S., Evans, N.D. and Stevens, M.M. (2009). Complexity in biomaterials and other challenges in the translation of tissue engineering, *Nat. Mater.*, **8**, 457–470.

9. Hench, L.L., Splinter, R.J., Greenlee, T.K., *et al.* (1971). Bonding mechanisms at the interface of ceramic prosthetic materials, *J. Biomed. Mater. Res. Symp.*, **2**, 117–141.

10. Hench, L.L. and Clark, A.E. (1982). "Adhesion to Bone", in Williams, D.F. and Winters, G.D. (eds), *Biocompatibility of Orthopaedic Implants*, CRC Press, Boca Raton, FL.

11. Piotrowski, G., Hench, L.L., Allen W.C., *et al.* (1975). Mechanical studies of the bone–bioglass interfacial bond, *J. Biomed. Mater. Res. Symp.*, **9**, 47–61.

12. Weinstein, A.M., Klawitter, J.J. and Cook, S.D. (1980). Implant–bone interface characteristics of bioglass dental implants, *J. Biomed. Mater. Res.*, **14**, 23–29.

13. Hench, L.L. (1981). "Stability of Ceramics in the Physiological Environment", in Williams D.F. (ed.), *Fundamental Aspects of Biocompatibility*, CRC Press, Boca Raton, FL.

14. Hench, L.L. and West, J.K. (1996). Biological applications of bioactive glasses, *Life Chem. Rep.*, **13**, 187–241.

15. Hench, L.L. (1998). Bioceramics, *J. Am. Ceram. Soc.*, **81**, 1705–1728.

16. Hench, L.L. and Wilson, J. (eds) (1993). *An Introduction to Bioceramics*, World Scientific, London.

17. Nakamura, T., Yamamuro, T., Higashi, S., *et al.* (1985). A new glass–ceramic for bone replacement: evaluation of its bonding to bone tissue, *J. Biomed. Mater. Res.*, **19**, 685–698.

18. Kangasniemi, K. and Yli-Urpo, A. (1990). "Biological Response to Glasses in the $SiO_2$-$Na_2O$-$CaO$-$P_2O_5$-$B_2O_3$ System", in Yamamuro, T., Hench L.L. and Wilson, J. (eds), *Handbook of Bioactive Ceramics*, vol. 1, CRC Press, Boca Raton, FL, pp. 97–108.

19. Oonishi, H., Hench, L.L., Wilson, J., *et al.* (2000). Quantitative comparison of bone growth behaviour in granules in bioglass®, A–W glass ceramic, and hydroxyapatite, *J.Biomed. Mater. Res.*, **51**, 37–46.

20. Wheeler, D.L., Eschbach, E.J., Hoellrich, R.G., *et al.* (2000). Assessment of resorbable bioactive material for grafting of critical-size cancellous defects, *J. Orthop. Res.*, **18**,140–148.

21. Lobel, K. (1996). "Ossicular Replacement Prostheses", in Hench L.L. and Wilson, J. (eds), *Clinical Performance of Skeletal Prostheses*, Chapman & Hall, London.

22. Wilson, J. and Low, S.B. (1992). Bioactive ceramics for periodontal treatment: comparative studies, *J. Appl. Biomater.*, **3**, 123–129.
23. Hench, L.L., Polak, J.M., Xynos, I.D., *et al.* (2000). Bioactive materials to control cell cycle, *Mater. Res. Innovat.*, **3**, 313–323.
24. Xynos, I.D., Hukkanen, M.V.J., Batten, J.J., *et al.* (2000). Bioglass® 45S5 stimulates osteoblast turnover and enhances bone formation *in vitro*: implications and applications for bone tissue engineering, *Calcif. Tiss. Int.*, **67**, 321–329.
25. Hench, L.L., Day, D.E., Holand, W., *et al.* (2010). Glass and medicine, *Int. J. Appl. Glass Sci.*, **1**, 104–117.
26. Xynos, I.D., Edgar, A.J., Buttery, L.D., *et al.* (2000). Ionic dissolution products of bioactive glass increase proliferation of human osteoblasts and induced insulin-like growth factor II mRNA expression and protein synthesis, *Biochem. Biophys. Res. Comm.*, **276**, 461–465.
27. Xynos, I.D., Edgar, A.J., Buttery, L.D., *et al.* (2001). Gene expression profiling of human osteoblasts following treatment with the Ionic dissolution products of Bioglass® 45S5 dissolution, *J. Biomed. Mater. Res.*, **55**, 151–157.
28. Hench, L.L. (2003). Glass and genes: the 2001 W.E.S. Turner Memorial Lecture, *Glass Tech.*, **44**, 1–10.
29. Bielby, R.C., Christodoulou, I.S., Pryce, R.S., *et al.* (2004). Time and concentration-dependent effects of dissolution products of 58s sol-gel bioactive glass on proliferation and differentiation of murine and human osteoblasts, *Tissue Eng.*, **10**, 1018–1026.
30. Christodoulou, I., Buttery, L.D.K., Saravanapavan, P., *et al.* (2005). Characterization of human fetal osteoblasts by microarray analysis following stimulation with 58S bioactive gel-glass Ionic dissolution products, *J. Biomed. Mater. Res.*, **77B**, 431–446.
31. Jones, J.R., Tsigkou, O., Coates, E.E., *et al.* (2007). Extracellular matrix formation and mineralization on a phosphate-free porous bioactive glass scaffold using primary human osteoblast (HOB) cells, *Biomaterials*, **28**, 1653–1663.
32. Hench, L.L. and West, J.K. (1990). The sol-gel process, *Chem. Rev.*, **90**, 33–72.
33. Saravanapavan, P., Jones, J.R. and Hench, L.L. (2003). Bioactivity of gel-glass powders in the CaO-SiO2 system: a comparison with

ternary (CaO-P2O5-SiO2) and quaternary glasses (SiO2-CaO-P2O5-Na2O), *J. Biomed. Mater. Res.*, **66A**, 110–119.

34. Gough, J.E., Jones, J.R. and Hench, L.L. (2004). Nodule Formation and mineralisation of human primary osteoblasts cultured on a porous bioactive glass scaffold, *Biomaterials*, **25**, 2039–2046.

35. Ilharreborde, B., Morel, E., Fitoussi, F., *et al.* (2008). Bioactive glass as a bone substitute for spinal fusion in adolescent idiopathic scoliosis: a comparative study with Iliac crest autograft, *J. Ped. Orthoped.*, **28**, 347–351.

36. Thomas, D.J., Griffiths, H.J. and Perrin, J.C. (2010). "The Use of a Calcium Phosphosilicate in Lower Lumbar Spinal Fusion", in *Clinical Report from the Department of Orthopedic Surgery, Sewickly Hospital at Pittsburgh, PA* (report available from NovaBone® Products, Jacksonville, FL).

37. Gillam, D.G., Tang, J.Y. Mordan, N.J., *et al.* (2002). The effects of a novel Bioglass® dentifrice on detine sensitivity: a scanning electron microscopy investigation, *J. Oral Rehabil.*, **29**, 305–313.

38. Gorustovich, A.A., Roether, J.A. and Boccaccini, A.R. (2010). Effect of bioactive glasses on angiogenesis: a review of *in vitro* and *in vivo* evidences, *Tissue Eng. B*, **16**,199–207.

# Nanoparticles for Novel Healthcare Therapeutics



## CHAPTER 4

Sudipta Seal, William Self, James McGinnis and A.S. Karakoti

## 4.1. Introduction

Nano-sized particles in various shapes and sizes are being considered for therapeutic applications due to their unique surface chemical characteristics. One of the nano-materials with great therapeutic potential is nano-cerium oxide ($CeO_2$), also known as nano-ceria — a class of rare earth oxides. The feature of nano-ceria that makes it attractive for medical applications is its unique redox properties. Multiple valence surface redox states make nano-sized ceria (3–5 nanometers diameter) nanoparticles (NP) capable of free-radical scavenging and thereby useful for treatment of medical disorders caused by reactive oxygen intermediates (ROIs).

Another important characteristic of nano-ceria is its auto-catalytic ability to regenerate free-radical scavenging surface sites under a wide range of environmental conditions. This feature makes single dosage treatments possible. An overview of the use of nano-ceria as an antioxidant in medical applications has been published.[1]

It is important to recognize that cellular responses to nano-sized materials are not necessarily beneficial. There is the possibility of toxicity depending upon a vast array of physical variables, such as surface area, particle size, size distribution, particle aspect ratio (length/diameter ratio), net surface charge, distribution of surface charges, zeta potential, and rates of dissolution. Solution pH can affect many of these variables and alter the biochemistry of cellular interactions with nano-materials. These topics are reviewed in the July 2006 issue of the *Journal of Materials*.[2] The review includes the important topics of classification of nano-materials; mechanisms behind the toxicity of nano-materials, surface chemistry of nano-materials, NP entry routes into human bodies, and others. Table 4.1 discusses various published work on the antioxidant activity of nano-ceria in imparting protection against diseases caused by the ROIs.

Because of the importance of potential toxicity of nano-materials, toxicity tests have been conducted on nano-ceria particles synthesized in various chemical environments. The results show that nano-ceria can be synthesized in a manner to make the material non-toxic and to possess the positive attributes of free-radical scavenging.[1] Although not a complete list, some of the potential therapeutic applications of nano-ceria are summarized in the following sections and in Table 4.1.

## 4.2. Use of Vacancy Engineered Nano-Ceria to Control Oxidative Micro-Chemical Environment of Cells During Regeneration of Tissues

Nano-size (3–5 nm) particles of cerium oxide (nano-ceria) have been shown to exhibit impressive, positive effects on cells of the central nervous system (CNS).[11,12] Nano-ceria has been shown to prevent retinal degeneration induced by intracellular peroxides. Potential applications of nano-ceria also include the treatment of spinal cord injury and other CNS degenerative diseases, as described below. Recent findings also show effectiveness of nano-ceria in a synergistic role with controlled release of ionic stimuli in stimulating osteogenesis.

## 4.3. Effects of Nano-Ceria Nanoparticles on Neuronal and Central Nervous System Biology

We discovered for the first time intrinsic therapeutic activity of a single dose of an engineered rare earth oxide (nano-ceria)

**Table 4.1.** Use of nano-ceria anti-oxidant activity to protect against diseases caused by reactive oxygen intermediates (ROIs).

| Disease or mode of injury | Model system studied | Dosage | Results | Ref. |
| --- | --- | --- | --- | --- |
| Retinal and macular degeneration of eye (blindness) | Rat retina (*in vivo*) light damage model | 1 nM to 1 μM | Nano-ceria prevented light-induced damage to retina | 3 |
| Radiation damage during radiotherapy during treatment of cancer | CRL8798 normal breast epithelial cell line and MCF-7 breast carcinoma cell line | 10 nM | Nano-ceria imparted protection against radiation damage to normal cells | 4 |
| Neurodegeneration/spinal cord injury/ischemic events | Adult mice spinal cord | 100 mM | Nano-ceria imparted protection to the spinal cord neurons against injury | 5 |
| Chronic inflammation (rheumatoid arthritis, multiple sclerosis, atherosclerosis, and heart disease) | J774A.1 murine macrophages | 1 nM to 10 μM | Nano-ceria imparted protection to cells against chronic inflammation. Nano-ceria suppressed the production of iNOS and COX-2 type protein mediators. | 6 |
| Oxidative stress | HT22 nerve cells | 10 ng to 50 μg/ml | Cerium oxide protected the cells against glutamate-induced stress and toxicity | 7 |

*(Continued)*

**Table 4.1.**   (*Continued*)

| Disease or mode of injury | Model system studied | Dosage | Results | Ref. |
| --- | --- | --- | --- | --- |
| Diseases caused by ROI (Alzheimer's, Parkinson's Huntington's disease, atherosclerosis and stroke) | Radical scavenging capability using SOD mimetic assay | 15–60 nM | Nano-ceria demonstrated higher rate constant for SOD activity than CuZn SOD. Nano-ceria can act as an effective antioxidant in a single dose | 8 |
| Cardiac dysfunction, ER stress, myocardial oxidative stress | MCP-1 transgenic mice and wild type FVB/N mice | 15 nM | Nano-ceria inhibited myocardial oxidative stress and markedly reduced the expression of pro-inflammatory cytokines, tumor necrosis factor-$\alpha$, and interleukin (IL)1$\beta$ and IL-6 | 9 |
| Oxidative stress | Murine insulinoma $\beta$TC-tet cells | N/A | Cerium oxide nanoparticles reduced the cellular oxidative stress and inhibited cell apoptosis upon the exposure to free radical inducer such as hydroquinone | 10 |

NPs as an extremely strong free radical scavenger (FRS). This finding opens a new frontier of engineered active NP-based drug discovery research. However the current understanding of biological interaction of these NPs with human cells and tissue is limited. Correlation of the material, size, and surface properties with the biological activity is required in order to ensure safe clinical usage.

A single dose of nano-ceria NPs in serum-free cell culture media is effective in scavenging free-radicals to reduce damage to neurons. The NPs prolong the brain cell longevity by 2–3 fold or more,[11,12] and can reduce hydrogen peroxide ($H_2O_2$) and UV light-induced cell injury by more than 60%. Nano-ceria NPs also conferred radiation protection to a normal human breast cell line, CRL 8798, but not to a human breast tumor line, MCF-7. These results suggest that nano-ceria NPs could serve as selective radio-protectants for normal tissues in cancer radiotherapy.[4]

Seal, Self and McGinnis have demonstrated that nano-ceria exhibits superoxide dismutase mimetic activity and can reduce the reactive oxygen intermediates (ROI) in primary cell cultures of rat retina. These promising preliminary data reveal the potential use of engineered NPs in preventing many diseases including glaucoma, published in *Nature Nanotechnology*.[3] The present work is bound to have a much broader impact, as engineered rare earth metal oxide NPs may be effective in inhibiting the progression of radical induced cell deaths leading to macular degeneration, retinitis pigmentosa, and other blinding diseases, as well as the death of other cell types involved in diabetes, Alzheimer's disease, atherosclerosis, stroke, and others.

The unique *therapeutic benefit* of single-dose ceria NPs (existing antioxidants require multiple doses), promoting cell longevity, and decreasing oxidative stress, originates in the electronic transitions between the higher to the lower oxidation states of Ce in select biological environments. The co-existence of both $Ce^{3+}$ and $Ce^{4+}$ in the surface of nano-ceria NPs plays a critical role in its antioxidant behavior and makes the material a highly effective and biologically active ROS scavenging NP. Experimental data show that the reverse transition of oxidation states also occurs over time in cell cultures.[1] Therefore, ceria NP acts as a catalyst similar to the naturally occurring enzyme, called superoxide dismutase (SOD).[8]

## References


1. Karakoti, A.S., Monterio-Riviere, N.A., Aggarawl, R., *et al.* (2008). Nano-ceria as antioxidant: synthesis and biomedical applications, *JOM-J. Min. Met. Mat. S.*, **60**, 33–37.
2. Karakoti, A.S., Hench, L.L. and Seal, S. (2008). The potential toxicity of nanomaterials — the role of surfaces, *JOM-J. Min. Met. Mat. S.*, **58**, 77–82.
3. Chen, J., Patil, S., Seal, S., *et al.* (2006). Rare earth nanoparticles prevent retinal degeneration induced by intracellular peroxides, *Nature Nanotech.*, **1**, 142–150.
4. Tarnuzzer, R.W., Colon, J., Patil, S., *et al.* (2005). Vacancy engineered ceria nanostructures for protection from radiation-induced cellular damage, *Nano Lett.*, **5**, 2573–2577.
5. Das, M., Patil, S., Bhargava, N., *et al.* (2007). Autocatalytic ceria nanoparticles offer neuroprotection to adult rat spinal cord neurons, *Biomaterials*, **28**, 1918.
6. Alili, L., Sack, M., Karakoti, A.S., *et al.* (2011). Combined cytoxic and anti-invasive properties of redox-active nanoparticles in tumor-stroma interactions, *Biomaterials*, **1**, 1–12.
7. Schubert, D., Dargusch, R., Raitano, J., *et al.* (2007). Cerium and yttrium oxide nanoparticles are neuroprotective, *Biochem. Bioph. Res. Co.*, **342**, 86–91.
8. Korsvik, C., Patil, S., Seal, S., *et al.* (2007). Superoxide dismutase mimetic properties exhibited by vacancy engineered ceria nanoparticles, *Chem. Commun.*, **10**, 1056–1058.
9. Niu, J., Azfer, A., Rogers, L.M., *et al.* (2007). Cardioprotective effects of cerium oxide nanoparticles in a transgenic murine model of cardiomyopathy, *Cardiovasc. Res.*, **73**, 549.
10. Tsai, Y.Y., Oca-Cossio, J., Atkinson, M.A., *et al.* (2007). Inhibiting cellular apoptosis with CeO2 nanoparticles as free radical scavengers, *Free Radical Bio. Med.*, **43**, S180–S180.

11. Callahan, P., Colon, J., Merchant, S., *et al.* (2003). Deleterious effects of microglia activated by *in vitro* trauma are blocked by engineered oxide nanoparticles, *J. Neurotraum.*, **20**, 1057.
12. Rzigalinski, B.A., Bailey, D., Chow, L., *et al.* (2003). Cerium oxide nanoparticles increase the lifespan of cultured brain cells and protect against free radical and mechanical trauma, *Faseb J.*, **17**, A606.

# Glass Microspheres for Cancer Treatment



## CHAPTER 5

Delbert E. Day

## 5.1. Introduction

Radiotherapy and chemotherapy are common treatments for cancer, and are aimed at killing the cells in malignant tumors. Unfortunately they also damage other cells, causing varying degrees of side effects. This chapter documents the steps involved in going from the concept of delivering high levels of local radiation to kill cancer cells in the liver to achieving a successful glass delivery system for the radiation. The success of this revolutionary approach to cancer therapy is saving the lives of thousands of patients and provides a methodology to target other debilitating diseases that affect the quality of life of aging people.

## 5.2. Destroying Malignant Tumors

The technological innovation described in this chapter is the development and application of radioactive yttrium aluminosilicate glass (hereafter abbreviated as YAS) microspheres that are being used to treat patients with a generally inoperable and deadly form of primary liver cancer; called hepatocellular carcinoma (HCC).[1] The incidence of this deadly disease is increasing worldwide and is considered the sixth most common cancer in the world (one million new cases annually), ranking third as the cause of cancer-related deaths (500,000 deaths per year). In the United States, the National Cancer Institute estimates there were 19,160 new cases of HCC and 16,780 deaths in 2007. Life expectancy for patients diagnosed with HCC is measured in months.

Malignant tumors are typically treated either surgically, by chemotherapy, or with radiation. Unfortunately, none of these methods has been effective in treating HCC and the 5-year survival rate for patients with HCC is less than 7%.[2] External beam radiation is employed to treat many forms of cancer, but the maximum dose that can be delivered to HCC tumors is limited by the unavoidable damage done to nearby healthy tissue, and is too small to be effective (therapeutic). A potential solution to this problem is to place the radiation source inside the diseased tumor, a treatment called intra-arterial therapy, so that a larger and therapeutic dose of localized radiation can be delivered to the tumor(s) *in situ* without damaging nearby healthy tissue.

### 5.2.1. *In situ irradiation of tumors*

For a material such as glass to be used for the *in situ* delivery of radiation to tumors in the liver, or in other organs, it must satisfy the following criteria:

1.  It must be biocompatible and non-toxic in the body.
2.  It must be chemically resistant to the body fluids so that none of the radioisotope is released within the body (blood stream). The approach taken was to incorporate the radioisotope into a chemically durable glass matrix, which confines it to the target site.
3.  If a glass is to be made radioactive by neutron activation, then it must be free of any other elements that form unwanted radioisotopes during neutron activation, other than the desired element being activated. Neutron activation eliminates the need to handle radioactive glass, but it limits the composition that can be used.
4.  The concentration of the element in the glass that can be activated by the neutrons must be high enough to provide the level of specific activity required for the desired treatment. This

concentration varies with the element being activated, the organ being treated, and the neutron flux and activation time.

5.  It must be capable of being formed into particles of the desired size and shape. The size and shape of the radioactive particles depend upon the characteristics of the organ being treated, and the method of delivering the particles to the target organ.

Radioactive particles (microspheres) can be delivered to a target organ using either the blood supply to that organ or by direct injection into the tumor. In the case of liver tumors, the aim is to inject radioactive microspheres into the hepatic artery, since it is the main blood supply to the malignant tumor(s) in the liver. In the early 1960s, attempts were made to irradiate tumors in the liver *in situ*, using polymer microspheres coated with radioactive Y-90, but the radioactive Y-90 could not be confined so the procedure was discontinued.[1]

### 5.2.2.  *Developing glass microspheres*

The key task was to develop a glass composition which fulfilled the preceding requirements. Yttrium-90 was the radioisotope of choice for this application, since it could be produced by neutron activation of naturally abundant Y-89, had an acceptable half-life of 64.2 hours, and emitted beta radiation, which had an average range in soft tissue of only 2.5 mm. This short-range radiation minimized the amount of radiation reaching healthy liver tissue. The desire to use neutron activation, along with the requirement for an extremely high chemical durability, eliminated common glasses that contain elements such as Na, K, Ca, and Ba, since they form undesirable radioisotopes during neutron activation. Since elemental oxygen, aluminum, and silicon do not form objectionable radioisotopes during neutron activation, an investigation was undertaken of the glass-forming tendency and selected properties of $Y_2O_3$-$Al_2O_3$-$SiO_2$ compositions.

A wide field of glass formation was found[3] for yttria–alumina–silica compositions which melted below 1600°C (a requirement for practical glass production) and which contained sufficient yttrium for this application, up to 55 wt% $Y_2O_3$. Glasses with an excellent chemical durability were identified and glass microspheres of the desired size (25–35 microns) were readily made by flame spheroidization.[4]

Laboratory and animal experiments were conducted to determine the safety of using YAS microspheres in humans, followed by clinical studies in Canada, Scotland, and the United States.[1] An initial major concern was the potential leaching of Y-90 from the glass microspheres, as had occurred in earlier studies where polymer beads coated with Y-90 were used. Experiments quickly showed that no detectable amount of Y-90 was released from a YAS glass with a nominal composition of $40Y_2O_3$-$20Al_2O_3$-$40SiO_2$ wt%. Microspheres of this YAS glass, which the US Food and Drug Administration considers a medical device (Table 5.1), have been used in hundreds of patients with no instances of unwanted release of Y-90 from the glass.

**Table 5.1.**   Time line for the development of YAS-90 glass microspheres.

| Year | Event |
| --- | --- |
| 1982 | Work begins at the University of Missouri-Rolla (now the Missouri University of Science and Technology) and the University of Missouri–Columbia |
| 1985 | First animal (dogs) experiments at University of Michigan |
| 1986 | Glass microsphere technology licensed by University of Missouri |
| 1986 | First human trials in Canada |
| 1987 | First human trials in United States at University of Michigan |
| 1988 | First patent issued: US 4,789,501 |
| 1991 | Microspheres approved for use in Canada |
| 1999 | US Food and Drug Administration (FDA) grants humanitarian device exemption (HDE) status to YAS microspheres for treating HCC or as bridge to transplantation |
| 2002 | Microspheres approved for commercial use in United States by FDA |
| 2005 | Microspheres granted CE mark and approved for use in European Union |
| 2010 | Patients currently being treated with Y-90 glass microspheres in approximately 100 sites world-wide |

## 5.3. Clinical Results of *In Situ* Radiation Delivery

Liver cancer, especially HCC, has been and still is a difficult disease to treat, but the availability of Y-90 glass microspheres has opened a new avenue for treating liver cancer, with a technique called radioembolization. This refers to the combined effect of the radiation and the embolization of the capillaries in a malignant tumor(s). The Y-90 microspheres not only safely deliver a much larger dose of radiation to the tumor(s), but the roughly 2–8 million glass microspheres in a typical injection become lodged (embolized) in the capillaries, thereby reducing the blood (nourishment) flow to the malignant tumor(s). These two synergetic effects help shrink and destroy the tumor(s).[1]

Typically, the procedure is done on an out-patient basis with the patient receiving the equivalent of an injection of several million radioactive Y-90 glass microspheres through a catheter inserted in the femoral/hepatic artery. The microspheres are released into the blood stream as close as possible to the tumor(s) in order to maximize the number of microspheres deposited in the malignant tumor(s). The microspheres are sized so that they are small enough to enter the capillaries of the tumor(s), but are too large to pass through the capillaries. Typically, only a small percentage of the radioactive microspheres, 2–10%, reach healthy tissue.

The patient is discharged after a brief period of observation, if there are no complications. The inoperable HCC tumor(s) in the patient's liver continue to be irradiated *in situ* by the localized Y-90 beta radiation as the patient returns to a normal routine. In about 4 weeks the microspheres are no longer radioactive.

Initially, patients received only one injection of Y-90 glass microspheres since the radiation dose was much larger than doses used previously. As confidence in the safety of *in situ* irradiation increased and the benefits became better known, patients received multiple injections when necessary. The *in situ* irradiation of HCC tumors with radioactive YAS glass microspheres is increasingly recognized as having many advantages.[1] With Y-90 radioembolization, it is possible to irradiate multiple, small tumors at much higher doses, 5–50 times that delivered by an external beam. The localized beta radiation destroys tumors regardless of the tumor origin, and the relatively low toxicity of the beta radiation towards healthy tissue permits repeated injections. Compared with chemotherapy and other radiation procedures, the side effects are small; a small fraction of patients report flu-like symptoms such as fatigue, a slight fever, or abdominal pain, for a few days.

## 5.4. Survival Data

Of greatest interest are the survival data for patients treated with the Y-90 glass microspheres. For patients with inoperable HCC, the stage of the disease at the time of treatment is especially important to survival. The survival data for a group of 229 patients with HCC who received no specific treatment provides a reference point.[5] The median survival time for this group was 51 days — 21 days for patients at an advanced stage and 249 days for patients with an early stage disease. For a group of 42 patients treated with Y-90 glass microspheres, the median survival was 236 days (advanced stage) and 660 days (early stage).[6] In another group of 150 patients with inoperable HCC, the median survival time for the entire group was 800 days (2.2 years) after treatment with the Y-90 glass microspheres.[7] Finally, the 1-, 2-, and 5-year survival rates for a group of 20 patients, ages 42–82, with inoperable HCC and treated with Y-90 glass microspheres, were recently reported to be 100%, 75%, and 46%, respectively.[8] These survival rates are encouraging.

## 5.5. Summary

The radioembolization of HCC tumors with radioactive YAS glass microspheres is recognized as a safe and effective way of treating this deadly form of cancer. Glass technology and science have made an important contribution to the success of this new procedure for treating patients with inoperable liver cancer. There are other opportunities where microspheres of YAS and other rare earth alumino-silicate glasses can be used as *in situ* radiation devices to destroy malignant tumors in other organs in the body.[9]

## Acknowledgments

This chapter is an edited version of a section of a review article, "Glass and Medicine," by Larry L. Hench, Delbert E. Day, Wolfram Hohland and Volker M. Rheinberger (2010), in the *International Journal of Applied Glass Science* **1**(1), 104–117.

## References


1. Salem, R. and Thurston, K.G. (2006). Radioembolization with Yttrium-90 microspheres: a state-of-the-art brachytherapy treatment for primary and secondary liver malignancies, part 3: comprehensive literature review and future direction, *J. Vasc. Interv. Radiol.*, **17**, 1571–1594.
2. El-Serag, H.B., Mason, A.C. and Key, C. (2001). Trends in survival of patients with hepatocellular carcinoma between 1977 and 1996 in the United States, *Hepatology*, **33**, 62–65.
3. Hyatt, M.J. and Day, D.E. (1987). Glass properties in the yttria-alumina-silica system, *J. Am. Ceram. Soc.*, **70**, C283–C287.
4. Erbe, E.M. and Day, D.E. (1993). Chemical durability of $Y_2O_3$-$Al_2O_3$-$SiO_2$ glasses for the *in vivo* delivery of beta radiation, *J. Biomed. Mater. Res.*, **27**, 1301–1308.
5. Okuda, K., Ohtsuki,T., Obata, H., *et al.* (1985). Natural history of hepatocellular carcinoma and prognosis in relation to treatment: study of 850 patients, *Cancer*, **56**, 918–928.
6. Keppke, A.L., Salem, R., Reddy, D., *et al.* (2007). Imaging of hepatocellular carcinoma after treatment with yttrium-90 microspheres, *Am. J. Roentgenology*, **188**, 768–775.
7. Kulik, L.M., Atassi, B., Holsbeeck, L.V., *et al.* (2006). Yttrium-90 microspheres (TheraSphere™) treatment of unresectable hepatocellular carcinoma: downstaging to resection, RFA and bridge to transplantation, *J. Surg. Oncol.*, **94**, 572–586.

8.  Gaba, R.C., Lewandowski, R.J., Kulik, L., *et al.* (2009). Radiation lobectomy: preliminary findings of hepatic volumetric response to lobar yttrium-90 radioembolization, *Ann. Surg. Oncol.*, **16**, 1587–1596.
9.  White, J.E. and Day, D.E. (1994). Rare earth aluminosilicate glasses for *in vivo* radiation delivery, *Key Eng. Mater.*, **94–95**, 181–208.

# Bioactive Glasses and Angiogenesis



## CHAPTER 6

Alejandro A. Gorustovich, Luis A. Haro Durand
and Aldo R. Boccaccini

## 6.1. Introduction

Angiogenesis, the formation of new blood vessels from the endothelium of the existing vasculature, plays a pivotal role in wound healing and tissue engineering (TE). In response to the injury, microvascular endothelial cells degrade the extracellular matrix (ECM), migrate into the perivascular space, proliferate, and align themselves into patent blood vessels to participate in provisional granulation tissue formation. This complex process is regulated at molecular level by growth factors, ECM proteins, membrane receptors, and signalling molecules that tightly modulate the formation of new blood vessels, leading to stabilization and eventually regression and involution of the newly formed vasculature as newly formed tissue remodels.[1,2]

New concepts to develop TE constructs based on engineered biomaterials for the replacement of diseased or damaged tissue are needed that take angiogenesis into consideration. Several approaches are being proposed to induce rapid vascular in-growth such as gene and/or protein delivery of angiogenic growth factors, and *ex vivo* culturing of scaffolds with endothelial cells alone or in combination with other cell types.[3,4] In order to utilize the number of cells available, cells need to be surrounded by a suitable environment, usually provided by a three-dimensional porous biomaterial scaffold that will promote cell viability and the formation of functioning pre-vascular structures, and that connects the implanted construct to the host tissue.

Enhancement of the angiogenic potential of implantable engineered tissue scaffolds is the focus of considerable research efforts in TE strategies. Neovascularization represents a significant element contributing to the success of regenerating and growing new tissue. Blood vessels provide growing cells with oxygen and nutrients necessary for their survival, with the notable exception of cartilage, which relies on diffusion through gel matrix of high water content. If TE scaffolds have the ability to induce neovascularization, the viability of native or transplanted cells within scaffolds will be increased, which will enhance the possibility of engineering larger volumes of new tissues. The angiogenic potential of most synthetic and natural materials used to fabricate TE scaffolds is limited or absent, thus numerous attempts have been made to enhance angiogenesis associated with TE constructs, either by supplementation with angiogenic factors, or by changing the physicochemical properties of the materials.

Neovascularization may be enhanced through controlled delivery of certain bioactive molecules, such as specific angiogenic growth factors, including vascular endothelial growth factor (VEGF). Several approaches are being explored to incorporate relevant growth factors into synthetic biomaterial scaffolds for use in TE. However, materials processing techniques used to fabricate such engineered scaffolds typically involve high temperatures (in the case of bioceramics and glasses) or the use of organic solvents, conditions that are expected to denature proteins present during the process. Due to the importance of angiogenesis in new tissue formation, there is a continuous need to develop alternative means to induce angiogenesis in TE constructs, in particular for the regeneration of highly vascularized tissues such as bone.

Investigations in the bone tissue engineering field demonstrate that bioactive silicate glasses may improve vascularization and bone regeneration in both healthy and highly compromised experimental models of bone healing. As discussed in Chapter 3, bioactive silicate glass of composition (in wt %): 45% $SiO_2$, 24.5%

Na$_2$O, 24.5% CaO, and 6% P$_2$O$_5$ (45S5 Bioglass®) was the first man-made inorganic material developed to bond to bone. Several other types of silicate glasses have been developed, based on that original silicate composition, with the goal to improve degradation behavior, mechanical properties or processibility, such as the ability to form components of different shapes and microstructure. The bioactivity of these inorganic materials is attributed to: (i) surface reactions taking place at the material–tissue interface providing an ideal environment for colonization, proliferation, and differentiation of osteoblasts to form new bone; and ii) the direct effect of glass ionic dissolution products on osteogenesis (see Chapter 3). These significant attributes make bioactive glasses, in particular the composition 45S5 Bioglass®, the biomaterials of choice for development of scaffold materials for bone tissue engineering. In this context, the chemical composition of bioactive glasses can be varied to tailor their rate of dissolution in the biological environment. Moreover, the structure and chemistry of glasses, in particular sol-gel derived glasses (see Chapter 9 for a discussion of sol-gel processing of bioactive glasses), can be tailored at a molecular level by varying either composition or processing parameters. It is possible to design glasses with degradation properties and bio-reactivity specific to a particular TE application. This chapter discusses the role of bioactive glasses to stimulate new bone growth and to serve as an angiogenic factor inducing increased vascularization when incorporated in bone TE constructs.[5]

## 6.2. Bioactive Glass Stimulates Fibroblasts to Secrete Angiogenic Growth Factors

*In vitro* studies have shown that bioactive glasses in particulate form, notably the composition 45S5 Bioglass®, stimulate a significant increase in the secretion of angiogenic growth factors from fibroblasts, such as VEGF and basic fibroblast growth factor (bFGF).[6–11] Culture medium collected from rat and human fibroblasts grown for 24 hours on polystyrene surfaces coated with

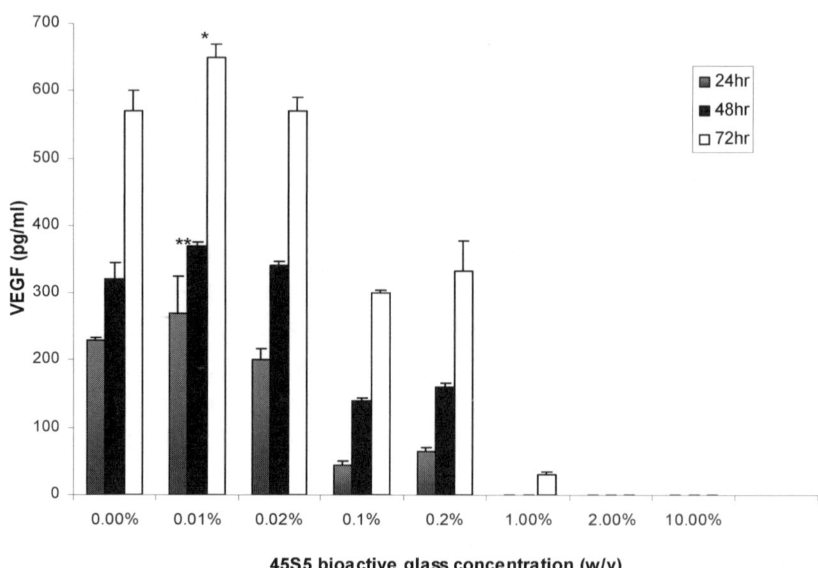


**Figure 6.1.**  VEGF secretion by 208F fibroblasts grown on 45S5 Bioglass®-coated cell culture plates. Assessment in conditioned culture medium collected after 24, 48, and 72 hours. Data are the mean values of triplicate experiments. The vertical bars are the standard deviations (p < 0.05, **p) (adapted from).[6]

low concentrations (0.01% wt/vol) of 45S5 Bioglass® (micrometer-size range) particles was found to contain significantly higher concentrations of VEGF as compared with control fibroblasts grown without the glass. Moreover, fibroblasts grown on surfaces coated with 45S5 Bioglass® also contained increased amounts of VEGF mRNA relative to unstimulated cells. Figure 6.1 shows an example of the reported results,[6] with the remarkable VEGF secretion by 208F fibroblasts grown on 45S5 Bioglass®-coated cell culture plates after 24, 48, and 72 hours.

It has been also reported[11] that micro-porous spheres of poly(lactide-*co*-glycolide) (PLGA) containing 10% (w/w) 45S5 Bioglass® particles stimulated a significant increase in VEGF secretion from human fibroblasts consistently over a ten-day period compared with the neat PLGA (no Bioglass®) micro-porous spheres. It was described that mouse fibroblasts cultured on the surface of PLGA discs with 0.01, 0.1, and 1 w/v % 45S5 Bioglass® particles secreted increased amounts of VEGF compared with cells cultured on PLGA alone. The most consistent significant increase in VEGF secretion occurred after culturing for 48 and 72 hours with the spheres containing 0.1 w/v % Bioglass®. The PLGA reduces the dissolution of the glass compared to the results shown in Fig. 6.1.

Several studies have shown that concentrations of 45S5 Bioglass® greater than 0.1 w/v % inhibited secretion of VEGF from fibroblasts possibly due to cytotoxic effects related to either the increased concentration of ionic dissolution products, or to the increased pH of the culture medium associated with higher concentrations of Bioglass®. In a related investigation,[8] human fibroblasts were encapsulated in alginate beads containing 0.01% and 0.1% (w/v) 45S5 Bioglass® particles with the final aim of incorporating cells and bioactive glass particles into a three-dimensional system that could be used as a bioreactor to deliver angiogenic growth factors for therapeutic angiogenesis. It was observed that fibroblasts secreted increased quantities of VEGF compared with cells encapsulated in alginate alone, or with 1% (w/v) 45S5 Bioglass® particles.

In addition, 45S5 Bioglass® has also been used to stimulate the bFGF production from human fibroblasts. Basic FGF secretion was significantly increased when the cells were grown on surfaces coated with 0.1–2% (w/v) 45S5 Bioglass® for periods of 24 and 48 hours. Table 6.1 summarizes a list of references which have investigated *in vitro* growth factor secretion with respect to various Bioglass substrates.

## 6.3. Bioactive Glass Stimulates Endothelial Cell Proliferation and Endothelial Tube Formation

Endothelial cells are those that form the tubular structure of blood vessels. Endothelial cell proliferation was significantly increased by conditioned medium collected from human fibroblasts encapsulated in alginate beads containing 0.1% (w/v) 45S5 Bioglass® particles, and from human fibroblasts grown for 72 hours on surfaces coated with 0.1% (w/v) 45S5 Bioglass® particles.[7,8] These results

**Table 6.1.**   A summary of the presented results, providing an overview of the *in vitro* evidence of the effect of bioactive glass stimulating secretion of relevant angiogenic growth factors, including the material used in each case (incorporating bioactive glass in particulate form).

| Angiogenic indicator | Cell line | Material | Ref |
|---|---|---|---|
| *Secretion of* VEGF | Rat fibroblasts (208F) | BG-coated cell culture plates | 6 |
| | Mouse fibroblasts (L929) | PLGA/BG composite discs | 9 |
| | Human fibroblasts (CCD-18Co) | BG-coated cell culture plates | 7 |
| | | BG/alginate beds | 8 |
| | | BG/PLGA porous microspheres | 11 |
| | Human endothelial (HMVEC) | Collagen/BG scaffolds | 12 |
| *Secretion of* bFGF | Human fibroblasts (CCD-18Co) | BG-coated cell culture plates | 7,10 |

indicate that the Bioglass® stimulated the fibroblasts to secrete the angiogenic factors (e.g. VEGF and bFGF) into the conditioned medium, and that this medium was capable of inducing endothelial cell proliferation. A small amount of ZnO addition into the glass composition was also beneficial for bovine aortic endothelial cell attachment[13] compared to the 45S5 composition, but larger amounts were detrimental. This may be because small amounts of Zn crosslink the silica network, as Zn acts as a network intermediate, reducing dissolution rate. Too much Zn may cause too much Zn ion release, causing toxicity.

A significantly enhanced mitogenic stimulation of endothelial cells in the presence of 3D porous scaffolds made from poly(lactide-*co*-glycolide) with approximately 0.5 mg 45S5 Bioglass® particle coating has been also reported.[14] In related experiments, endothelial cells cultured in indirect contact with Bioglass®-loaded collagen sponges exhibited a dose-related proliferative response to the soluble products of the scaffolds.

45S5 Bioglass® dosages of 0.6, 1.2, and 6 mg resulted in enhanced proliferation, with the greatest proliferative response achieved with collagen sponges loaded with 1.2 mg of Bioglass® particles. In contrast, a substantial inhibition of endothelial cell proliferation was observed with sponges containing the highest Bioglass® mass (12 mg). Future studies at cellular, molecular, and proteomic level are required to investigate the response of endothelial cells to ionic dissolution products of bioactive glasses. It can be anticipated that the correct design of silicate bioactive glass compositions will follow from understanding the direct effect of dissolution products on the up-regulation mechanisms of angiogenic growth factors.

Several authors explored the proangiogenic potential of 45S5 Bioglass® by examining its capacity to promote the generation of endothelial tubules, as reviewed elsewhere.[5] For example, it has been reported that conditioned medium from human fibroblasts grown for 72 hours on surfaces coated with a slurry of 0.1% (w/v) 45S5 Bioglass® particles induced a significant increase in the formation of anastomosing networks of newly formed endothelial tubules.[7] Moreover, a dose-related response of tubule formation to 45S5 Bioglass® was observed[12] using co-cultures of endothelial cells and fibroblasts stimulated with conditioned medium from 45S5 Bioglass®-treated rat aortic rings. The greatest average number of tubules was generated using 1.2 mg 45S5 Bioglass®, while other masses (0.12, 0.6, and 6 mg) failed to produce enhanced tubule formation.

## 6.4. Bioactive Glass Stimulates Neovascularization of Tissue-Engineered Scaffolds

The first evidence of the relationship between a bioactive glass porous matrix and *in vivo* vascular development was achieved using porous-ball scaffolds based on a fibrous form of bioactive glass composed (in mol%) of 41% $SiO_2$, 32.24% $CaO$, 9.26% $P_2O_5$, and 17.5% $Al_2O_3$.[15] Two receptors for VEGF (Flt-1 and KDR) were expressed in the scaffolds at both two and four weeks after subcutaneous implantation in rats. Similarly, vascularization, as determined by hemoglobin (Hb) content extracted from implants, was higher in sol-gel bioactive glass-coated collagen scaffolds placed subcutaneously in comparison to the glass-free (control) group.[16] One important concomitant finding was that glass-coated collagen samples did not present inflammatory cells associated with the presence of Hb as observed for the control group. Experimental evidence supports the anti-inflammatory properties of bioactive glasses. Results using the mesenteric-window assay in adult mice showed that the intraperitoneal administration of the ionic dissolution products of melt-derived 45S5 Bioglass® and 45S5 containing 2 wt% $B_2O_3$ (45S5.2B) did not elicit an inflammatory response in the test tissue. It is known that inflammation causes angiogenesis. In this context, it is interesting to note that bioactive glasses are able to induce an angiogenic response without inflammation.

It has been described that bioactive glass scaffolds prepared from a melt-derived glass powder of composition (in wt%) 43.70% $SiO_2$, 19.20% CaO, 5.46% $P_2O_5$, 9.40% $B_2O_3$, and 22.24% $Na_2O$ allowed well-formed vascularization toward the implant block three months after implantation in a bone defect performed in the lateral aspect of diaphysis of the radius bone of black Bengal goats.[17] Using the same experimental model, it was demonstrated by angiography that there was well-organized trans-implant angiogenesis and establishment of vascular supply across the bone defects treated with bioactive glass porous struts of 35–40 vol. % porosity, prepared from a bioactive glass powder of composition (in wt%) 58.60% $SiO_2$, 23.66% CaO, 3.38% $P_2O_5$, 3.78% $B_2O_3$, 1.26% $TiO_2$, and 9.32% $Na_2O$.

Previous and recent *in vivo* studies have evaluated the effect on the angiogenic response in both soft connective and bone tissues (as reviewed elsewhere)[5] of the incorporation of relatively small quantities of 45S5 Bioglass® (in particulate form) into tissue-engineered scaffolds. For example, a biodegradable poly(glycolic) acid (PGA) mesh coated with Bioglass® particles was developed and implanted subcutaneously into rats. It was shown that the composite scaffolds became infiltrated by a significantly increased number of blood vessels compared with uncoated control scaffolds.[6] On the other hand, polymer (PLGA) foam composites containing 0.1 w/v % Bioglass® particles implanted subcutaneously into mice did not produce a significant increase in the number of blood vessels counted in the granulation tissue surrounding the scaffolds.[9] Similarly, a quantitative assessment of the number of blood vessels infiltrating the voids inside micro-porous spheres of PLGA containing 10% (w/w) 45S5 Bioglass®, placed into a subcutaneous wound model in rats, revealed no significant difference between the PLGA micro-porous spheres with or without Bioglass® after two weeks of implantation.[11] In other tests, 45S5 Bioglass®-derived glass-ceramic scaffolds did not produce an angiogenic response when they were placed on the chorioallantoic membrane of chick embryos. It remains to be determined whether this effect is the result of suboptimal concentration of Bioglass® or if it is related to the physical properties of the scaffold, such as pore dimension, pore-size distribution, interconnectivity, and pore orientation.

In critical-sized cranial bone defects in rats, a VEGF-releasing PLGA porous scaffold coated with 0.5 mg of 45S5 Bioglass® particles was capable of enhancing neovascularization. In addition, greater neovascularization and bone regeneration was found in irradiated critical-sized calvarial defects filled with collagen sponges loaded with 1.2 mg of Bioglass® than in controls at two weeks post-implantation in rats.[18]

The evidence from *in vivo* investigations to support the effect of bioactive glasses to stimulate neovascularization of TE scaffolds is still limited and more research is needed. Recently, employing the quail embryo chorioallantoic membrane as the experimental model, preliminary data shows that there is a significant effect on angiogenesis of the ionic dissolution products (IDPs) from 45S5 Bioglass® containing 2 wt% $B_2O_3$ (45S5.2B). The 45S5.2B IDPs have proangiogenic potential as confirmed by increased endogenous expression levels of integrin $\alpha_v\beta_3$, a marker of angiogenic vascular tissue, 48 hours after treatment and an enhanced of number of blood vessels as early as five days after treatment (Fig. 6.2).

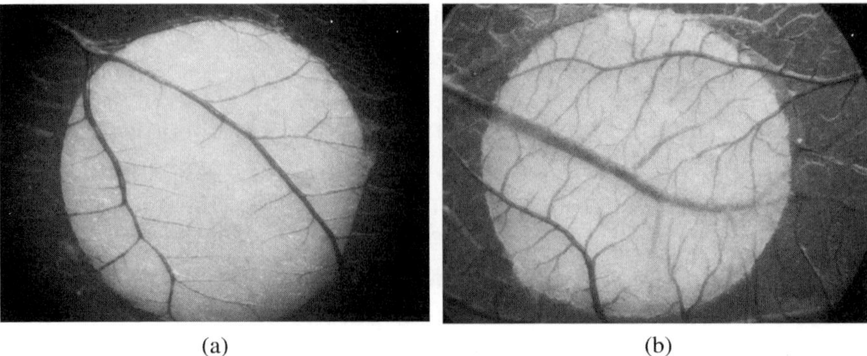

(a)                                                      (b)

**Figure 6.2.**    Angiogenic response of the quail embryo chorioallantoic membrane (CAM) to ionic dissolution products from 45S5 Bioglass® containing 2 wt% $B_2O_3$ (45S5.2B). An increase in the blood vessel density five days after treatment (b) compared to untreated CAM (a) (original magnification ×25).

## 6.5.  Summary and Outlook

Based on the emerging evidence in the literature about the angiogenic potential of bioactive glass, the incorporation of this material into tissue engineering scaffolds — in particular the composition 45S5 Bioglass® — can be widely beneficial in regenerative medicine strategies. *In vitro* studies have demonstrated increases in angiogenic indicators through both direct and indirect contact of cells with 45S5 Bioglass® particles or with their dissolution products. Moreover *in vivo* studies have confirmed the ability of certain bioactive glasses to stimulate neovascularization. Further research should concentrate on assessing the specific effect on angiogenesis of particular ion dissolution products from bioactive glasses, and in particular their relative concentration. In addition, the possible effect of scaffold morphology on neovascularization must be investigated, including volume fraction of porosity, pore size distribution, interconnectivity of pores, and pore orientation. The recent availability of nano-sized bioactive glass particles, and the successful incorporation of such nanoparticles in biodegradable polymers to form TE scaffolds, provides an opportunity to assess the effect on angiogenesis and neovascularization of the enhanced surface bioreactivity and degradation rate exhibited by nanoparticles.

# References


1. Lovett, M., Lee, K., Edwards, A., *et al.* (2009). Vascularization strategies for tissue engineering, *Tissue Eng. Pt. B-Rev.*, **15**, 353–370.
2. Phelps, E.A. and Garcia, A.J. (2009). Update on therapeutic vascularization strategies, *Regen. Med.*, **4**, 65–80.
3. Patel, Z.S., Young, S., Tabata, Y., *et al.* (2008). Dual delivery of an angiogenic and an osteogenic growth factor for bone regeneration in a critical size defect model, *Bone*, **43**, 931–940.
4. Tsigkou, O., Pomerantseva, I., Spencer, J.A., *et al.* (2010). Engineered vascularized bone grafts, *Proc. Natl. Acad. Sci. USA*, **107**, 3311–3316.
5. Gorustovich, A., Roether, J.A. and Boccaccini, A.R. (2010). Effect of bioactive glasses on angiogenesis: a review of *in vitro* and *in vivo* evidences, *Tissue Eng. Pt. B-Rev.*, **16**, 199–207.
6. Day, R.M., Boccaccini, A.R., Shurey, S., *et al.* (2004). Assessment of polyglycolic acid mesh and bioactive glass for soft-tissue engineering scaffolds, *Biomaterials*, **25**, 5857–5866.
7. Day, R.M. (2005). Bioactive glass stimulates the secretion of angiogenic growth factors and angiogenesis *in vitro*, *Tissue Eng.*, **11**, 768–777.
8. Keshaw, H., Forbes, A. and Day, R.M. (2005). Release of angiogenic growth factors from cells encapsulated in alginate beads with bioactive glass, *Biomaterials*, **26**, 4171–4179.
9. Day, R.M., Maquet, V., Boccaccini, A.R., *et al.* (2005). *In vitro* and *in vivo* analysis of macroporous biodegradable poly(D,L-lactide-co-glycolide) scaffolds containing bioactive glass, *J. Biomed. Mater. Res.*, **75A**, 778–787.
10. Moosvi, S.R. and Day, R.M. (2009). Bioactive glass modulation of intestinal epithelial cell restitution, *Acta Biomater.*, **5**, 76–83.
11. Keshaw, H., Georgiou, G., Blaker, J.J., *et al.* (2009). Assessment of polymer/bioactive glass-composite microporous spheres for tissue regeneration applications, *Tissue Eng. A*, **15**, 1451–1461.
12. Leu, A. and Leach, J.K. (2008). Proangiogenic potential of a collagen/bioactive glass substrate, *Pharm. Res.*, **25**, 1222–1229.
13. Aina, V., Malavasi, G., Fiorio, P.A., *et al.* (2009). Zinc-containing bioactive glasses: surface reactivity and behaviour towards endothelial cells, *Acta Biomater.*, **5**, 1211–1222.

14. Leach, J.K., Kaigler, D., Wang, Z., *et al.* (2006). Coating of VEGF-releasing scaffolds with bioactive glass for angiogenesis and bone regeneration, *Biomaterials*, **27**, 3249–3255.
15. Mahmood, J., Takita, H., Ojima, Y., *et al.* (2001). Geometric effect of matrix upon cell differentiation: BMP-Induced osteogenesis using a new bioglass with a feasible structure, *J. Biochem.*, **129**, 163–171.
16. Andrade, A.L., Andrade, S.P. and Domingues, R.Z. (2006). *In vivo* performance of a sol-gel glass-coated collagen, *J. Biomed. Mater. Res. B*, **79B**, 122–128.
17. Ghosh, S.K., Nandi, S.K., Kundu, B., *et al.* (2008). *In vivo* response of porous hydroxyapatite and $\beta$-tricalcium phosphate prepared by aqueous solution combustion method and comparison with bioglass scaffolds, *J. Biomed. Mater. Res. B*, **86B**, 217–227.
18. Leu, A., Stieger, S.M., Dayton, P., *et al.* (2009). Angiogenic response to bioactive glass promotes bone healing in an irradiated calvarial defect, *Tissue Eng. A*, **15**, 877–883.

## CHAPTER 7

# Clinical Applications of Bioactive Glasses for Maxillo-Facial Repair

Ian Thompson

## 7.1. Introduction

An improved quality of life and increased life expectancy have created new medical challenges, age-related disease, and an increased number of trauma victims. Replacement of body tissues that have been worn out due to prolonged use, and the regeneration of weakened or damaged bone structures are of increasing importance world-wide. This chapter summarizes the use of one specific class of materials, bioactive glasses, which are being clinically applied in the reconstruction and regeneration of bone following maxillo-facial surgery.

## 7.2. Clinical Requirements

Surgeons face a number of options when determining a treatment plan for a patient requiring bone grafting. The first decision is to determine the source of the graft material. Several options are possible, each with advantages and limitations. They are:

*Autogenous*: Use of bone from the patient's own (autograft) skeleton

*Allogenic*: Use of bone donated from a (allograft) human source

*Xenogenic*: Use of bone obtained from another (xenograft) species

*Alloplastic*: Use of a synthetic graft material (synthetic graft)

### 7.2.1. *Autografts*

Autogenous bone used as an autograft is generally regarded as the "gold standard", due to its characteristic of being both *osteoconductive* (provides a surface along which bone cells migrate) and *osteoinductive* (stimulates undifferentiated stem cells to change their phenotype to that of bone cells). In other words they can regenerate bone naturally.

However, autografts have two major drawbacks: donor site morbidity, and unpredictable rates of resorption due to biomechanical factors. Therefore, in order to overcome the donor site problems many surgeons have used allogenic bone from bone banks. A danger here is immunorejection and disease transmission.

### 7.2.2. *Allogenic bone grafts*

The work by Urist *et al.* helped the development of a clinically safe allogenic bone for bone banks. The resultant material was known as antigen-extracted autodigested allogenic bone, or so-called AAA bone. This allows the bone to be antigen-low while maintaining high levels of morphogenetic proteins.[1] The Urist method for producing AAA bone is shown in Table 7.1.

**Table 7.1.**   Production of AAA bone (from[1]).

| 1.  Harvest of bone | Rodent, for experimental purposes only; bovine or human, for implantation |
|---|---|
| 2.  1:1 chloroform methanol<br>4 hours @ 25°C | Extract lipids |
| 3.  0.1M phosphate buffer<br>72 hours @ 37°C | Autodigestion of antigens |
| 4.  0.6N hydrochloric acid<br>24 hours @ 2°C | Extract acid soluble proteins |
| 5.  Freeze-drying<br>24 hours @ −72°C | Dehydration, thus preserving residual proteins |
| 6.  Double plastic envelope<br>+ vacuum-sealed glass containers | Prevent rehydration |

Recent papers suggest that AAA bone has variable osteoinductive characteristics and often may only exhibit osteoconductive properties. The reduction of osteoinduction is due to the sterilization process that removes or denatures the bone morphogenetic proteins.[2,3] The AAA bone process is also described as demineralized freeze-dried bone (DFDB). The use of xenografts as DFDB materials has led to a commercial xenogenic product formed from a bovine source, Bio-Oss®.

Both allogenic and xenogenic bone are available but they are expensive to produce, have minimal osteoinductive properties, and both are generally regarded as osteoconductors. They are used sparingly in our surgeries due to the possible infection risk by the transfer of latent virus or prion particles, such as HIV or CJD.

## 7.3. Comparison of Types of Grafts

The following summary of research papers shows the extensive variability of performance of different types of bone graft, when compared with factors such as graft sites, material, techniques, recipients, and recipient sites.

1. Healing in tibial and mandibular defects in rabbits showed that a mixture of AAA bone and hydroxyapatite produced better results than just AAA bone alone, which was better than hydroxyapatite (HA) alone.[4]
2. Healing of critical-sized defects at six months in primate calvarium[5] showed the following:

| | |
|---|---|
| Best | Autografts |
| ↑ | AAA bone |
| ↓ | Bovine BMP in a polylactide carrier |
| Worst | Control site with no graft |

A critical-sized defect is defined as the smallest size of an intraosseous wound that will not heal spontaneously during the lifetime of the individual.[6,7]

3.  Bone augmentation around titanium fixtures in dog mandibles[8] showed the following:

| | |
|---|---|
| Best | Autografts + membrane (Gore-tex®) |
| ↑ | Membrane alone |
| | Autogenous bone alone |
| ↓ | Allograft |
| Worst | Control site with no graft |

4.  Bone healing after sinus lifts in humans showed that the use of autogenous bone had good results with no implant failures within the grafted areas at eight months. Use of demineralized freeze-dried allogenic bone (DFDB) had poor quality bone at 16 months with 25% of the implants lost.[9]

## 7.4.  Limitations of Autogeneous Bone Grafts

Although autografts are the ideal source of bone for surgical repair there are many donor site problems, including:

- Blood loss
- Infection
- Pain
- Loss of mobility, as muscles are removed from the donor site, particularly with iliac crest sites
- Nerve damage, particularly with mandibular and iliac crest donor sites
- Pneumothorax, specifically with ribs as donor sites

### 7.4.1.  *Source of autogeneous bone for grafting*

There are a considerable number of sources of autograft material from a patient:

- iliac crest (anterior or posterior), ribs, calvarium, or mandible (chin or ramus);
- cancellous or cortical bone; or
- membranous (e.g. mandible and calvarium), or endochondral (hip and rib) bone.

The donor site is usually the iliac crest of the pelvis. Another implant (usually synthetic) is needed to fill the space, and recovery time can be lengthy and painful. Rapid resorption of the transplanted bone can occur because its structure is that of pelvic bone, not that of the defect site, and therefore osteoclasts may resorb it. Each of the above grafting alternatives has positive and negative effects on healing rates and ultimate clinical outcome of the bone graft. Despite the above concerns, autografts are regarded as the better option for bone grafts when compared with the various xenografts and allografts cited above. All autograft procedures require a donor site to harvest the bone graft, which results in pain and discomfort for the patient, an increased risk of infection, a second surgical team, and increased costs associated with all aspects of the operation. Thus, any bone grafting procedure has room for improvement with much recent effort being given to alloplastic, synthetic materials.

## 7.5. Need and Criteria for Alloplastic Bone Grafts

Dr Oonishi was one of the first surgeons to recognize the potential value of alloplastic materials in bone grafting as an alternative to autogeneous bone, and pioneered a series of studies that quantified the rate of bone regeneration in the presence of alloplastic particles.[10,11] His research group in Osaka, Japan, studied a large range of alloplastic materials as bone graft substitutes. Professor T. Yamamuro at Kyoto University also recognized the need for alternatives to autogeneous bone in grafting large defects, as discussed in Chapter 8. These materials are designed to provide a base for new bone to grow on, while providing a rapid return of clinical function. The goal is to stop fibrous (scar) tissue growing into a site where bone formation is desirable, such as at a proposed implant site.

These studies have led to a set of design criteria for alloplastic bone grafting materials. The material should:

- Provide an osteoconductive surface.
- Be entirely resorbable within 6–12 months.

- Be dimensionally stable in a bony defect.
- Be dimensionally stable while it is resorbing.
- Be compatible (as should its resorption products) with the host tissues.
- Stimulate formation of new bone with an architecture equivalent to the bone being replaced.

The following features will also be required of the material:

- Ease of application.
- Ability to be sterilized.
- Possible use as a carrier system for an osteoinductive medium.

There are a number of alloplastic materials used clinically in maxillo-facial reconstructions:

- Dense hydroxyapatite (HA), and porous HA, in particles or solid blocks.
- Bioactive glass.
- Collagen meshes.
- Resorbable polymers.

Synthetic hydroxyapatite ($Ca_{10}(PO_4)_6OH_2$) particles and blocks have been routinely used as alveolar ridge augmentation materials.[12,13] They are popular because HA has a similar composition to the apatite in bone and is osteoconductive. However, they have shown a number of problems: particle misplacement, postoperative particle migration into undesirable areas, implant flattening, and excessive post-operative ridge width.[14–16] A further clinical problem noted has been nerve dysesthesia, a change in the ability of the nerve to conduct the impulse which may result in a "pins and needles" sensation or numbness. The dysesthesia has been attributed to particles moving into the mental foramen and pressurizing the nerve.[17]

In an attempt to stop particle migration a number of binding agents have been used in conjunction with the alloplastic materials: plaster of paris, collagen, and fibrin. Another experimental approach has been to use degradable tubes filled with particulate hydroxyapatite. The results in 12 patients show that the tubes become "clinically solid" within 8–16 weeks.[18] It is stated that "the ridges are desirable for denture construction. Although the augmentation becomes clinically solid, it is doubtful if the HA is interdispersed with actual bone."[19]

Due to these variable results the decision was made *not* to use HA based materials for maxillo-facial reconstruction in the clinical trials summarized below. The material chosen for the clinical trials was bioactive glass, because this bioactive material meets all of the above-listed design criteria.

## 7.6. Applications of Bioactive Alloplastic (Synthetic) Powders

A range of bioactive glass and bioactive glass-ceramic powders have been used in clinical practice for over 20 years, and all of these have had generally good clinical and commercial success. The literature shows two established types of bioactive glass and bioactive glass-ceramics that vary in composition, particle size, surface chemistry, and origin of manufacture. Bioactive glass of the original 45S5 (46.1 mol% $SiO_2$, 24.4 mol% $Na_2O$, 26.9 mol% CaO, and 2.6 mol% $P_2O_5$) composition is marketed as Perioglas® and NovaBone® by NovaBone Corp. (Alachua, FL) Cerabone is the trademarked name of AW-GC apatite–wollastonite glass ceramic discussed in Chapter 8 (Table 7.2).

The change in chemical composition of the alloplastic material gives rise to differing rates of osteo-stimulation and mechanical properties; typically, the higher the silica content the slower is the dissolution of the glass structure once implanted into the body, with a resultant lower level of bioactivity and osteo-stimulation. The most bioactive composition, 45S5 Bioglass®, releases a critical concentration of ionic dissolution products as it resorbs. The biologically active soluble silica and calcia ions result in up-regulation of seven families of genes in osteoprogenitor cells and give rise to rapid bone regeneration.[20]

The sites for application of particulate bioactive glass are numerous, from simply filling extracted tooth sockets, thereby alleviating alveolar ridge resorption, to use in the reconstruction of complex

**Table 7.2.**    Bioglass® and A-W glass-ceramic properties compared with human bone.

| Material | Density (g/cm$^3$) | Hardness (Vickers, HV) | Bending strength (MPa) | Fracture toughness ($K_{1C}$) (MPa m$^{1/2}$) | Young's Modulus (GPa) |
|---|---|---|---|---|---|
| Cortical bone | 1.6–2.1 | — | 50–150 | 2–12 | 7–30 |
| Cancellous bone | 1.0 | — | 10–20 | 0.1 | 0.05–0.5 |
| Bioglass® (45S5) | 2.6572 | 458 ± 9.4 | 42 (tensile) | 0.6 | 35 |
| A-W glass-ceramic | 3.07 | 680 | 200–220 | 2.0 | 100–150 |

orthopedic fractures and spinal fusion, as reviewed by the author previously.[21] However, the particulate systems lack some dimensional stability when applied as a large volume of powder and when first placed into the surgical site. Therefore, the powders are typically used in intra-bony cavities, as the "containing" walls of the cavity reduce the creep of the glass particles around the local surgical site due to blood flow and patient locomotion. Histology has shown the glass particles form a biological bond to collagen fibers within 24 hours. After this initial interaction of glass surface and collagen fibers the particulate material becomes very stable; the literature has not reported migration of glass particles after this initial biological bond formation.

A number of composite and paste systems have been developed to minimize the movement of the glass particles within the surgical site.[22] All of the above applications have been approved by governmental regulatory bodies based upon a comprehensive series of *in vitro* and *in vivo* tests. One of the most important *in vivo* tests is the Oonishi model. This test is relatively inexpensive and has the highly important characteristic of being able to quantify comparisons of the rate of osteo-regeneration of differing materials. This model was used in a series of studies by the authors to quantify bone growth response to several alloplastic materials systems.[23] The positive results from these *in vivo* studies provided the basis for ethical committee approval to conduct the clinical trials described herein.

## 7.7. The Oonishi Model: A Critical Size Defect in the Rabbit Femoral Condyle

Holes 6 mm in diameter are drilled bilaterally in the femoral condyles of mature rabbits. Immediately afterward, granules of the alloplastic material to be tested are placed in sufficient amounts to fill the holes in the femoral condyles. Five femoral condyles are used for each time period, and for or a detailed investigation nine time periods are studied: 2, 3, 5 days and 1, 2, 3, 6, 12, and 24 weeks. A less costly preliminary comparison of materials can be done by reducing time periods to one, two, three, and six weeks. Most key differences between materials show within these time periods. Following euthanasia, non-decalcified specimens of the femoral condyles are observed by light microscopy and backscattered scanning electron microscopic (SEM) imaging. Rates of bone in growth from the periphery of the defect to the interior are measured quantitatively using high performance image processing. Details are given in the pioneering references where the Oonishi model is used to compare various bioactive materials.[10] An important finding from these studies is the extensive network of bone bridges that form between bioactive glass particles. The architecture of bone that results is equivalent to that of normal trabecular bone present in the femoral condyles.[11] The mechanical properties of the regenerated bone are also equivalent as shown quantitatively by Wheeler *et al.*'s studies using the Oonishi model.[24]

## 7.8. Clinical Cases

### 7.8.1. *Bioactive glass alloplastic bone grafts for maxillo-facial reconstruction*

One of the more common applications of bioactive glass is the filling of cystic cavities in mandibles. Cystic cavities as large as 15 cm$^3$ are frequently filled with bone chips harvested from the iliac crest plus some additional bone graft substitute material such as 45S5 Bioglass® bioactive glass. The surgical steps used to fill a cystic cavity are as follows.

The site is incised and the cyst is removed from the mandible (Fig. 7.1). The cyst is sent for pathology examination (Fig. 7.2). The

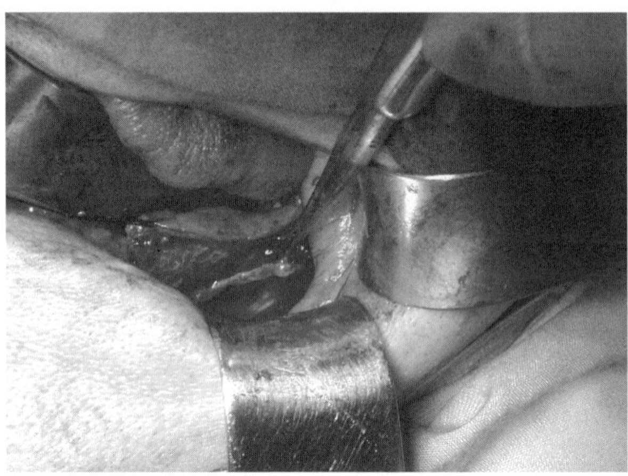

**Figure 7.1.**   Lower left jaw, open surgical site after removal of cyst. Intact nerve.

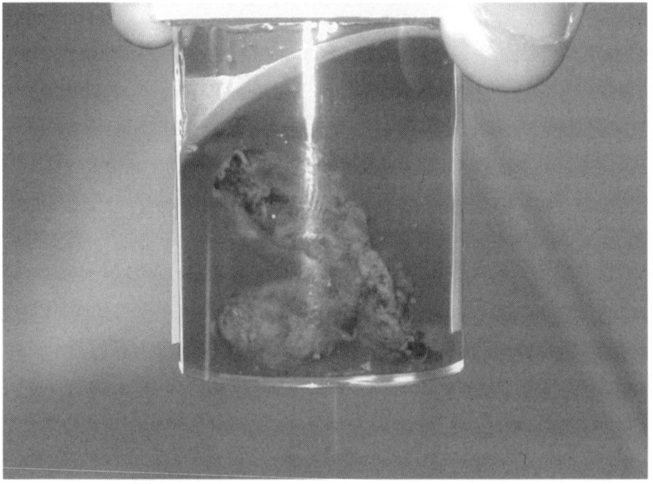

**Figure 7.2.**   Removed cyst (approx. 12cm).[3]

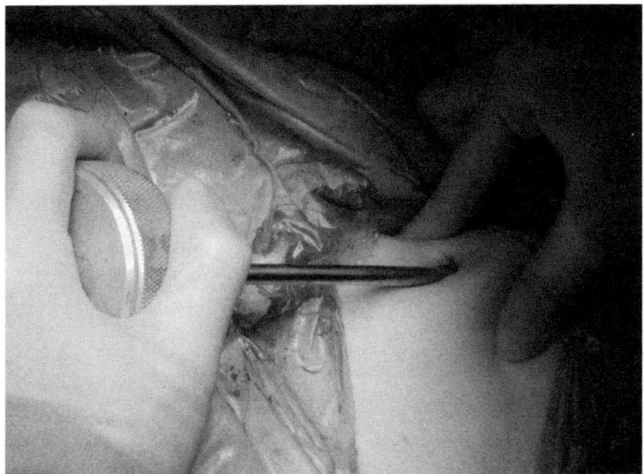

**Figure 7.3.**   Trephines of bone are removed from the iliac crest.

internal cavity is fenestrated with a small burr to allow a small amount of blood to flow back into the cavity. Bone chips are harvested from the iliac crest using a trephine (Fig. 7.3). The chips are crushed using surgical instruments and mixed with bioactive glass particles. A small amount of blood is added to wet the glass particles, thus making the graft material slightly cohesive and easier to place into the site (Fig. 7.4). The glass powder, bone chip, and blood material is placed into the surgical site (Fig. 7.5). The site is closed with sutures (Fig. 7.6).

The immediate post-operative X-ray shows how the cyst has deformed the macroscopic shape of the mandible (Fig. 7.7). After six months of healing the mandible has regained its natural anatomic shape and begun to regenerate the internal trabecular structure (Fig. 7.8).

Clinical comparisons of bioactive glass with or without autogenous bone in a similar model to that described above has shown there is no reduction in healing time when only the alloplastic material is used. Thus, grafting clinical sites with only bioactive glass particles does not require additional autogenous bone to heal a bony defect. This is an important finding since it eliminates problems of second site morbidity, cost, time, and pain.

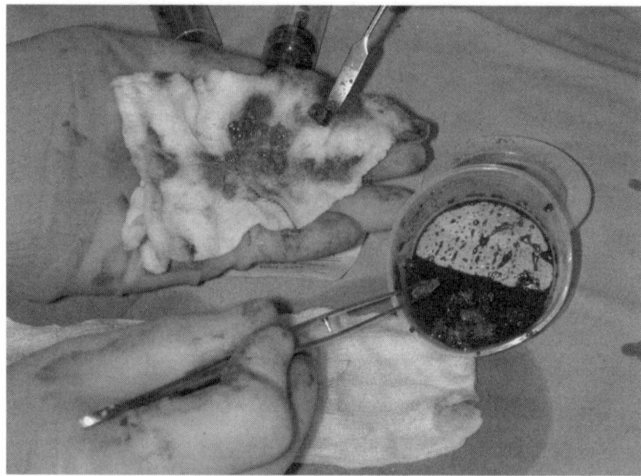

**Figure 7.4.**    Trephines of bone are crushed, mixed with bioactive glass, and blood.

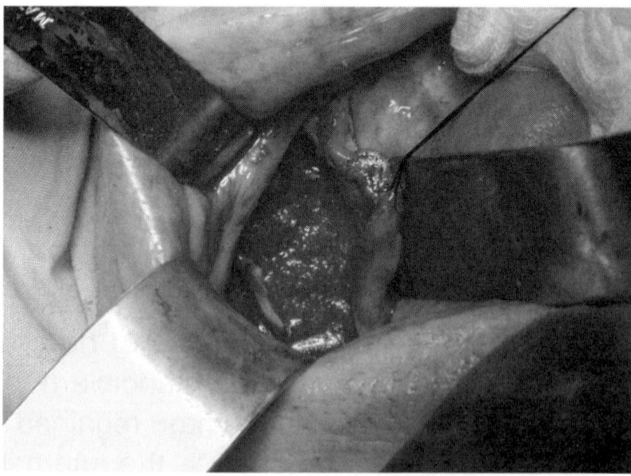

**Figure 7.5.**    The graft materials are placed into the surgical site.

The positive findings of successful use of bioactive glass particulate alone in a graft site indicates that if there is a sufficient supply of blood then there will be a cascade of healing inductive factors such as TGF-β delivered to the site, thereby eliminating the need for bone morphogenetic proteins and other inductive factors being

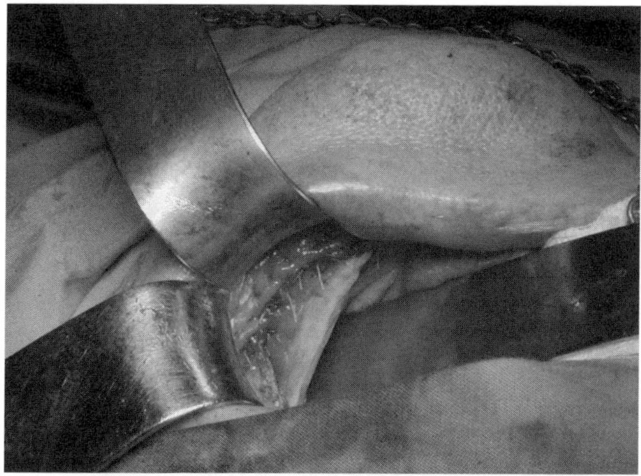

**Figure 7.6.** Site is closed with sutures.

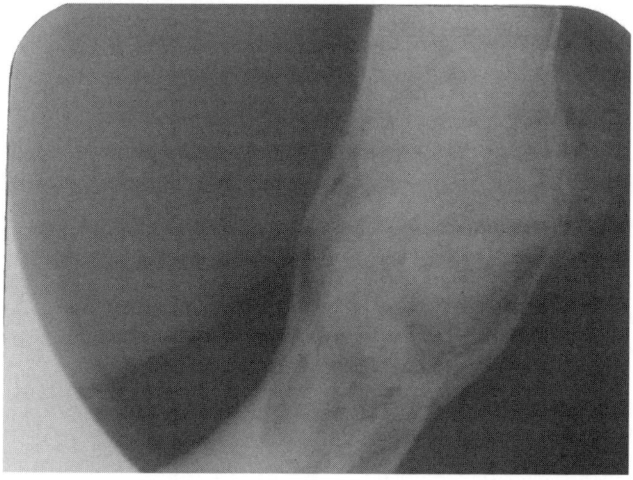

**Figure 7.7.** Post-operative X-ray: cystic activity has deformed the mandible.

released from an autogenous bone graft. These clinical conclusions are consistent with the *in vitro* findings of gene up-regulation of osteoprogenitor cells induced by bioactive glasses (Chapter 3).

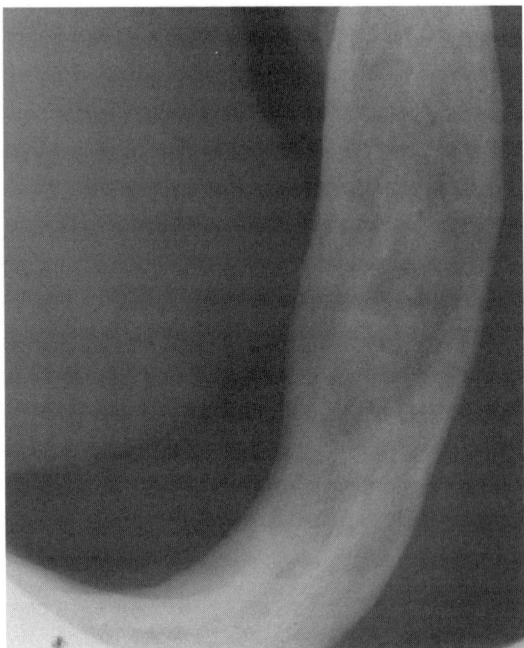

**Figure 7.8.**    Six months post-operative X-ray: remodelling of mandible.

## 7.8.2. *Monolithic bioactive glass implants combined with bioactive alloplastic powders*

Bulk pieces of bioactive glass have been clinically successful when used as middle ear prosthesis, where they were used to replace the small bones (ossicles) in the middle ear to restore hearing; in dental ridge maintenance implants; and in non-load bearing facial skeletal reconstruction. See Chapter 3 for details and references.

Clinicians have generally moved away from using bulk bioactive glass implants due to the difficulty in altering the shape of the implant to fit the patient during surgery. Recently the work of one of the authors at Guy's Hospital London has shown that prefabrication of bulk bioactive glass implants using computer tomography (CT) scans to create accurate molds has led to a good clinical outcome and makes the implants much easier for surgeons to implant without modification of the device. The CT scans of the patient provide

3D images of the defect site. The images can be used as data input into computer aided design (CAD) files that can be used to machine custom molds for the casting of the Bioglass®. The dimensional accuracy leads to improved contact between the implant and host tissue resulting in more rapid site stability post surgery. There is also less surgical trauma to the patient when the implant is custom made to fit the surgical site.

The following describes a case of a patient who suffered an orbital floor trauma due to a road traffic accident. The resulting hypoglobus left the patient with blurred vision, which would increase to permanent blindness if not relieved. Initial attempts to use an autogeneous bone graft from the iliac crest failed as the material resorbed away after six months. The clinician's previous experiences led porous polyethylene to be considered to be an

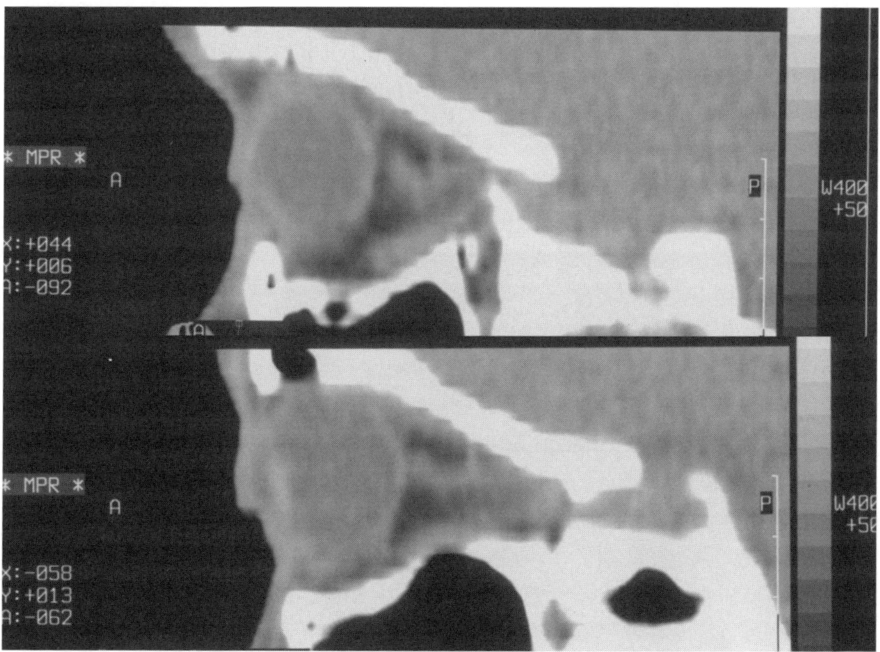

**Figure 7.9.** (Top) Collapsed orbital floor, globe has fallen back into skull so kinking optic nerve. (Bottom) Normal orbital floor.

infection risk. Consequently local ethical permission was granted to cast a bioactive glass implant from rapid prototyped molds. Figure 7.9 shows the CT data of the affected site and the unaffected site (note the change in orbital floor height and contour). Using CT data a dimensionally accurate mold was created that allowed the implant to be cast with sufficient contour to mimic the correct orbital floor anatomy. The molten 45S5 glass implant was cast and subsequently annealed (reheated) to remove any thermal stresses. Samples were taken for mechanical and chemical testing.

The surgical site was opened and the implant positioned onto the collapsed orbital floor (Fig. 7.10).

Powdered bioactive glass was packed around the implant to improve the contact between the collapsed bony orbital floor and the implant's lower surface. The implant was secured into the site with degradable sutures. A postoperative X-ray shows the location of the glass implant and the regained symmetry between the left and right orbital floor positions (Fig. 7.11). A recent five-year follow-up has shown the patient has regained full movement of the eye and lost all complications from blurred vision. The patient is also happy with the aesthetic look of his face due to correction of the hypoglobus.

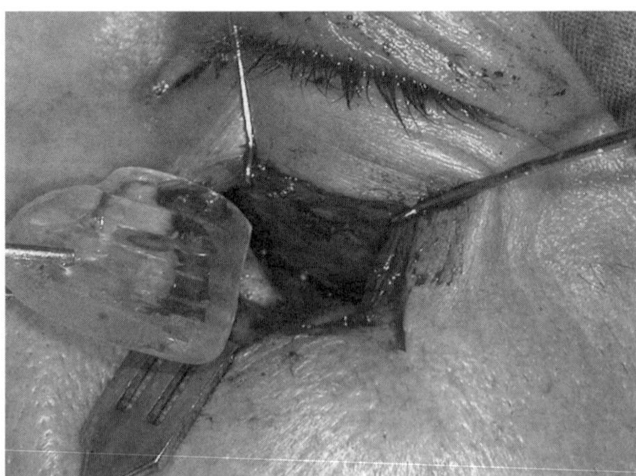

**Figure 7.10.**    Placement of bioactive glass monolith in surgical site.

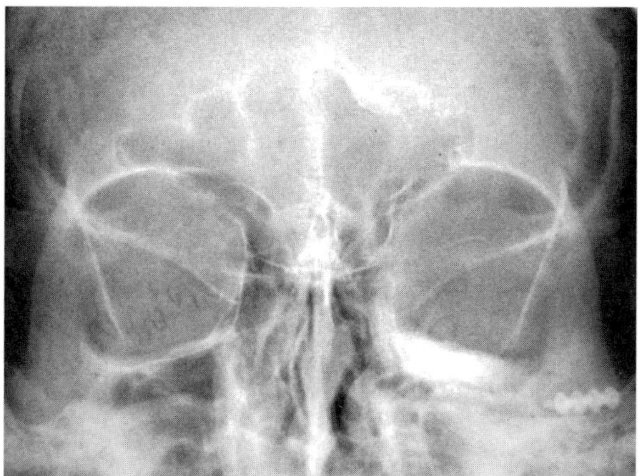

**Figure 7.11.**   Post-operative X-ray, showing bioactive glass implant. Note: new orbital floor height is symmetrical to unaffected side.

This case is part of a 30-patient trial at Kings College/Guy's Hospital, London. The trial's preliminary findings show zero post-operative infection, and zero loss of implants from migration or any other factors. The patients are happy with the regained function of their damaged orbital floors, and also with correction of their facial symmetries.

## 7.9. Conclusion

Bioactive glass has been and continues to be a significantly effective bone graft substitute material when used as a particulate. The published clinical data support the long-standing claims that the bioactive glass is more than just a simple osteoconductive material. Its use in a range of clinical applications has only been compromised when the glass is over-packed into a site, thus making it difficult for in-growing bone to colonize the mass of glass particulate. The glass also has a number of applications when cast as a block or in a monolithic shape, and is particularly effective when used in combination with particulate grafts. Use of CT scans to create highly dimensionally accurate molds for casting bulk bioactive glass implants makes it possible to achieve rapid surgical reconstruction of large-scale defects. Particulate bioactive glass particles used in conjunction with the accurate castings are especially effective clinically, have lower costs, and minimize pain and trauma to the patient.

## References


1. Urist, M.R., Mikulski, A. and Boyd, S.D. (1975). A chemosterilized antigen extracted autodigested alloimplant for bone banks, *Arch. Surg.*, **110**, 416–427.
2. Sennerby, L. and Lundgren, S. (1999). "Histologic Aspects of Simultaneous Implant and Graft Placement", in Jensen, O.T. (ed.), *The Sinus Bone Graft*, Quintessence Publishing, Chicago, IL, pp. 95–105.
3. Aspenberg, P., Kalebo, P. and Alberktsson, T. (1988). Rapid bone healing delay by bone matrix implantation, *Int. J. Oral Maxillofac. Implants*, **3**, 123–127.
4. Oberg, S. and Rosenquist, J.B. (1994). Bone healing after implantation of hydroxyapatite Granules and Blocks (Interpore 200®) combined with autolyzed antigen extracted allogenic bone and fibrin glue, *Int. J. Oral. Maxillofac. Surg.*, **23**, 110–114.
5. Hollinger, J.O., Schmitz, J.P., Mark, D.E., *et al.* (1990). Osseous wound healing with xenogenic bone implants with a biodegradable carrier, *Surgery*, **107**, 50–54.
6. Schmitz, J.P. and Hollinger, J.O. (1986). The critical size defect as an experimental model for craniomandibulofacial nonunions, *Clin. Orthop. Rel. Res.*, **205**, 299–308.
7. Hollinger, J.O. and Kleinschmidt, J.C. (1990). The critical size defect as an experimental model to test bone repair materials, *J. Craniofac. Surg.*, **1**, 60–68.
8. Jensen, O.T., Greer R.O., Johnson, L., *et al.* (1995). Vertical guided bone grafts augmentation in a new canine mandibular model, *Int. J. Oral Maxillofac. Implants*, **7**, 62–71.
9. Nishibori, M., Betts, N.J., Salama, H., *et al.* (1994). Short term healing of autogenous and allogenic bone grafts after sinus augmentation: a report of 2 cases, *J. Periodontotol.*, **65**, 958–966.

10. Oonishi, H., Hench, L.L., Wilson, J., *et al.* (1999). Comparative bone growth behaviour in granules of bioceramic materials of various sizes, *J. Biomed. Mater. Res.*, **44**, 31–43.

11. Oonishi, H., Hench, L.L., Wilson, J., *et al.* (2000). Quantitative comparison of bone growth behaviour in granules in Bioglass®, A-W glass ceramic, and hydroxyapatite, *J. Biomed. Mater. Res.*, **51**, 37–46.

12. Kent, J.N., Quinn, J.H., Zide, M.F., *et al.* (1982). Correction of alveolar ridge deficiencies with nonresorbable hydroxyapatite, *J. Am. Dent. Assoc.*, **105**, 993–1001.

13. Griffiths, G.R. (1985). New hydroxyapatite ceramic materials: potential use for bone induction and alveolar ridge augmentation, *J. Prosthet. Dent.*, **53**, 109–114.

14. Silverberg, M., Singh, M., Gans, B., *et al.* (1986). Polyglycolic acid mesh contained hydroxylapatite for augmentation of bone in the rat, *Proc. 9th Int. Conf. Oral. Maxillofac. Surg.*, **145**, 52–56.

15. Gongloff, R.K., Montgomery, C.K., Lee, R., *et al.* (1986). Collagen tubes: role in subperiosteal contour augmentation, *Int. J. Oral. Maxillo. Surg.*, **15**, 669–674.

16. Barsan, R.E. and Kent, J.N. (1985). Hydroxylapatite reconstruction of alveolar ridge deficiency with an open technique — a preliminary report, *Oral Surg.*, **59**, 113.

17. Gongloff, R.K. (1988). Use of collagen tube contained implants of particulate hydroxyapatite for ridge augmentation, *J. Oral and Maxillofac. Surg.*, **46**, 641–647.

18. Collins, T.A. (1989). Use of collagen tubes containing particulate hydroxlapatite for augmentation of the edentulous atrophic maxilla: a preliminary report, *J. Oral Maxillofac. Surg.*, **47**, 137–141.

19. Blonk, M.S., Kent, J.N., Ardoin, R.C., *et al.* (1987). Mandibular augmentation in dogs with hydroxylapatite combined with demineralized bone, *J. Oral Maxillofac. Surg.*, **45**, 414–420.

20. Hench, L.L. (2003). Glass and genes: the 2001 W.E.S. Turner memorial lecture, *Glass Tech.*, **44**, 1–10.

21. Thompson, I. and Hench, L. (2000). "Medical Application of Composites", in Kelly, A. and Zweben, C. (eds), *Comprehensive Composite Materials*, vol. 6, Elsevier Science, Oxford, UK, pp. 727–753.

22. Thompson, I. and Hench, L.L. (1998). Mechanical properties of bioactive glasses, glass–ceramics and composites, *P. I. Mech. Eng. H.*, **212**, 127–136.
23. Chan, C., Thompson, I., Robinson, P., *et al.* (2002). Evaluation of a Bioglass®–dextran composite as a bone graft substitute, *Int. J. Oral Maxillofac. Surg.*, **31**, 73–77.
24. Wheeler, D.L., Eschbach, E.J., Hoellrich, R.G., *et al.* (2000). Assessment of resorbable bioactive material for grafting of critical-size cancellous defects, *J. Orthop. Res.*, **18**, 140–148.

# Clinical Applications of Bioactive Glass-Ceramics



## CHAPTER 8

Takao Yamamuro

## 8.1. Introduction

This chapter reviews my 30 years experience as an orthopedic surgeon in the development and clinical use of bioactive glass-ceramics, especially A-W (apatite-wollastonite) bioactive glass-ceramic. My first clinical use of a synthetic bone graft occurred in 1981. The patient was a 27-year-old female who had developed a large giant cell tumor in the right ilium (pelvis). At operation, the tumor was resected through a window made in the lateral wall of the ilium. The bone defect remaining in the pelvic bone was too large to be filled completely with autogeneous bone. In order to fill the defect, a mixture of porous and granular synthetic hydroxyapatite ceramic (HA) was prepared and mixed with fibrin glue and autogeneous bone chips. The mixture of autogeneous and synthetic graft was compactly filled into the bone defect. A radiograph taken post surgery showed a radio-lucent line between the implanted graft and the pelvic bone. A radiograph taken after eight months showed that the radiolucent line persisted. At ten months a biopsy of the implanted mass was performed and histology showed that the implanted HA granules had united with newly formed bone. A radiograph taken at 12 months postoperatively showed no radiolucent line between graft and pelvic bone. The patient was permitted to bear full body weight on the affected side.

## 8.2. Clinical Conclusions: HA Bone Graft

The clinical conclusions are:

1. It took one year for complete filling of the gap between implant and bone, and bone bonding with the granules of HA. A radiograph taken after 15 years showed a stable graft of HA incorporated into the living bone. The patient has continued to live a normal life for more than 30 years.
2. HA does bond to bone but with a low level of osteoconductivity.
3. A graft of HA granules is too weak mechanically to be used for conditions of loading of full body weight.
4. A more bioactive and stronger bone substitute than synthetic HA ceramic is needed for orthopedic applications.

## 8.3. Development of Bioactive Apatite–Wollastonite Glass-Ceramic (AW-GC)

A new bioactive glass-ceramic composed of apatite crystals (A) for bioactivity and wollastonite (W) crystals for strength was developed at Kyoto University.[1–3] The material is prepared by hot pressing powders to form a very high strength dense polycrystalline ceramic with glassy phase bonding. See Chapter 7 for a table of composition of AW-GC. Chapter 5 of reference[2] by Professor T. Kokubo documents development of the processing method and physical properties and *in vitro* bioactivity testing of AW-GC. Chapter 6 of reference[2] by the author (Professor T. Yamamuro) summarizes clinical applications of AW-GC up to the year 1993.

*In vitro* studies in simulated body fluid showed that AW-GC developed a surface layer of hydroxycarbonate apatite (HCA) from the Ca and P ions in the body fluid within seven days, whereas it took nearly 30 days for a similar layer to form on a synthetic HA ceramic. A comparison of 45S5 Bioglass® in the same *in vitro* test model showed formation of an HCA layer from the ions of the body fluid within 24 hours, as described in Chapter 3. Also, an *in vivo* study in rabbits at Kyoto University showed that bone bonding strength of AW-GC at 8 weeks after implantation was equivalent to synthetic HA ceramic at

**Table 8.1.** Mechanical properties of A-W glass-ceramic compared to synthetic HA and cortical bone.

| Material | Bending strength (MPa) | Compressive strength (MPa) | Young's modulus (GPa) |
|---|---|---|---|
| Cortical bone | 30–190 | 90–230 | 7–30 |
| Synthetic HA | 110–170 | 500–900 | 35–120 |
| A-W glass-ceramic | 200–220 | 1000 | 350–380 |

25 weeks. Bone bonding to 45S5 Bioglass® was more rapid than to AW-GC but the mechanical strength of Bioglass® was considered too weak to be used as a bone substitute in clinical applications.

The bending strength, compressive strength, and fracture toughness of AW-GC are all significantly higher than those of human cortical bone, as illustrated in Table 8.1. In contrast, the mechanical properties of synthetic HA ceramic are lower than human cortical bone.

The successful mechanical properties and bone bonding ability of AW-GC led to the decision to make bone spacers of the material that would be suitable for weight-bearing clinical applications. In order to establish both safety and efficacy an animal model was designed and used for testing of the prototype AW-GC device. The model was a vertebral prosthesis used to replace L3 and L4 lumbar vertebrae of sheep. Specimens harvested at six months showed bone bonding to trabeculae all the way around the prosthesis, as shown in Fig. 8.1. These findings confirmed earlier studies of bone bonding and high strength of bonding to bone in a rabbit model, as summarized in references.[1–3] Based upon these findings summarized in references,[2–7] vertebral prostheses were made of AW-GC in different sizes for clinical use. Figure 8.2 shows photographs of these devices.

## 8.4. Case Histories

### 8.4.1. *Vertebral prostheses*

During the period from 1983–1994, when the author retired from Kyoto University, the vertebral prosthesis made of AW-GC had

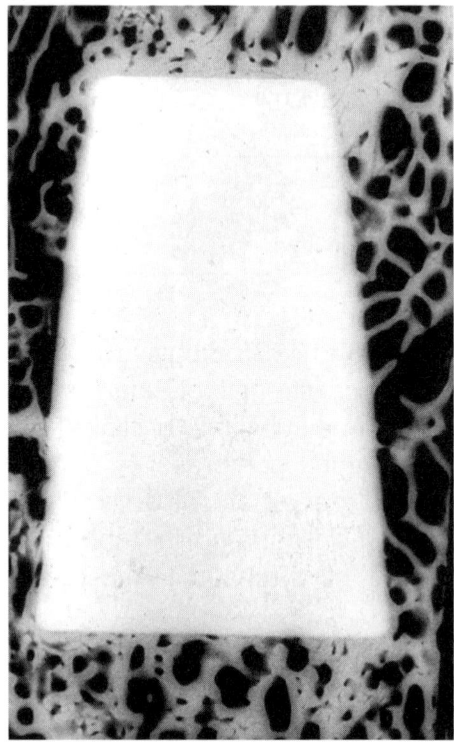

**Figure 8.1.**    AW-GC vertebral prosthesis tested in lumbar vertebrae of a sheep. Contact X-ray micrograph shows bonding of bone trabeculae to the prosthesis by six months.

been used in 1,070 cases in Japan. The device was used to treat bone tumors, burst fracture, and fracture dislocation of the spine. By 2010 the AW-GC vertebral prosthesis had been used to treat approximately 3,000 cases. There have been no serious complications reported (such as loosening, collapse, or breakage of the prosthesis). Figure 8.3 illustrates the progressive growth of bone around an AW-GC vertebral prosthesis.

## 8.4.2. *Intervertebral spacers*

During the period from 1989 to 1994 intervertebral spacers made of AW-GC were used for 1,005 lumbar spine fusion cases. By 2010

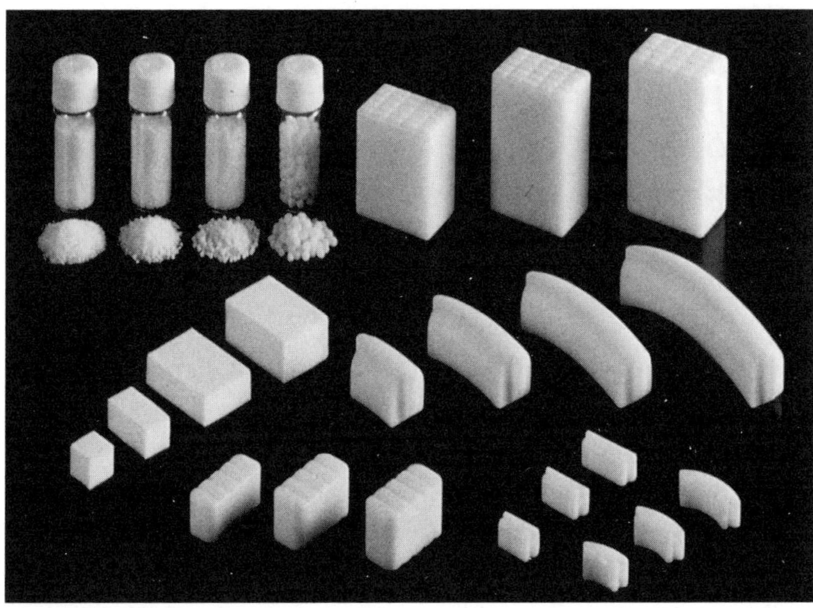

**Figure 8.2.**   AW-GC prostheses and granules used for various orthopedic applications.

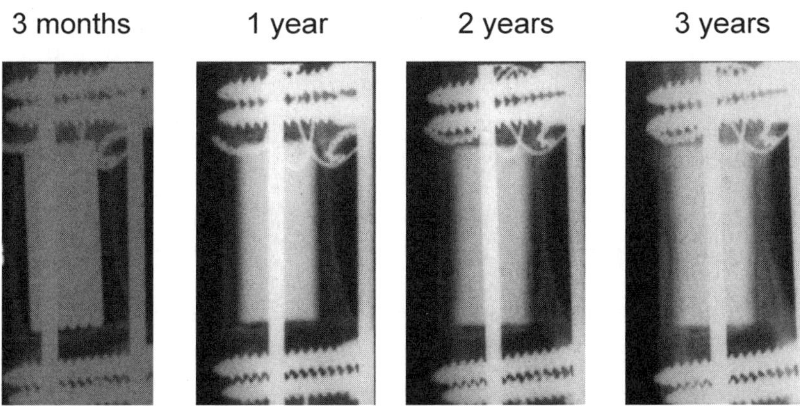



**Figure 8.3.**   Time course of radiological changes around an AW-GC vertebral prosthesis observed in a young female. The patient was treated for a metastasis of alveolar soft part sarcoma in L3 vertebra, which was removed prior to implantation of AW-GC. No recurrence has been observed for 20 years, and the bone trabeculae around the prosthesis became thicker with time.

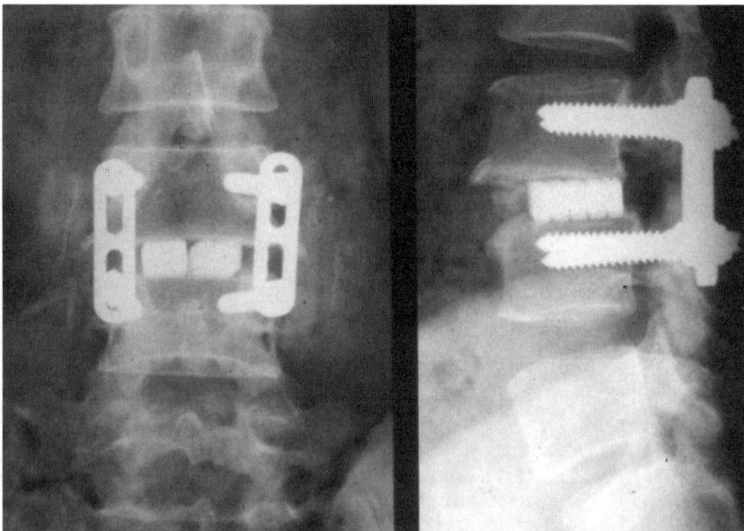

**Figure 8.4.**   Intervertebral spacers of AW-GC used to treat a 40-year-old patient with isthmic spondylolisthesis causing severe low back pain and neuralgia in the leg. The spacers had an osteoconductive effect.

the number of cases had reached approximately 5,000. The AW-GC spacer has an osteoconductive effect as shown by radiological follow-up studies. No case has shown late narrowing of the inter-vertebral space, often observed when bone autograft or allograft is used for the interbody fusion. Figure 8.4 shows a clinical case of use of AW-GC intervertebral spacers.

### 8.4.3.   *Laminoplasty spacer*

Laminoplasty spacers were made of AW-GC to be used for lamino-plastic enlargement in cases of multiple spinal canal stenoses of the cervical spine. Typically, four to five laminae are enlarged and bone defects remaining in each lamina are filled with the spacer that is fixed to the spinous process with threads. In most cases, bone bond-ing of the spacer was observed on CT images within six months of surgery resulting in excellent stability of the spacer. The advantages of using the AW-GC spacer in laminoplastic enlargement are: 1) no

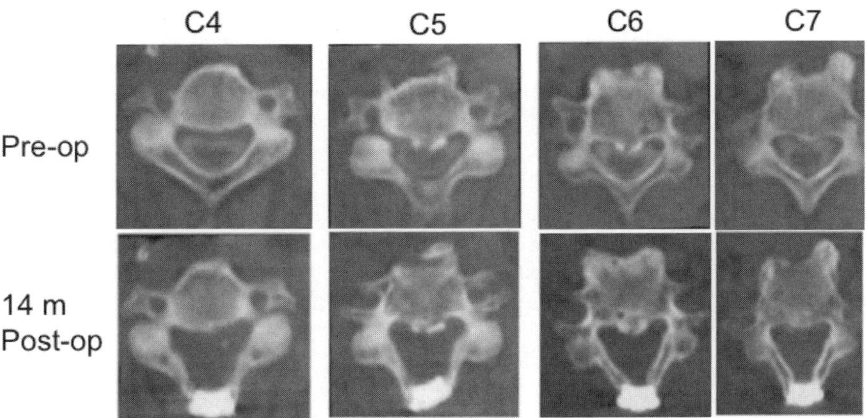

**Figure 8.5.**   AW-GC used as laminoplasty spacer to enlarge the spinal canal for treatment of a 75-year-old male suffering from tetraplegia due to ossification of the posterior longitudinal ligament from C4 to C7. All of the spacer has united to the adjacent laminae by 14 months postoperatively.

surgical intervention for harvesting an autograft is required; and 2) late collapse of the implant never occurs, as can be the case with autograft.

From 1988 to 1994 the laminoplasty spacer made of AW-GC was used in 778 cases in Japan. By 2010 the number of cases had reached 12,000. This operation is considered to be the best indicated for myelopathy caused by multi-level spinal canal stenosis of the cervical spine. Figure 8.5 illustrates use of the AW-GC laminoplasty spacer in a 75-year-old male with tetraplegia.

### 8.4.4. *Iliac crest prosthesis*

Harvesting of autograft from the iliac crest is often necessary. The large bone defect remaining following the removal of the bone can create various cosmetic and neurological problems, especially chronic pain. Iliac crest prostheses made of AW-GC were used to fill the bone defect in 4,113 cases during the period from 1987 to 1994. By 2010 the number of cases had reached approximately 20,000. It is reported that 97% of the patients are satisfied with

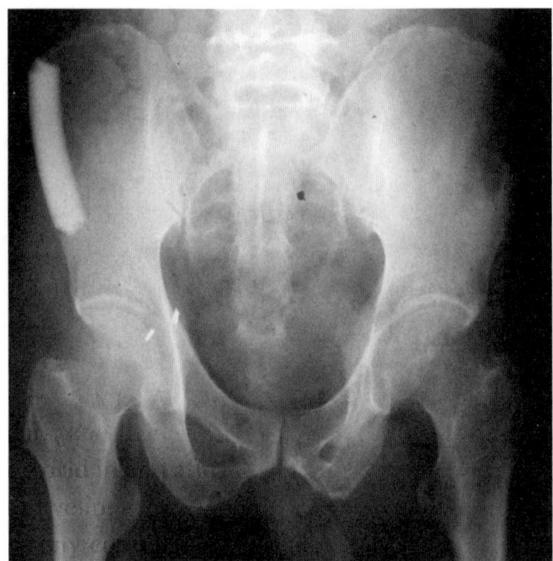

**Figure 8.6.** Replacement of a large bone defect in the iliac crest with an AW-GC prosthesis after harvest of an autograft.

the iliac crest prosthesis. Figure 8.6 shows a typical clinical case using AW-GC as iliac crest prosthesis.

### 8.4.5. *Bone defect spacers*

Removal of large bone tumors developed in weight-bearing locations of long bones requires a strong, tough bone bonding material to replace the lost bone. AW-GC large bone spacers have been used to fulfill this need. Clinical results after four years show satisfactory weight bearing and function of the long bone. These findings show that AW-GC can be used clinically as a bone substitute in locations where synthetic HA implants cannot be used.

### 8.5. Conclusion

Twenty-seven years of clinical use of bioactive apatite-wollastonite glass-ceramic (AW-GC) in thousands of patients has demonstrated excellent success. The concept of bioactive bone bonding of orthopedic prostheses has revolutionized the long-term treatment of many ailments.

## References


1.  Kokubo, T., Ito, S., Sakka, S., *et al.* (1986). Formation of a high-strength bioactive glass–ceramic in the system $MgO$-$CaO$-$SiO_2$-$P_2O_5$, *J. Mater. Sci.*, **21**, 536–540.
2.  Yamamuro, T. (1993). "A/W Glass-Ceramic: Clinical Applications", in Hench, L.L. and Wilson, J. (eds), *An Introduction to Bioceramics*, World Scientific, Singapore, pp. 89–104.
3.  Hench, L.L. (1998). Bioceramics, *J.Am.Ceram. Soc.*, **81**, 1705–1728.
4.  Yamamuro, T., Shikata, J., Kakutani, Y., *et al.* (1988). "Novel Methods for Clinical Applications of Bioactive Ceramics", in Ducheyne, P. and Lemons, J.E. (eds), *Bioceramics: Material Characteristics Versus In Vivo Behavior*, New York Academy of Science, New York, NY, pp. 107–114.
5.  Yoshii, S., Kakutani, Y., Yamamuro, T., *et al.* (1988). Strength of Bonding Between A-W glass–ceramic and the surface of bone cortex, *J. Biomed. Mater. Res.*, **22**, 327–338.
6.  Kitsugi, T., Yamamuro, T. and Kokubo, T. (1989). Bonding behavior of a glass-ceramic containing apatite and wollastonite in segmental replacement of the rabbit tibia under load-bearing conditions, *J. Bone Joint Surg. Am.*, **71**, 264–272.
7.  Neo, M., Kotani, S., Nakamura, T., *et al.* (1992). A comparative study of ultrastructure of the interfaces between four kinds of surface-active ceramic and bone, *J. Biomed. Mater. Res.*, **26**, 1419–1432.

# Hierarchical Porous Scaffolds for Bone Regeneration

CHAPTER **9**

Julian R. Jones

## 9.1. Introduction

Regenerative medicine is the future for bone repair, but in some ways it is already the present. Bone can heal itself if the defect is small, but otherwise it needs assistance. In Chapters 1–3 stem cell therapy and using materials to guide stem cell behavior are discussed. Annually, orthopedic and neurological surgeons perform half a million bone graft operations in the United States and 300,000 operations in Europe. In the majority of those procedures they are treating bone defects caused by trauma, tumor removal or non-union of fractures, using temporary templates that stimulate the body's own natural regenerative processes. However, the scaffold is the patient's own bone, taken from the pelvis; in other words, an autograft. Actually, it is a ready-made tissue-engineered living construct of a porous scaffold (a composite of hydroxycarbonate apatite and collagen) containing osteoprogenitor cells and other adult stem cells.

Unfortunately there are many drawbacks to using the patient's bone, some of which are obvious, some a little less so. The most important is that there is only so much bone you can take from places that do not really need it. The body is not wasteful in bone production. In most places in the body it is needed for bearing loads such as body weight. If bones are not loaded, the body tends to take the bone away through remodelling by osteoclast action. The pelvis is the most common site that can be a bone donor. This requires surgeons to carry out two operations to heal one defect; in fact surgeons usually work in pairs, with one operating on the pelvis and one working on the original defect. This doubles the resources needed for the operation and of course the cost. More importantly, removing bone from the pelvis creates another defect that needs to be healed (without the help of more bone). The healing of this defect is extremely painful and lengthy. Patients generally feel more pain at the donor site than at the original defect and one in four patients will experience complications at the defect site even 12 months after the operation. Some will require further surgical intervention, which is not ideal for the patient, the health-care service or the economy.

Artificial bone graft materials are needed that can regenerate bone defects without the need for graft operations so that patients can heal in a pain-free manner and return to their normal lives more quickly, including those still of a working age who can return to work more rapidly. Bone can also be sourced from bone banks, but the bone in these banks is from cadavers and has to be irradiated to remove biological molecules that would otherwise transmit disease or trigger immunorejection by the patient, and the irradiation reduces the mechanical properties of the bone. An artificial bone graft that can replace the need for grafts would have a massive impact on the global economy. The device market itself is thought to be worth US\$2 billion/year without taking into account the economic impact of reduced operating costs and faster recovery times. However there are differences in the engineering design criteria for an artificial bone graft and a scaffold for bone regeneration. This is because many artificial bone grafts are designed to augment the bone and stay there for a long time rather than regenerate the bone to its original state and function. Depending on the design and the claims a company can make when marketing the materials, there are also differences in the regulatory procedures that have to be undertaken.

## 9.2. An Ideal Scaffold

An ideal scaffold to regenerate a bone defect and leave no trace of an implant must possess the following characteristics:[1]

1. be biocompatible and bioactive, promoting osteogenic cell attachment and bone formation, and bond to the bone without scar tissue sealing it off from the body;
2. act as a template for bone growth and therefore have an interconnected porous structure that can allow bone in-growth and vascularization;
3. resorb safely in the body and have a controllable degradation rate;
4. exhibit mechanical properties similar to that of the host bone;
5. have a fabrication process which allows the scaffold to be shaped to fit a range of defect geometries and be up-scalable for mass-production; and
6. be sterilizable and meet the regulatory requirements for clinical use.

Current products are far from this ideality. Bone defect fillers need not be scaffolds; particles and granules are often preferred by orthopedic surgeons for their ease of handling. If the material is designed to augment bone for long periods it should not degrade but can be remodelled by osteoclasts, which will take many months or perhaps even years. Currently there are no tissue-engineering constructs available as commercial products, largely due to the amount of time and investment needed to take the products to market. Surgeons are very aware of the potential benefits of tissue-engineering strategies, such as the fact that osteoprogenitor cells cultured on a template would enhance bone production rate, so they do what they can by aspirating blood and bone marrow from within the bone defect and its surroundings. They then mix the blood and marrow with synthetic bone graft particles or granules to form a putty-like mixture that can be press-fitted into bone defects. The aim is that the putty can be easily handled and the blood and bone marrow may contain osteoprogenitor cells that will activate when re-implanted.

## 9.3. Commercially Available Synthetic Bone Grafts

There are numerous synthetic bone graft materials as commercial products, but none that fulfill all the criteria listed above. Bioceramics are usually chosen because they can be bioactive, but they are brittle materials. Some porous metallic constructs are available (e.g. titanium alloys), but they are not bioactive or degradable. Polymers can be tough but they have poor compressive strength, and are not bioactive unless expensive biological growth factors are incorporated; thus, there are no polymeric synthetic bone grafts. Composites would provide a combination of bioactivity and compressive strength from bioceramics and toughness from biodegradable polymers; however, it is difficult to synthesize composites in such a way that the bioceramics are not masked by the polymer. Consequently, most commercial bone graft substitutes are bioactive ceramics.

### 9.3.1. *Commercial bioactive glasses*

Chapter 3 shows that 45S5 Bioglass® has considerable use in bone regeneration because it is arguably the most bioactive material,[2] bonding to bone rapidly and stimulating bone growth by its dissolution products. However there are relatively few clinical products based on this material. What are the reasons? Some reasons are related to business and marketing, others are due to scientific factors. Considering the scientific factors, all the processing techniques for making porous scaffolds from particles involve sintering, the fusion of particles at high temperature. Raising the temperature above the glass transition temperature $(T_g)$ causes local flow of the glass and allows particles to fuse. However, to maintain the amorphous glass structure and properties, the temperature must not be heated above the crystallization temperature $(T_{c\ onset})$. Unfortunately for the original Bioglass® composition (46.1% $SiO_2$, 24.4% $Na_2O$, 26.9% $CaO$, and 2.6% $P_2O_5$, in mol %) the low silica content causes $T_g$ and $T_{c\ onset}$ to be too close together, so it is not possible to sinter the glass without crystallizing, which leads to formation of a glass-ceramic and reduction in bioactivity.

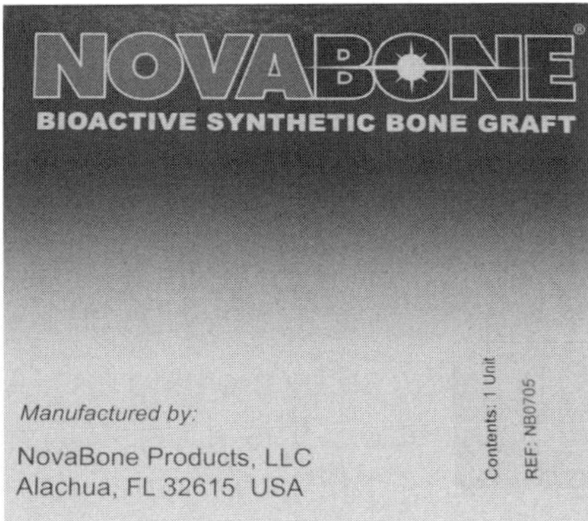

**Figure 9.1.**   NovaBone® (a) packaging, (b) electron microscope image of Bioglass® particles.

Therefore, only Bioglass® particles are available commercially as Perioglas® and NovaBone® from NovaBone Products LLC (Alachua, FL). Perioglas® comprises particles used for filling bone defects in the jaw following periodontal disease, whereas NovaBone® is an orthopedic bone defect filler (Fig. 9.1). The particle size range is 90–710 µm. Biogran®, also made of Bioglass® granules, was marketed by Orthovita Inc. Finer ($< 5$ µm) particles are produced by NovaMin LLC (Alachua, FL) for use in toothpaste (e.g. Sensodyne Repair and Protect) to promote mineralization of microtubules in enamel, with the aim of reducing tooth sensitivity, discussed in Chapter 32.

Other slight variations on the Bioglass® composition have been released as commercial products, such as BonAlive® ($SiO_2$ 53 wt%, $Na_2O$ 23%, $CaO$ 20%, and $P_2O_5$ 4%) by BonAlive Biomaterials Ltd (Finland) and StronBone® (44.47 mol% $SiO_2$, 27.26% $Na_2O$, 21.47% $CaO$, 2.39% $SrO$, and 4.42% $P_2O_5$) by RepRegen Ltd (London, UK). The potential advantage of StronBone® is the inclusion of small amounts of strontium which is thought to help fight osteoporosis.

### 9.3.2. *Commercial synthetic hydroxyapatites*

As bone mineral is a hydroxyapatite, an obvious choice of material for a synthetic bone graft is a synthetically produced apatite. Many synthetic bone grafts are made from synthetic hydroxyapatite (HA), chemical formula $Ca_{10}(PO_4)_6OH_2$. However, bone mineral is a carbonated hydroxyapatite, approximated by $(Ca,X)_{10}(PO_4,HPO_4,$ $CO_3)_6(OH,Y)_2$, where X is a cation, either a magnesium, sodium or strontium ion that can substitute for the calcium ions, and Y is an anion (a chloride or fluoride ion) that can substitute for the hydroxyl group.[3]

Conventional synthetic HA (sHA) is produced by creating an HA powder by solution chemistry, which is sintered. It is a crystalline ceramic and stoichiometric HA (Ca/P ratio of 1.67) commonly produced by precipitation by reacting calcium hydroxide with orthophosphoric acid solution (at pH > 9) at temperatures of between 25 and 90°C.[4] The first clinical use of sHA was in 1978 when dense sintered hydroxyapatite cylinders were used as immediate dental root implants after tooth extraction.[5] Chapter 8 by Professor T. Yamamuro reports early use (1981) of synthetic HA in orthopedic surgery.

Several methods have been used to produce porous sHA. The simplest way to generate porous scaffolds from ceramics, such as HA, is to sinter spherical particles. Porosity is often increased by adding sacrificial porogens, which are often particles that will be removed after compaction or during sintering. The sacrificial particles can either be soluble, such as salt or sucrose,[6] or combustible, for example PMMA microbeads. However, these methods give rise to a heterogeneous pore distribution and the interconnectivity of the pores is low. To improve interconnectivity, open-celled polyurethane foams can be immersed in slurries of sHA. The foams are then heated at 250°C to burn out the organic components (pyrolysis) and sintered at 1350°C for 3 hours, producing a scaffold with 300 μm interconnected pore diameters.[7] This method is successful, but leaves the foam struts hollow and is difficult to up-scale for mass production.

ApaPore®, a successful porous sHA product (Apatech Ltd, Elstree, UK), is a porous hydroxyapatite with an interconnected macroporosity and some microporosity. It has been successfully used in impaction grafting for the cemented revision of failed

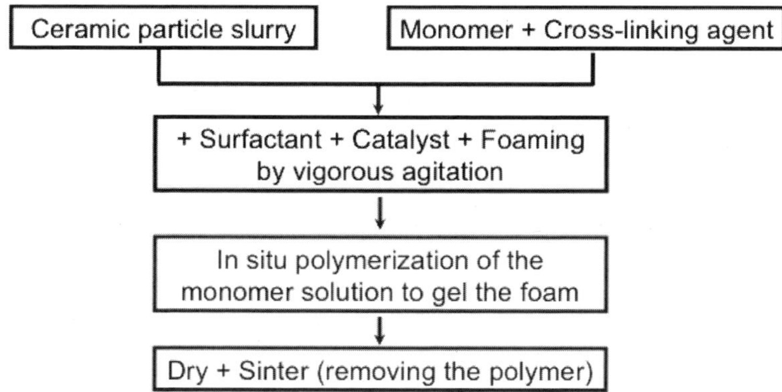


**Figure 9.2.**   Schematic of the gel-cast foaming process.

total joint arthroplasties, spinal fusions and bone defect treatment. Although Apatech has not disclosed the process, the morphology indicates that it is similar to the gel-cast foaming process, which has been the most successful published method for producing an interconnected porous ceramic.[8] In the gel-cast foaming process, suspensions of sHA particles and organic monomers are foamed with the aid of a surfactant (Fig. 9.2). The surfactant is critical to obtaining an interconnected pore network. Once the foam is formed the monomers are polymerized and the porous network sets. The polymer is removed by a sintering process, which also fuses the ceramic particles together and provides strength.

Hi-Por Ceramics Ltd (Sheffield, UK) markets a foamed competitor, and Ossatura™ (Isotis Orthobiologics US, TX) is a granular macroporous (75%) material composed of approximately 80% sHA and 20% β-TCP (tricalcium phosphate). TCP undergoes dissolution in aqueous solutions and therefore is incorporated in the product so that it will degrade more rapidly than the HA, leaving pores for bone to grow into.

Very popular HA bone graft materials have also been derived from special species of corals (Porites[9]) and from bovine bone.[10] These can contain some of the minor and trace elements originally present in the coral or in the bone. Bio-Oss® (Osteohealth, Shirley,

New York, NY) is a porous bovine bone.[10] Osteograf-N™ (CeraMed Co, Denver, CO), and Endobon™ (Merck Co, Darmstadt, Germany) are sintered materials.

Porous coralline HA is sold as Interpore™ and Pro-Osteon™, which are manufactured by Interpore International, Inc., Irvine, CA.[9] Hydroxyapatite grafts derived from coral or coralline hydroxyapatite are prepared by hydrothermal conversion of coral, which is primarily $CaCO_3$, at 260°C and 15,000 psi in the presence of ammonium phosphate.[9] A secondary phase of $\beta$-tricalcium phosphate ($\beta$-TCP) also forms during the hydrothermal conversion.

Bovine-derived apatites are produced by removal of the organic matrix. The resulting material is then either left unsintered or sintered above 1,000°C. The unsintered bone mineral consists of small crystals of HCA, whereas the sintered bone mineral consists of much larger apatite crystals without $CO_3$ ions.[3] Coralline hydroxyapatite contains traces of Mg, Sr, $CO_3$, and F. Bovine bone-derived apatite contains Mg, Na, and $CO_3$ ions.[3] The advantage of these materials is that they have an interconnected macroporous network that is conserved from the coral and bone, without the need for porogens or foaming processes. A disadvantage is the lack of supply of the coral, and ethical and environmental issues regarding coral harvesting.

Although sHA has excellent biocompatibility and osteoconductivity, it does not resorb in the body and therefore will not fulfill the criteria of an ideal tissue scaffold. Therefore, it is not an ideal material for regenerative medicine applications, but is an excellent material as permanent bone defect filler or permanent grafting material (bone substitute).

The rate of dissolution of HA materials depends on porosity and crystallinity.[11] Although stoichiometric sHA has very low degradation rates, degradation can be increased by substituting other components that are found in biological apatites. The type and extent of substitution affects the rate of dissolution. Bone has 5–8 wt% $CO_3^{2-}$; the carbonate substitution (nominally for $PO_4$) contributes to the most soluble apatite (HCA), however F substitution (for OH) decreases the solubility to lower than sHA.[12] Higher dissolution also leads to higher bioactivity.

Synthetic carbonated apatite (HCA) is synthesized by adding calcium nitrate into the solution reaction and sintering in a $CO_2$ atmosphere to prevent carbonate loss.[13,14] However, Apatech decided to opt for silicon substituted (nominally for $PO_4$) apatite.[15] The concept of using Si is based upon the composition of bioactive glasses that are more osteogenic than apatites and the main difference is that bioactive glasses release soluble silica, as discussed in Chapter 3. Very small amounts of Si were introduced into the HA composition. Cell culture tests, using human bone cells, and *in vivo* tests in rabbits indicated that 0.8 wt% Si produced the best quality bone.[16,17]

The improved bioactivity of the Si-HA compared to sHA is likely to be due to a combination of the Si producing defects in the crystal structure, increasing dissolution rate of the HA and the release of soluble silica. Small structural differences were observed between sHA and Si-HA, with Si-HA having a unit cell with a shorter a axis and larger c axis and a lower number of OH groups. The incorporation of silicon in the HA lattice also caused an increase in the distortion of the $PO_4$ tetrahedra. Si-HA is marketed as various products under the name Actifuse® (Fig. 9.3), which is used as a synthetic graft for spinal fusion operations. It is a granular material that the surgeons tend to mix with blood and insert into titanium alloy or polyethyl ethylketone (PEEK) cages, which are used to replace herniated intervertebral discs.

The Actifuse® actively fuses the vertebrae together in an operation that would commonly use bone from the pelvis or bone spurs harvested from other regions of the spine. It is 80% porous Si-HA (0.8 wt%), produced by a similar method as their ApaPore® product, where a cage or screw fixation device is used to relieve the graft site from physiological loads. It is applied as a granular structure that is combined with localized blood or with blood marrow aspirate, in a ratio of 1–1.5:1 (blood/BMA to Actifuse) to allow for optimal handling and placement.

The limitation with all these materials is that they cannot be used in load-bearing applications because of their low fracture strength.[12]

**Figure 9.3.**   Actifuse™, a porous silicon substituted hydroxapatite artificial bone graft material.

## 9.4.  Latest Developments in Bioactive Glass Scaffolds

In terms of commercial products, bioactive glass lags behind apatites somewhat. However, bioactive glasses have the potential to outperform all the commercial products currently available. The delay in the development of bioactive glass scaffolds has been because the commercially available (and FDA approved) composition (Bioglass®) cannot be made into porous scaffolds without it crystallizing. However, NovaBone Products now has a porous 45S5 Bioglass® product in the market that has a minimal amount of crystallization. There are therefore two options: (1) change the composition in order to create a melt-derived bioactive glass that can be sintered and ensure the new composition is still bioactive; or (2) make the glass by an alternative route.

There are two processing routes to produce bioactive glasses; the traditional melt-derived approach and the sol-gel process, each yielding very different glasses.

### 9.4.1. *New melt-derived glasses*

Bioglass® cannot be made into scaffolds or even easily drawn into fibers without it crystallizing, because of its random 2D amorphous structure. It has less than 50% silica content so that the silica network is disrupted enough to be degradable in aqueous solutions (i.e. body fluid). This means it must have greater than 50% of its other components, the majority of which are calcia (CaO) and soda ($Na_2O$), which act as network modifiers. This composition means that Bioglass® has a network connectivity (number of bridging oxygen bonds, -Si-O-Si-, bonds per silicon atom) of two. If the network connectivity increases above 2.4, bioactivity is significantly reduced, which is why bioactivity is lost in melt-derived glasses that have silica content greater than 60 mol%.

Recently, new compositions have been devised that do not crystallize on drawing and sintering. One is 13–93 (54.6 mol% $SiO_2$, 6% $Na_2O$, 22.1% CaO, 1.7% $P_2O_5$, 7.9% $K_2O$, and 7.7% MgO), which was developed in Finland. Porous scaffolds have been produced, using the polymer foam replication technique,[18] but this composition takes seven days to form an HCA layer in simulated body fluid tests; that is, it has Class B bioactivity. In contrast, 45S5 Bioglass® particles formed the same layer within eight hours in the same test: Class A bioactivity. This is because the network connectivity is higher in glass composition 13–93 compared to 45S5 Bioglass®, due to the increased silica content.

In order to obtain a similar result without compromising bioactivity, ICIE16 (49.46% $SiO_2$, 36.27% CaO, 6.6% $Na_2O$, 1.07% $P_2O_5$, and 6.6% $K_2O$, in mol%) was developed by Elgayar *et al.*[19] Also, to improve on the pore morphology, the gel-cast foaming process that was used so successfully on HA ceramics was adapted to produce porous bioactive glass scaffolds. Prior to this, however, the sol-gel process was used to produce porous scaffolds.

### 9.4.2. *Sol-gel-derived bioactive glasses*

For a melt-derived glass to bond to bone, the silica content has to be 60 mol% or lower. However, HCA layer formation and bone bonding can be achieved for glasses with up to 90 mol% silica if the glass is sol-gel-derived.[20] This is because sol-gel glasses have lower network connectivity than their nominal composition indicates.[21] This is due to OH being incorporated into the silica network during the process and because sol-gel glasses have inherent nanoscale porosity, which increases their specific surface area (> 100 $m^2g^{-1}$ compared to 2 $m^2g^{-1}$ for melt-derived glasses), and thereby increases their degradation rate.[21]

The first sol-gel-derived bioactive glasses were developed in the early 1990s.[20,22] The 58S (60 mol% $SiO_2$, 36 mol% CaO, and 4 mol% $P_2O_5$) composition was found to form the HCA surface layer more rapidly than similar compositions of melt-derived glass.[23]

The sol-gel process involves the hydrolysis of alkoxide precursors to create a sol. In the case of silicate-based bioactive glasses, the silicate precursor would be an alkoxide such as tetraethyl orthosilicate (TEOS) or similar. If other components apart from silica are required in the glass composition they are added to the sol either as other alkoxides or as salts. In the case of 58S, phosphate is incorporated by adding triethyl phosphate (TEP) and calcium by adding nitrate tetrahydrate. The sol can be considered as a solution of silica species that can undergo polycondensation to form silica nanoparticles which then coalesce during gelation. Water and ethanol are by-products of the condensation reaction, which must be evaporated by using carefully controlled low heating rates. The final step is to heat the dried gel to at least 600°C in order to remove the nitrates from the calcium nitrate. During the thermal processing, the coalesced nanoparticles sinter together, leaving interstitial nanoporosity. The nanopores are usually in the range of 1–30 nm diameter and can be tailored during processing, by controlling the pH of the catalyst,[24] the nominal composition,[25] and the final temperature.[26] It is however difficult to produce large crack-free monoliths (greater than 10 mm thickness) because the driving off of water, organics, and nitrates initiates capillary stresses that cause cracking.

Although sol-gel glasses can degrade more rapidly and be more bioactive than the melt-derived glasses, perhaps the biggest benefit of using the sol-gel route over melt-derived is that porous scaffolds with interconnected macro-pores suitable for tissue engineering applications can be easily produced by introducing a foaming step.[26] The sol-gel foaming process has some similarities to the gel cast foaming process in that a surfactant is added and the sol is foamed by vigorous agitation as the viscosity rapidly increases (Fig. 9.4). However in the sol-gel process, no polymer is

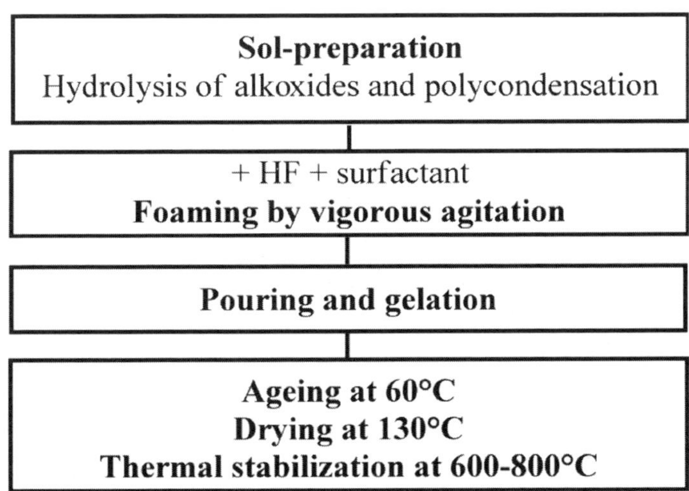


**Figure 9.4.**   The sol-gel foaming process.

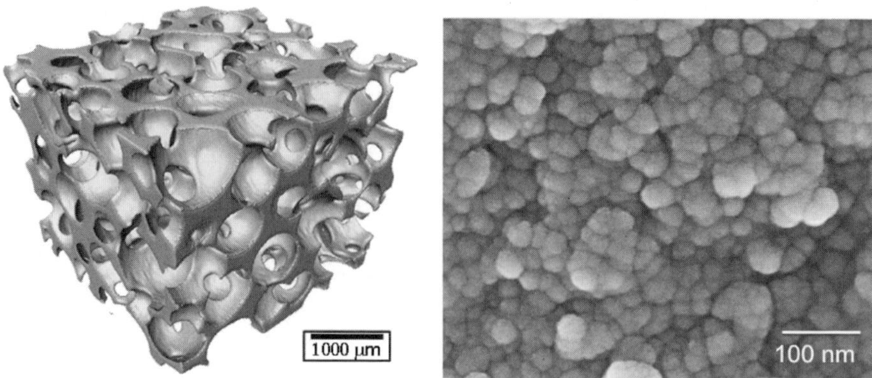

**Figure 9.5.**   Sol-gel foam bioactive glass scaffolds (a) X-ray microtomography image of the macroporosity (reprinted with permission from Jones *et al., Biomaterials,* 2007:28: 1404–1413); and (b) electron microscopy image of the nanoporosity.

added to gel the system, as the sol gels spontaneously. On gelation, the spherical bubbles become permanent in the gel, and as drainage occurs in the foam struts, the gel shrinks; the bubbles merge and interconnects open up at the point of contact between neighboring bubbles.

The sol-gel foam scaffolds have a hierarchical structure of interconnected macro-pores (Fig. 9.5a), which mimic the porous structure of cancellous bone and allow the scaffold to act as a 3D template for tissue growth, and a nanoporous porosity that allows control of degradation (Fig. 9.5b). Fig. 9.5a shows an X-ray micro computer tomography ($\mu$CT) image of a bioactive glass foam scaffold. Cell response studies on the bioactive glass foam scaffolds have found that primary human osteoblasts lay down mineralized immature bone tissue, without additional signalling species. This occurred in scaffolds of both the 58S composition[27] and the 70S30C composition,[28] which showed that phosphate is not required in the glass composition for bone matrix production and mineralization to occur. Compressive strengths of 2.4 MPa have been obtained while maintaining modal interconnect diameters above 100 $\mu$m.[1] The strength values are similar to those of clinically used porous hydroxyapatite.[10] Although their compressive strength may be suitable, bioactive scaffolds suffer similar problems to sHA for brittleness.

### 9.4.3. *Glass-ceramics*

In addition to glasses and ceramics, a third class of materials has been developed that is in widespread use in Japan: glass-ceramics, particularly the apatite-wollastonite glass-ceramics (AW-GC). Traditional glass-ceramics are polycrystalline materials produced by controlling nucleation and crystallization of fine-grained crystals. The most successful glass-ceramic has been A/W (apatite/wollastonite) bioactive glass-ceramic. It has a very fine-grained apatite (A) and wollastonite ($W = CaSiO_3$) crystals bonded by a bioactive glass interface produced by hot-pressing powders.[29] Mechanical strength, toughness, and stability of A/W glass-ceramics (AW-GC) in physiological environments are excellent, and bone bonds to A/W-GC implants with high interfacial bond strengths, as discussed in Chapter 8. The bioactivity of this glass ceramic was attributed to apatite formation on its surface in the body, brought about by the dissolution of calcium and silicate ions from the glass ceramic.[30] The AW-GC material was approved for orthopaedic applications in Japan with particular success in vertebral replacement and spinal repair.[31] It is used clinically as artificial vertebrae and in iliac crest reconstruction. No A/W scaffolds have been developed. Chapters 8 and 14 in this book by Professor T. Yamamuro summarize the development and clinical applications of this biomaterial.

## 9.5. The Future: Bioactive Nanocomposites

As no present bioceramic or biopolymer fulfills all of the criteria for an ideal scaffold listed above, new materials must be developed. There have been several attempts to combine bioactive glasses with biodegradable polymers to create a scaffold material with degradability, bioactivity, and toughness. These attempts commonly used biodegradable polymers that are already approved for clinical use for some applications such as the polyesters poly(lactides) or poly(glycolides) (and their co-polymers). Composite scaffolds with interconnected pore networks have been produced by incorporating bioactive glass particles into polylactide foams. However, their application in bone regeneration is flawed as the bioactive particles are generally covered by the polymer matrix. The host bone will therefore not come into contact with the glass. This may be rectified as the polymer phase begins to degrade and the glass is exposed. However, it is not as simple as that because of the way the polymer degrades. Polyesters degrade by hydrolysis (chain scission by reaction with water). Initially, degradation is slow and depends on rate of water uptake (diffusion) into the polymer. Once the chain scission begins, molecular weight will decrease. The pH will then drop locally due to the acidic degradation products (e.g. lactic or glycolic acid), which will catalyze the degradation process (self-catalysis). This will make the degradation rate become extremely rapid, causing the scaffold to break down and rapidly lose mechanical strength, possibly leaving the defect unhealed. The autocatalysis can even cause thicker sections of polymer to degrade faster than thin sections. Co-polymerization of poly(lactides) with poly(glycolides) can help tailor the degradation rate of the polymers, but the degradation rate will not be linear. Nanoscale composites may allow a closer relationship between the glass and the polymer and overcome this problem.

The aim of creating nanocomposites is to have a nanoscale interaction between the bioactive inorganic phase and the organic phase, creating a tough material. This intimate interaction should allow bone cells to come into contact with both phases at the same time, and the material should degrade congruently, in a more linear fashion than conventional polyesters or their composites. Nanocomposites can be divided into two classes. One is a nanoscale version of a conventional composite, where nanoparticles are dispersed in a polymer matrix. The second is where the inorganic and organic phases are covalently bonded together at a molecular scale during processing. These materials are termed *hybrids* here

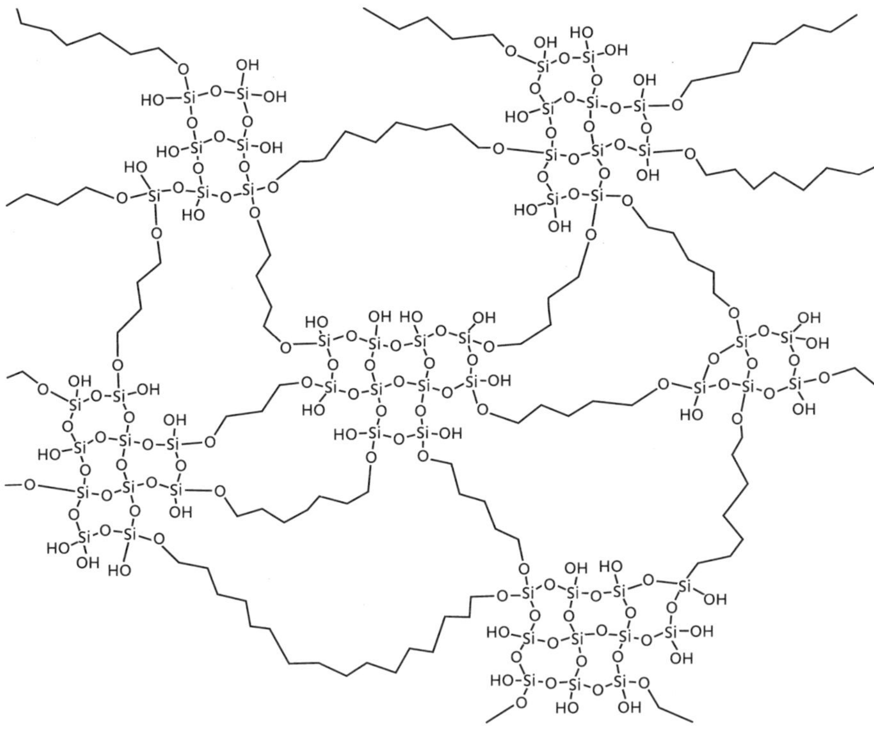

**Figure 9.6.**   Schematic of a hybrid material. Courtesy of Esther Valliant, Imperial College, London.

(Fig. 9.6), having also previously been termed ceramers and ormosils, and have the greatest potential of combining the desired properties of the constituent materials for bone regeneration. In hybrids, due to the fine scale interaction, the organic and inorganic components are not distinguishable from each other at the micron scale.

The sol-gel foaming process yielded ideal pore networks for bioactive glass scaffolds, so a logical step is to introduce a polymer phase into this process. Polymer chains can be incorporated into the sol while the inorganic chains are forming, so that the polymer network forms at the same time as the silica-based nanoparticles assemble. However, there are complex chemistry challenges associated with this procedure including which polymer to use and how to incorporate calcium.

Traditionally calcium nitrate has been used as a precursor and donator of calcium into the inorganic network. However, temperatures of at least 600°C are needed to drive off the nitrate by-products that are toxic to cells. The temperature must also reach 450°C before calcium nitrate dissociates and calcium is incorporated into the network.[21] Hybrids cannot be heated to high temperatures, as the polymer will be damaged, so another calcium precursor is needed. The best method for introducing calcium is yet to be found.

Many bioresorbable polymers, for example poly(lactides), cannot be simply introduced into the sol as they are not soluble in it. Bioactive glass/poly (vinyl alcohol) (PVA) hybrid scaffolds were produced using the sol-gel foaming technique.[32] PVA was chosen because it is soluble in water and could be added to a typical sol used to synthesize bioactive glass. The scaffolds produced had pore networks very similar to the bioactive glass foams, with macro-pore diameters of up to 500 µm. Compression testing on these foams demonstrated that polymer addition increased with strain to failure of up to 8% strain, compared to ~1% without the polymer added. A problem with the silica/PVA nanocomposites is that the PVA was not covalently bonded to the inorganic phase, therefore the scaffold is likely to break up quite rapidly in body fluid.

Although conventional polyesters are insoluble in water, they can be functionalized so that not only are they incorporated in the sol-gel process, they can also form covalent bonds with the silica network, creating a true hybrid material. The functionalization of the polymer involves the introduction of coupling agents.

One example is silica/poly (ε-caprolactone) (PCL) hybrids.[33] Hydroxyl groups at either end of poly (ε-caprolactone diol) polymer chains can be reacted with 3-isocyanatopropyl triethoxysilane (IPTS). This process results in a polymer end capped with a triethoxysilyl group. When the end-capped PCL is introduced into a sol the siloxane groups hydrolyze and then Si-OH groups from the end-capped polymer condense with the Si-OH groups from the hydrolyzed TEOS in the sol to yield an interconnected PCL-silica network. The Young's modulus and tensile strength of the bioactive glass/PCL hybrid with 60 wt% polymer were 600 and 200 MPa respectively, which is in the range of cancellous bone. However, the mechanical properties were measured on dense materials and would be dramatically lower if the materials were processed into porous scaffolds. The molecular weight of the PCL was low, at 6,693, indicating that long-term mechanical properties and stability in body fluid may also be low. Polymer chains must entangle with each other if the material is to have toughness, and the molecular weight should be two orders of magnitude higher for entanglement to happen. Polymers that can be functionalized using side groups of the chain rather than end-capping would therefore be preferable.

Alternatives to conventional polyesters are natural polymers. A popular strategy is to try to mimic the structure of bone as closely as possible, as its mechanical properties are superior to any biomedical composite produced to date. Reasons for bone's superior properties are that it has a hierarchical structure and is a nanocomposite containing collagen, which itself is a triple helix of polypeptides (amino acid chains). Collagen is a structural protein, with its triple helix structure analogous to that of a rope, giving it excellent mechanical strength.

Collagen would therefore be an ideal choice to use as a natural polymer in hybrid synthesis. Unfortunately the triple helix structure that gives collagen its properties makes it very insoluble. It will dissolve in acetic acid, but only in very low concentrations. Therefore it is not possible to produce a hybrid with significant amounts of polymer using collagen, so it would remain brittle. There is also the dilemma of where to source the collagen. Currently, it cannot be synthesized in any significant quantity so it must be sourced from animals. Although collagen is unlikely to be rejected by a patient's body (because the amino acid structure of collagen is similar for most species and therefore cannot be detected as foreign), patients may refuse an implant on religious or moral grounds due to the animal species — such as bovine or porcine — it originates from.

Natural polymers can also be functionalized. Gelatin has great potential as it is effectively denatured collagen and when collagen is dissolved in acetic acid, it denatures. The benefit of gelatin is that it is water-soluble. It also contains COOH groups (carboxylic acid) and $NH_2$ groups along its backbone that are available for functionalization. A similar process to that used for the PCL-diol is used, except that glycidoxypropyl trimethoxysilane (GPTMS) is used as the coupling agent.[34] The glycidol group (expoxy ring) can open and react with the functional groups on the polymer chain. The functionalized polymer has short molecules bonded to it with Si-OH groups on the end of them, ready to undergo condensation with other Si-OH groups from the silica in the sol, to form Si-O-Si bonds.

Another popular natural polymer is chitosan (2-amino-2-deoxy-2-Ducan), which is derived from crustacean shells.[35] The chitosan is reacted with methanesulphonic acid to form butyrylchitosan, which is then reacted with acryloxypropyl trimethoxysilane (APTMS) to form a silanated butyrylchitosan, which was then introduced into a sol of hydrolyzed TEOS.

The covalent coupling of polymers to the bioactive silica is necessary to be able to control the degradation or dissolution of the hybrids, especially if the polymers are water soluble. However, careful control of the chemistry should allow complete control of degradation rate and the mechanical properties. The future may well be the use of human recombinant proteins, but much development is needed to increase their yields.

## 9.6. Regulatory Issues

Once a promising new scaffold material has been developed and tested in the laboratory with cell culture tests and long-term mechanical tests as a function of degradation, it is time to translate the work from the bench to the bedside. This is not a trivial (or inexpensive) process, and takes considerable time and investment. Difficult decisions also have to be made. An important early decision is for which class of material regulatory approval should be sought. At the time of writing, regulatory bodies, such as the FDA and the EU are (quite rightly) making regulatory approval more difficult to achieve. Although it may disadvantage new companies with potentially important new products, safety must be paramount. The regulatory class depends on the claims made by a company. For example, in the EU, if a company claims that a scaffold will bond to bone, degrade over time, and stimulate bone regeneration, it will be a Class 3 medical device. If the company claims a new material will simply do the same as other bone grafts and fill space it will be a Class 2 device, but then the marketing or sales people in the company cannot claim performance superior to current products. Of course, to obtain regulatory approval of a Class 3 device takes more investment and may involve lengthy clinical trials. Claiming pharmaceutical properties of an implant may also mean that approval is need from pharmaceutical (rather than device) regulatory bodies such as the UK Healthcare products Regulatory Agency (MHRA).

## 9.7. Summary

Bioactive sol-gel glass foams have the potential to improve performance of current artificial scaffolds for bone regeneration, as they have a hierarchical pore structure, but no bioceramics will ever fulfill all the criteria for an ideal scaffold. Inorganic/organic hybrid nanocomposites have the potential to be tough bioactive and biodegradable scaffolds. A massive challenge is expanding this technology to regenerate load-bearing skeletal components such as the hip. Replacing metals as load-bearing devices is the ultimate, but perhaps unachievable, goal. Tissue-engineering approaches are not yet widely used in the clinic, except that surgeons often mix blood and bone marrow with implants, hoping to activate some stem cells already belonging to the patient.

## References

1. Jones, J.R., Ehrenfried, L.M. and Hench, L.L. (2006). Optimising bioactive glass scaffolds for bone tissue engineering, *Biomaterials*, **27**, 964–973.
2. Oonishi, H., Hench, L.L., Wilson, J., *et al.* (2000). Quantitative comparison of bone growth behavior in granules of Bioglass®, A-W

glass-ceramic, and hydroxyapatite, *J. Biomed. Mater. Res.*, **51**, 37–46.

3. LeGeros, R.Z. (2002). Properties of osteoconductive biomaterials: calcium phosphates, *Clin. Orthopaed. Rel. Res.*, **395**, 81–98.

4. Kato, K., Aoki, H., Tabata, T., *et al.* (1979). Biocompatibility of apatite ceramics in mandibles, *Biomater. Med. Dev.*, **7**, 291–297.

5. Denissen, H.W., Kalk, W., Veldhuis, A.A.H., *et al.* (1989). 11-year study of hydroxyapatite implants, *J. Prosth. Dent.*, **61**, 706–712.

6. Andrade, J.C.T., Camilli, J.A., Kawachi, E.Y., *et al.* (2002). Behavior of dense and porous hydroxyapatite implants and tissue response in rat femoral defects, *J. Biomed. Mater. Res.*, **62**, 30–36.

7. Zhang, C., Wang, J.X., Feng, H.H., *et al.* (2001). Replacement of segmental bone defects using porous bioceramic cylinders: a biomechanical and X-ray diffraction study. *J. Biomed. Mater. Res.*, **54**, 407–411.

8. Sepulveda, P., Binner, J.G.P., Rogero, S.O., *et al.* (2000). Production of porous hydroxyapatite by the gel-casting of foams and cytotoxic evaluation, *J. Biomed. Mater. Res.*, **50**, 27–34.

9. Roy, D.M. and Linnehan, S.K. (1974). Hydroxyapatite formed from coral skeletal carbonate by hydrothermal exchange, *Nature*, **247**, 220–222.

10. Valentini, P., Abensur, D., Wenz, B., *et al.* (2000). Sinus grafting with porous bone mineral (Bio-Oss) for implant placement: a 5-year study on 15 patients, *Int. J. Periodon. Rest. Dent.*, **20**, 245–254.

11. Koerten, H.K. and van der Meulen, J. (1999). Degradation of calcium phosphate ceramics, *J. Biomed. Mater. Res.*, **44**, 78–86.

12. de Groot, K. (1983). *Bioceramics of CaP*, CRC Press, Boca Raton, FL.

13. Merry, J.C., Gibson, I.R., Best, S.M., *et al.* (1998). Synthesis and characterization of carbonate hydroxyapatite, *J. Mater. Sci. Mater. Med.*, **9**, 779–783.

14. Gibson, I.R. and Bonfield, W. (2002). Novel synthesis and characterization of an ab-type carbonate-substituted hydroxyapatite, *J. Biomed. Mater. Res.*, **59**, 697–708.

15. Gibson, I.R., Best, S.M. and Bonfield, W. (1999). Chemical characterization of silicon-substituted hydroxyapatite, *J. Biomed. Mater. Res.*, **44**, 422–428.

16. Botelho, C.M., Brooks, R.A., Best, S.M., *et al.* (2006). Human Osteoblast response to silicon-substituted hydroxyapatite, *J. Biomed. Mater. Res. A*, **79A**, 723–730.

17. Hing, K.A., Revell, P.A., Smith, N., *et al.* (2006). Effect of silicon level on rate, quality and progression of bone healing within silicate-substituted porous hydroxyapatite scaffolds, *Biomaterials*, **27**, 5014–5026.

18. Fu, Q., Rahaman, M.N., Bal, B.S., *et al.* (2008). Mechanical and *in vitro* performance of 13–93 bioactive glass scaffolds prepared by a polymer foam replication technique, *Acta Biomater.*, **4**, 1854–1864.

19. Elgayar, I., Aliev, A.E., Boccaccini, A.R., *et al.* (2005). Structural analysis of bioactive glasses, *J. Non-Cryst. Solids*, **351**, 173–183.

20. Li, R., Clark, A.E. and Hench, L.L. (1991). An investigation of bioactive glass powders by sol-gel processing, *J. Appl. Biomater.*, **2**, 231–239.

21. Lin, S., Ionescu, C., Pike, K.J., *et al.* (2009). Nanostructure evolution and calcium distribution in sol-gel gerived bioactive glass, *J. Mater. Chem.*, **19**, 1276–1282.

22. Pereira, M.M., Clark, A.E. and Hench, L.L. (1994). Calcium-Phosphate formation on sol-gel-derived bioactive glasses *in vitro*, *J. Biomed. Mater. Res.*, **28**, 693–698.

23. Sepulveda, P., Jones, J.R. and Hench, L.L. (2002). *In vitro* dissolution of melt-derived 45s5 and sol-gel derived 58s bioactive glasses, *J. Biomed. Mater. Res.*, **61**, 301–311.

24. Brinker, J. and Scherer, G.W. (1990). *Sol-gel science: the physics and chemistry of sol-gel processing*, Academic Press, Boston, MA.

25. Arcos, D., Greenspan, D.C. and Vallet-Regi, M. (2002). Influence of the stabilization temperature on textural and structural features and Ion release in $SiO_2$-$CaO$-$P_2O_5$ Sol-Gel glasses, *Chem. Mater.*, **14**, 1515–1522.

26. Sepulveda, P., Jones, J.R. and Hench, L.L. (2002). Bioactive sol-gel foams for tissue repair, *J. Biomed. Mater. Res.*, **59**, 340–348.

27. Gough, J.E., Jones, J.R. and Hench, L.L. (2004). Nodule formation and mineralisation of human primary osteoblasts cultured on a porous bioactive glass scaffold, *Biomaterials*, **25**, 2039–2046.

28. Jones, J.R. ,Tsigkou, O., Coates, E.E., *et al.* (2007). Extracellular matrix formation and mineralization of a phosphate-free porous bioactive

glass scaffold using primary human osteoblast (HOB) cells, *Biomaterials*, **28**, 1653–1663.

29. Nakamura, T., Yamamuro, T., Higashi, S., *et al.* (1985). A new glass-ceramic for bone-replacement — evaluation of its bonding to bone tissue, *J. Biomed. Mater. Res.*, **19**, 685–698.

30. Kokubo, T. (1991). Bioactive glass-ceramics — properties and applications, *Biomaterials*, **12**, 155–163.

31. Neo, M., Kotani, S., Nakamura, T., *et al.* (1992). A comparative study of ultrastructures of the interfaces between four kinds of surface-active ceramic and bone, *J. Biomed. Mater. Res.*, **26**, 1419–1432.

32. Pereira, M.M., Jones, J.R., Orefice, R.L., *et al.* (2005). Preparation of bioactive glass-polyvinyl alcohol hybrid foams by the sol-gel method, *J. Mater. Sci. Mater. Med.*, **16**, 1045–1050.

33. Rhee, S.H. and Lee, S.J. (2007). Effect of acidic degradation products of poly(lactic-co-glycolic)acid on the apatite-forming ability of poly(lactic-co-glycolic)acid-siloxane nanohybrid material, *J. Biomed. Mater. Res. A*, **83A**, 799–805.

34. Mahony, O., Tsigkou, O., Ionescu, C., *et al.* (2010). *Silica-gelatin hybrids with tailorable degradation and mechanical properties for tissue regeneration*, *Adv. Funct. Mater.*, **20**, 3835–3845.

35. Zhu, A., Zhang, M. and Shen, R. (2003). Covalent immobilization of o-butyrylchitosan with a photosensitive hetero-bifunctional crosslinking reagent on biopolymer substrate surface and blood compatibility characterization, *J. Biomater. Sci. Polym. Ed.*, **14**, 411–421.

# Cartilage Regeneration

CHAPTER **10**

Adam C. Grochowski and Michael B. Fenn

## 10.1. Introduction

Cartilage damage is a massive problem for people of all ages, particularly the elderly suffering from arthritis and athletes who sustain cartilage injuries. Although most people will experience some form of arthritis as they age, ranging from joint stiffness to immobility and intense pain, arthritis can affect people of any age. Repair or restoration of cartilage is difficult because it has limited self-repair capabilities, and a complex cellular and acellular structure. Severely damaged cartilage does not regenerate like many other tissues, such as bone or skin, due mainly to the lack of innervations and vascularization.

This chapter covers:

1. modes by which cartilage fails;
2. regeneration processes of cartilage;
3. current methods of cartilage repair;
4. limits to current methods of repair; and
5. future applications of cartilage regeneration.

## 10.2. Anatomy and Physiology of Cartilage[1]

Cartilage is a type of connective tissue that is able to withstand both compressive and tensile forces. It is a tough yet flexible and elastic tissue that is composed of up to 80% water. A dense membrane called the perichondrium surrounds the cartilage. This membrane and the movement of the water within the flexible cartilage are what give this tissue its ability to provide support under compression. The water also serves as the medium by which the two types of cartilage cells, chondrocytes (mature cartilage cells) and chondroblasts (cartilage-forming cells) receive nutrients by diffusion from the vascularized perichondrium that encases the cartilage tissue. It is important to remember that cartilage tissue is avascular, and because it does not have a blood supply, nutrients reach the cells via diffusion.

There are three types of cartilage: hyaline, elastic, and fibrous. Hyaline cartilage, the most abundant, supports the respiratory tract, connects the ribs to the sternum and is found at the end of long bones, where it is called articular cartilage. Figure 10.1 shows a typical joint with the location of articular cartilage. Elastic cartilage is similar to hyaline cartilage except that it contains a much greater concentration of elastin fibers, making it more

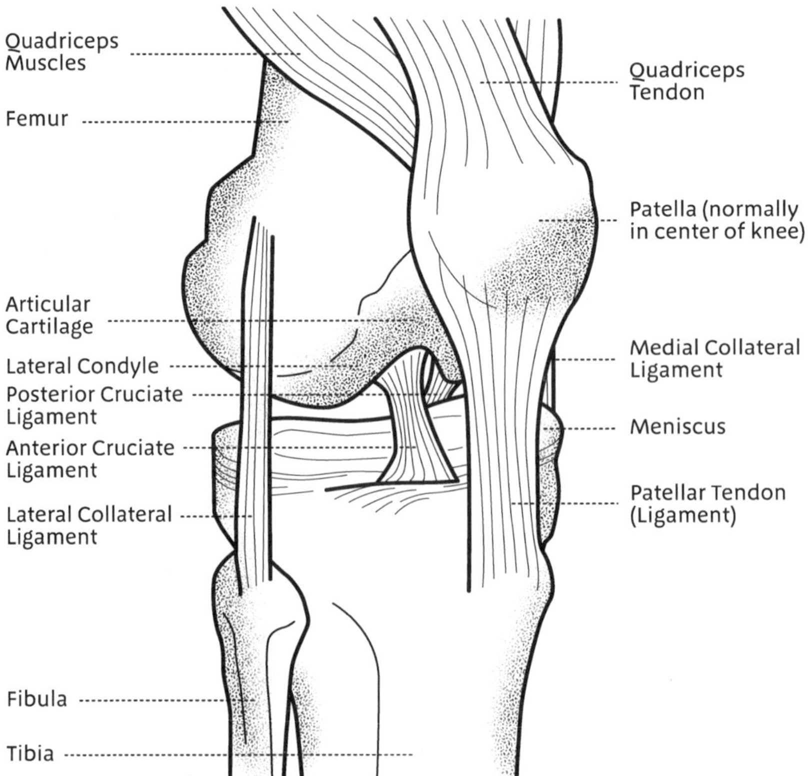


**Figure 10.1.**    Anatomy of a typical joint with articular cartilage.

flexible than articular cartilage. It is found in the ear, nose, cornea, larynx, and epiglottis. Fibrocartilage contains a regular array of collagen fibers bundled and aligned in the direction of the applied stresses and is found at sites that incur heavy loading and pressure. For example, fibrocartilage forms the intervertebral discs, the soft tissue that provides support, flexibility, and cushioning to the spine by filling the gap between the vertebral bodies. It is also found in the bones of the pelvis, and in the ball and socket joints of the shoulder and hip. The alignment of the collagen fibers, parallel to the loading direction, provides the cushioning properties.

Cartilage formation occurs very early in the human embryo, in about the fifth week of gestation, and provides a natural scaffold for embryonic skeletal development. Mesenchymal stem cells differentiate to become chondroblasts and begin to secrete hyaluran chondroitin sulphate and tropocollagen. Tropocollagen molecules are amino acid chains (polypeptides) that polymerize to form a triple helix, forming fine collagen fibers (fibrils) with excellent mechanical properties.

Articular cartilage, a type of hyaline cartilage, is composed of water and chondrocytes in an extracellular matrix consisting of proteins, mainly collagen, and proteoglycans containing mainly sulphated glycosominoglycans (GAGs) and hyluronic acid. The bulk of the composition of articular cartilage is fluid (~75%). The solid fraction consists of approximately 60% collagen, 30% GAGs, and 10% other proteins. GAGs consist of a protein backbone with highly charged polysaccharide side chains. The negative charges of the latter are what allow for the resistance to compression on the molecular level, due to bound water molecules. Collagen types present include collagen II, IX, and XI with type II dominating (>90%). Type II collagen contains many carbohydrate groups, allowing more interaction with water than other types. The fluid in cartilage contains water, gases, metabolites, and cations to balance the negatively charged GAGs. Articular cartilage is a specialized form of hyaline cartilage that can support the loads applied to it mainly due to the complex organization of the collagen fibrils.

Hyaline cartilage is distinguished by its homogenous matrix surrounding the lacunae of chondrocytes. Chondrocytes develop in the perichondrium, which surrounds the cartilage. As they mature, they are moved deeper into the cartilage, where they secrete the matrix and forming cartilage that consists of several layers. It is the low cellular content along with diffusion-limited nutrient transport that makes it difficult for cartilage to self-repair.

The composition of mature articular cartilage varies with depth and is composed of four structurally different zones. In the *superficial zone* cells are numerous, flattened, and parallel to the surface, producing a lubricating protein. The collagen fibers in the superficial zone are fine and run parallel to the surface. In the *transitional zone* chondrocytes are more spherical and collagen fibers are less oriented taking an oblique path through the matrix. In the deeper *radial zone* chondrocytes form radial columns and collagen fibers are orientated perpendicular to the articulating surfaces. The deepest *calcified zone* is adjacent to the subchondral bone and is partially mineralized. This complex hierarchical structure gives articular cartilage its unique mechanical properties that allow it to bear loads and cushion impacts. In turn it also makes articular cartilage very difficult to mimic. Figure 10.2 illustrates the zonal or layered structure of articular cartilage.

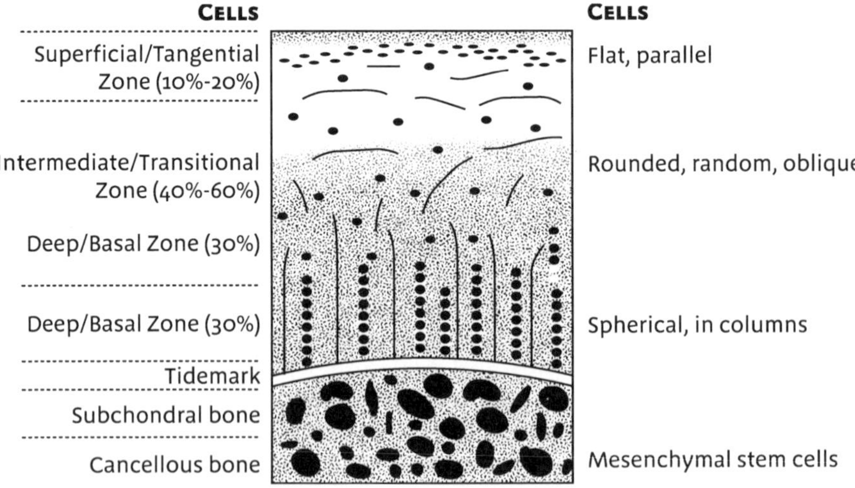


**Figure 10.2.**    The zonal structure of articular cartilage.

## 10.3. Cartilage Disease and Damage

Osteoarthritis (OA), also known as degenerative joint disease, is the most common form of arthritis and is the leading cause of chronic disability in the western world.[2] It is not only caused by aging but also may result from an underlying cause, such as trauma or metabolic irregularities. Hereditary and developmental factors can also contribute, and immune processes can accelerate the damage. Aging accelerates cartilage degeneration as a result of the decreasing water content of the tissue, which is due to a reduction in proteoglycan content, thus causing the cartilage to become less resilient and more prone to damage. When articular cartilage is lost, subchondral bone surfaces become less protected and can come in contact with one another, causing intense pain and bone damage. Current procedures for treatment consist of pain management with medication and prescribed physical therapy regimens. Surgical intervention is usually reserved as a last resort, only if the degeneration of the cartilage results in severe chronic pain or immobility. Surgical techniques range from minimally invasive arthroscopic procedures to total joint replacements. Total joint replacements have revolutionized surgery, giving pain-free mobility to many patients since the 1960s, but the implants have a limited lifetime (15–25 years), which is becoming too short as life expectancy is increasing. Surgical procedures usually provide the patient some degree of pain relief and restoration of mobility but the results vary considerably based on the patient, the extent of damage, surgical technique, and skill of the surgeon.

Rheumatoid arthritis (RA) is a systemic inflammatory disorder affecting around 1% of the population, in which the body's immune system attacks synovial tissue, causing painful swelling and deformity. Inflammation can also appear in other organs. Treatment typically consists of administering anti-inflammatory drugs and steroids, as currently there is no cure for RA.

Intervertebral discs can degenerate with age, a condition known as degenerative disc disease (DDD), which causes intense pain for the sufferer and leads to immobility. If the condition does not respond to anti-inflammatory drugs and physical therapy, surgery may be indicated. Current surgical treatments are based on a variation of spinal fusion with discectomy, in which the disc is removed and replaced by a titanium or polymeric cage and filled either with bone graft or bone graft substitute (e.g. synthetic hydroxyapatite ceramic, see Chapter 9). The graft hardens and bone in-growth occurs, fusing the spine at the affected level(s). Chapter 16, Degenerative Disc Disease in the Lumbar Spine, summarizes the treatment of DDD.

In terms of impact trauma, articular cartilage is usually at risk. Articular cartilage may sustain two types of injury, partial-thickness or full-thickness. Partial-thickness injuries are superficial lesions. After injury, there is a brief period of chondroblast proliferation and matrix synthesis, but regenerative potential is limited as progenitor cells cannot access the wound.[1] However, these lesions do not usually progress and rarely result in osteoarthritis.[2] Full-thickness injuries occur when the defect penetrates through to the subchondral bone. In this case a clot forms and progenitor cells can access the wound; healing of the injured cartilage can occur to some degree. In the first two weeks following the injury, the large cartilage defect is filled with irregular connective tissue. The cells of this irregular tissue then differentiate into cells that have the specific biochemical and morphological characteristics mimicking the characteristics of chondrocytes; these cells begin producing a matrix that contains collagen. The organization of the collagen fibrils does not return to the original state, with the cartilage having properties intermediate to fibrocartilage and hyaline cartilage, leading to reduced mechanical properties.

A method of regenerating cartilage is therefore needed due to the lack of self-regenerative capability of the tissue. Ideally a method for regenerating cartilage to its original form and function should be developed. There has been a great deal of research in tissue regeneration of cartilage. One of the major foci of the regenerative medicine approach is to develop materials that mimic the extracellular matrix of cartilage (biomimetic), and stimulate chondrogenesis, the formation of new cartilage (bioactive).

## 10.4. Regenerative Procedures

There are several procedures in clinical use for cartilage repair that do not involve implants.[4] A common treatment of cartilage degeneration is to create a channel that allows access of progenitor cells to migrate to the defect. The progenitor cells are located in the subchondral bone and can be accessed by puncturing the bone and therefore, creating a hole or defect which heals like that of a full-thickness defect. Several current clinical procedures for cartilage regeneration use an autologous cell-based therapy, in which healthy chondrocytes or progenitor cells harvested from the patient are injected into the defect. Results from these procedures have been variable, as some methods have had difficulty retaining cells in the defect site long enough to produce matrix.

Keyhole surgery is used to harvest slivers of undamaged cartilage, and cells are extracted and expanded in the laboratory. Retainment of the cells can be improved by suturing a periosteal flap over the defect as a barrier beneath which cultured autologous chondrocytes are injected.[5] The flap also encloses the implanted cells, but the need for a periosteal flap requires a donor site giving rise to increased morbidity, among other complications. The repair will typically restore function in 82% of patients within two years.[6] Hyaluronic acid gels have also been used to fill the defect and to support the cells by mimicking the natural extracellular matrix environment. The gel also acts as a delivery vehicle for the cells.

Other techniques used to regenerate cartilage are still at a trial stage. One such procedure that is currently being researched inhibits the inflammation-induced cartilage degradation process.[7] The strategy is to target the inflammatory triggers, such as the cytokines that provide signals between cells. Key cytokines are: Interleukin-1 (IL-1), which stimulates immune system cells that fight disease, and is involved in inflammatory responses; and tumor necrosis factor-alpha (TNF-$\alpha$), which is involved in systemic inflammation and acute response. Matrix metalloproteinases (MMPs) can also be targeted; these are enzymes involved in the degradation of the extracellular matrix. These enzymes and cytokines are targeted through gene delivery of secreted IL-1 receptor antagonist protein (IRAP) with soluble TNF receptors into the articular defect; this strategy has been shown to protect the articular cartilage from degradation.[7]

Another approach used to inhibit the inflammation and degradation of articular cartilage is through hindering the MMP via an antisense process.[8] Antisense technology is the inhibition of the translation of a specific mRNA by binding of complementary oligonucleotides. Cartilage degradation has been shown to be greatly reduced through the over-expression of MMP inhibitors via a gene-delivery process.[9] A different approach involves the targeting of the synovial cells that are responsible for the stimulation of cartilage degradation, and destroying these cells via the over-expression of pro-apoptotic factors.[10] Though these chondro-protective approaches have demonstrated the ability to decrease the growth of articular cartilage degradation, none have been able to regenerate cartilage tissue inside a cartilage defect.

## 10.5.  Scaffolds and Tissue Engineering

Scientists are currently researching the use of various scaffolds as mediums to promote cell growth and proliferation of articular cartilage. The scaffold must maintain the chondrocyte phenotype, which requires a 3D environment that mimics extracellular matrix (ECM). The scaffold must also mimic the physical and mechanical properties of native articular cartilage. A number of techniques for regenerating cartilage are currently in trial stages. Hydrogels are highly porous, usually consist of 80% or greater solvent, and have highly tuneable mechanical characteristics that could potentially mimic cartilage tissue. Scaffolds based on hydrogels can be made from a wide variety of materials, but are mainly based on two sub-categories: natural polymers, such as carbohydrates and proteins, or synthetic polymers, such as poly(ethylene gycol) and poly(vinyl alcohol). Chapter 11, Hydrogels for Tissue Engineering of Articular Cartilage, is devoted to a discussion of various hydrogels and their use in regenerative medicine.

Recently, autologous chondrocytes were expanded and embedded in a three-dimensional resorbable two-component gel-polymer scaffold called BioSeed-C and used for the treatment of trauma and osteoarthritic defects.[11] Results after two years showed improvement of the defect. Similar positive results were observed for collagen gel scaffolds containing autologous bone marrow stromal cells (progenitor cells harvested from the bone marrow of a patient and expanded in the laboratory) that were used in clinical trials in full-thickness articular cartilage defects in the medial femoral condyle.[12] The scaffolds were covered with a periosteal flap.

An alternative hydrogel that has received much attention is hyaluronic acid and its derivatives. Positive results were found for hyaluronan-based scaffolds (Hyalograft C) seeded with autologous chondrocytes, when compared to collagen gels, for the treatment of deep chondral lesions.[13] Drawbacks to natural polymers include difficulty of producing large quantities, a lack of reproducibility, difficulty in sterilization, and possible immune response and inflammation issues.

A wide variety of synthetic polymer scaffolds are under investigation, using polymers such as poly(lactic acid) (PLA), poly(glycolic acid) (PGA) and poly(urethane). Although none of these materials are in clinical use for cartilage repair, PLA, PGA, and associated copolymers have been approved for use as resorbable sutures. Using synthetic polymers negates any concerns about the presence of pathogens as observed in naturally derived materials and yields much greater reproducibility. Although degradable polyesters have been FDA approved for clinical use in specific applications, they have not been approved for use as regenerative scaffolds. A potential disadvantage of using polyesters in tissue scaffolds is due to their degradation mechanism via hydrolysis of the ester bonds. The resulting carboxylic acid groups can greatly reduce the PH, and a 'burst' of degradation products can be released. Although these products are normally safe, in a burst they can cause damage to tissue in confined environments due to a large reduction in pH.

For cartilage scaffolds, the polymers can be made into meshes to mimic the collagen matrix. The mesh is often made using electrospinning techniques, where an electric field is applied to a polymer solution, which accelerates a stream of the polymer towards a target. Electrospinning produces arrangements of fibers with diameters in the nanometer range, similar to that of naturally occurring collagen fibrils.[14] As articular cartilage is composed of a matrix of collagen fibrils containing proteoglycans, a combined approach has also been proposed using a woven fibrous polyester immersed in a hydrogel of agarose and fibrin to mimic the natural cellular and mechanical environment more closely.[15]

Severely damaged articular cartilage often causes damage to the underlying (subchondral) bone, so a tissue-engineering approach is required that can regenerate bone and cartilage simultaneously. A number of various approaches are being investigated to combat this complex challenge.[16] Such defects have a further level of complexity, since bone cells and cartilage cells prosper in different environments.[17] A promising strategy is to use scaffolds that have been engineered to have both chondrogenic and osteogenic regions, such as collagen-GAG-apatite scaffolds. Such a material is found in the Chondromimetic product produced by Orthomimetics (Cambridge, UK).

## 10.6.  Gene Therapy

Gene therapy has also been investigated as a means to repair cartilage damage. Due to the complex and dense matrix surrounding the defects in the articular cartilage, this is a very challenging technique. Direct application of recombinant adeno-associated virus (rAAV, a virus that depends on the adenovirus for replication) vectors to defects in articular cartilage have been shown to induce stimulation of gene expression that implied successful cartilage regeneration. Due to the difficulty of genetically modifying chondrocytes *in situ*, many scientists have focused their attention on gene delivery to the synovial tissue surrounding the targeted articular cartilage, since the synovium is easily accessible and allows for genetic modification by many gene characters. Adenovirus can be injected into the synorial tissue for a gene therapy approach to repair damaged articular cartilage.[18,19]

## 10.7. Recent Developments

Currently many cartilage injuries are treated by surgical means in which the goal is to replace the damaged tissue with a substitute, transplant or implant tissue of similar composition, thus requiring a donor site for harvesting healthy tissue. Such techniques have limitations due to donor site morbidity and limited sources of healthy tissue in the affected patient. The substitution of cartilage with other types of tissue yields poor mechanical properties when compared to naturally occurring tissue and often does not produce long-term relief.[20] Microfracture surgery or an osteotomy results in replacement of the damaged tissue with fibrocartilaginous tissue that has reduced mechanical properties.[20]

Considerable research has focused on improving the procedure in which healthy chondrocytes are harvested from the patient followed by *in vitro* expansion and finally re-implantation into the area of damaged tissue. This procedure is one of the few tissue-engineered approaches for cartilage replacement that is FDA-approved for clinical use.[21] Known as autologous chondrocyte implantation, this process is expensive and has a number of limitations besides the need for a donor site and poor loading characteristics of the newly formed tissue. Culturing of human chondrocytes *in vitro* is one of the major drawbacks of this procedure given that cultured human chondrocytes tend to grow in a monolayer as opposed to the three-dimensional growth observed *in vivo*.[22] Three-dimensional growth is critical to survival of the functioning articular chondrocytes or the cells will de-differentiate, lose their phenotype, and no longer produce type II collagen-rich tissue.[23]

Due to these limitations, research has focused on methods for creating a three-dimensional culture environment and simulating *in vivo* mechanical forces *in vitro* for chondrocyte cultures. A number of different bioreactor configurations have been investigated in attempts to grow three-dimensional *in vitro* cultures of explanted articular chondrocytes. The bioreactors allow for the cells to be cultured under conditions which provide appropriate mass transfer and physical stimuli.[24] Bone marrow-derived stem cells, differentiated with cartilage-inducing factor TGF-$\beta$ and grown in a rotating wall vessel (RWV), resulted in a three-dimensional articular cartilage tissue. These engineered tissues cultured in the RWV had improved mechanical properties and greatly reduced cell necrosis.[25] Controlling the biomechanical and biophysical forces exerted on the tissue while growing *in vitro* has been shown to have a profound influence on tissue growth and development.[26]

The use of stem cells for repair of articular cartilage has been investigated extensively. Most of the current research has focused on the use of mesenchymal stem cells (MSCs) mainly derived from bone marrow.[20] Adult MSCs have been differentiated into functional articular cartilage by a variety of methods, including inducing differentiation by biomechanical means and a number of different growth factors.[27–29] The use of MSCs for articular cartilage repair is an area that will have a major impact on future developments in this area of tissue engineering.

## 10.8. Summary

Developing a successful cartilage repair technique would improve the quality of life for millions of people. Articular cartilage is a difficult tissue to repair as it has a complex cellular and acellular structure and does not readily regenerate itself. Currently, cell therapies combined with surgical techniques are being used to heal lesions, with mixed results. The goal is to develop tissue-engineering techniques that supply a viable and fully functional cartilage tissue or tissue substitute which would provide long-term benefits to the patient. A combination of the appropriate bioactive factors, three-dimensional scaffolds materials, optimal mechanical properties, and the proper physical stimuli will be necessary to achieve a suitable tissue-engineering scaffold for articular cartilage therapy. Hydrogels possess many of the desired properties essential for articular cartilage tissue engineering, as will be discussed in the following chapter.


## References


1. Marieb, E.N. and Hoehn, K. (2007). *Human Anatomy and Physiology*, 7th ed., Pearson Education, Harlow, UK.
2. Felson, D.T., Lawrence, R.C., Dieppe, P.A, *et al.* (2000). Osteoarthritis: new insights, part 1: the disease and its risk factors, *Ann. Int. Med.*, **133**, 636–646.
3. Majithia, V. and Geraci, S.A. (2007). Rheumatoid arthritis: diagnosis and management, *Am. J. Med.*, **120**, 936–939.
4. Lu Y., Markel, M.D., Swain C., *et al.* (2006). Development of partial thickness articular cartilage injury in an ovine model, *J. Orthopaed. Res.*, **24**, 1974–1982.
5. Friedlaender, G.E. (1991). Bone allografts: the biological consequences of immunological events, *J. Bone Joint Surg. Am.*, **73**, 1119–1122.
6. Getgood, A., Brooks, R., Fortier, L., *et al.* (2009). Articular cartilage tissue engineering, *J. Bone Joint Surg. Br.*, **91B**, 565–576.
7. Gillogly, S.D., Myers, T.H. and Reinold, M.M. (2006). Treatment of full-thickness chondral defects in the knee with autologous chondrocyte implantation, *J. Orthopaed. Sports Phys. Ther.*, **36**, 751–764.
8. Peterson, L., Brittberg, M., Kiviranta, I., *et al.* (2002). Autologous chondrocyte transplantation biomechanics and long-term durability, *Am J. Sports Med.*, **30**, 2–12.
9. Knedla, A., Neumann, E. and Muller-Ladner, U. (2007). Developments in the synovial biology field 2006, *Arthritis Res. Ther.*, **9**, 209.
10. Rutkauskaite, E., Volkmer, D., Shigeyama, Y., *et al.* (2005). Retroviral gene transfer of an antisense construct against membrane type 1 matrix metalloproteinase reduces the invasiveness of rheumatoid arthritis synovial fibroblasts, *Arthritis Rheum.*, **52**, 2010–2014.
11. van der Laan, W.H., Quax, P.H.A, Seemayer, C.A., *et al.* (2003). Cartilage degradation and invasion by rheumatoid synovial fibroblasts

is inhibited by gene transfer of TIMP-1 and TIMP-3, *Gene Ther.*, **10**, 234–242.

12. Yao, Q., Wang, S., Gambotto, A., *et al.* (2003). Intra-articular adenoviral-mediated gene transfer of trail induces apoptosis of arthritic rabbit synovium, *Gene Ther.*, **10**, 1055–1060.

13. Ossendorf, C., Kaps, C., Kreuz, P.C., *et al.* (2007). Treatment of post-traumatic and focal osteoarthritic cartilage defects of the knee with autologous polymer-based three-dimensional chondrocyte grafts: 2-year clinical results, *Arthritis Res. Ther.*, **9**, R41.

14. Kuroda, R., Ishida, K., Matsumoto, T., *et al.* (2007). Treatment of a full-thickness articular cartilage defect in the femoral condyle of an athlete with autologous bone-marrow stromal cells, *Osteoarthr. Cartilage*, **15**, 226–231.

15. Tognana, E., Borrione, A., De Luca, C., *et al.* (2007). Hyalograft® C: hyaluronan-based scaffolds in tissue-engineered cartilage, *Cells Tissues Organs*, **186**, 97–103.

16. Yoshimoto, H., Shin, Y.M., Terai, H., *et al.* (2003). A biodegradable nanofiber scaffold by electrospinning and its potential for bone tissue engineering, *Biomaterials*, **24**, 2077–2082.

17. Moutos, F.T., Freed, L.E. and Guilak, F. (2007). A biomimetic three-dimensional woven composite scaffold for functional tissue engineering of cartilage, *Nat. Mater.*, **6**, 162–167.

18. Hiraide, A., Yokoo, N., Xin, K.Q., *et al.* (2005). Repair of articular cartilage defect by intraarticular administration of basic fibroblast growth factor gene, using adeno-associated virus vector, *Human Gene Ther.*, **16**, 1413–1421.

19. Martin, I., Miot, S., Barbero, A., *et al.* (2007). Osteochondral tissue engineering, *J. Biomech.*, **40**, 750–765.

20. Fritz, J.R., Pelaez, D. and Cheung, H.S. (2009). Current challenges in cartilage tissue engineering: a review of current cellular based therapies, *Curr. Rheumatol. Rev.*, **5**, 8–14.

21. Palsson, B.O. and Bhatia, S.N. (2004). *Tissue Engineering*, Pearson Prentice Hall, Upper Saddle River, NJ.

22. Benya, P.D. and Schaffer J.D. (1982). Dedifferentiated chondrocytes reexpress the differentiated collagen phenotype when cultured in agarose, *Cell*, **30**, 215–224.

23. Brittberg, M., Peterson, L., Sjögren-Jansson, E., *et al.* (2003). Articular cartilage engineering with autologous chondrocyte transplantation, *J. Bone Joint Surg.*, **85**, 109–115.
24. Blitterswijk, C.V. (2008). *Tissue Engineering*, Elsevier, Oxford, UK.
25. Ohyabu, Y., Kida, N., Kojima, H., *et al.* (2006). Cartilaginous tissue formation from bone marrow cells using rotating wall vessel (RWV) bioreactor, *Biotechnol. Bioeng.*, **95**, 1003–1008.
26. Butler, D.L., Goldstein, S.A., Guldberg, R.E., *et al.* (2009). The impact of biomechanics in tissue engineering and regenerative medicine, *Tissue Eng. B*, **15**, 447–484.
27. Richardson, S.M., Hoyland, J.A., Mobasheri, R., *et al.* (2010). Mesenchymal stem cells in regenerative medicine: opportunities and challenges for articular cartilage and interverterbral disc tissue engineering, *J. Cell. Phys.*, **222**, 23–32.
28. Nelson, L., Fairclough, J. and Archer, C.W. (2010). Use of stem cells in the biological repair of articular cartilage, *Expert Opin. Biol. Ther.*, **10**, 43–55.
29. Ahmed, T.A.E. and Hincke, M.T. (2010). Strategies for articular cartilage lesion repair and functional restoration, *Tissue Eng. B*, **16**, 305–329.

# Hydrogels for Tissue Engineering of Articular Cartilage



## CHAPTER 11

Javier S. Castro

## 11.1. Introduction

Tens of millions of people all over the world suffer from arthritis; approximately 30–40 million people in the United States alone suffer from it. Arthritis is a joint disease in which cartilage undergoes irreversible damage, limiting or in severe cases destroying the function of the joints. As a consequence, the quality of life of the patient decreases dramatically, suffering pain, limiting physical activity, or causing complete disability. It is one of the most common causes of physical disability among people older than 65.[1] Arthritis commonly affects the joints of the fingers, knees, hips, and spine, and less frequently affects the wrists, elbows, shoulders, and ankles. The most common risk factors for arthritis can be divided into genetic and non-genetic factors. Genetic factors include disorders of type II collagen, gender (most common in females after menopause), and race. The non-genetic factors include aging, trauma or injuries to joints (caused by accidents or sports injuries), lack of or excessive physical activity, and repetitive joint use and/or joint alignment.

Due to the poor self-regenerative capacity of cartilage, discussed in Chapter 10, current therapies treat the symptoms but are often unable to offer a long-term solution. On the other hand, surgical methods depend on the availability of healthy cartilage in the patient, which is generally not abundant. Another option is to transplant cartilage from external donor sources such as allografts or xenografts from human cadavers or animals. This carries with it intrinsic risks of infection, immune response, pathogen transmission, and graft viability. Possible solutions are to fabricate a synthetic cartilage with adequate biomimetic properties, or the development of scaffolds capable of triggering an accelerated regenerative process.

## 11.2.  The Hydrogel-Like Structure of Cartilage

Cartilage is a tough, translucent, elastic, and flexible connective tissue consisting of a matrix, mainly composed of collagen, proteoglycans, cartilage cells called chondrocytes scattered throughout the matrix, and a high content of water. The cells, chondroblasts and chondrocytes, are dispersed within the matrix and are responsible for cartilage proliferation and maintenance, and occupy less than 2% of the tissue volume. In addition to these components, cartilage is covered by a dense fibrous membrane of connective tissue called the perichondrium, which contains undifferentiated cells that can form chondrocytes or chondroblasts as needed for cartilage repair. The structure is described in detail in Chapter 10. It is of importance to note that articular cartilage lacks nerves and blood vessels, and the transport of nutrients to the cells is carried out by diffusion through the matrix. Both of these characteristics make regeneration and repair of damaged cartilage slow and difficult. Due to its low capacity for self-repair, diffusion-limited nutrient transport, and the potentially lower risk of implant rejection by the immune system, cartilage is one of the primary areas of interest in tissue-engineering research, along with man-made skin, and ahead of liver, heart or kidney tissue.[2] The material of choice for substitute or regeneration of cartilage has been hydrogels, composed of a polymeric matrix filled with water, resembling a structure similar to that of cartilage. Its unique swelling properties, highly hydrated structures, good biocompatibility, and biodegradability make hydrogels especially desirable for biomedical and pharmaceutical applications.[3] Hydrogels have been investigated for a wide range of medical applications such as contact lenses, corneal implants, biosensors, drug delivery systems, and artificial organs, as well as tissue engineering. In tissue engineering some of the potential applications are substitutes for skin, bone, tendons, ligaments, and cartilage.

## 11.3. Current Methods of Treating Cartilage Diseases

There are two main types of cartilage diseases: rheumatoid arthritis (RA) and osteoarthritis (OA). Rheumatoid arthritis is an autoimmune disease in which the immune system of the body attacks its own tissues and is characterized by inflammation of the joint, causing swelling, pain, and loss of function.[4] When cartilage is destroyed, a fibrous tissue grows and later ossifies, fusing the joint and making it immovable. A very common example is observed in the distortion of the fingers of people with RA. Osteoarthritis is a degenerative joint disease in which joint cartilage is gradually lost as a result of ageing, irritation of the joints, wear, and abrasion. As articular cartilage deteriorates, new bone forms, decreasing the space of the joint cavity and restricting the joint movement. Osteoarthritis first afflicts the larger joints (knees and hips), whereas rheumatoid arthritis first strikes smaller joints.[4]

Because cartilage has a very limited capacity for regeneration, there are no current treatments for curing or preventing the diseases. In both OA and RA, non-invasive treatments focus on pain relief and maintaining quality of life. A wide variety of analgesic and anti-inflammatory drugs are used for controlling pain. Intra-articular injections can be used when pain is not controlled by oral or topical approaches. Surgery is the ultimate option when non-invasive methods are unable to provide an acceptable level of comfort and quality of life to the patient. There are a variety of different surgical options including those in which damaged cartilage is taken out and replaced by an autograft or allograft cartilage implant.[5] In severe cases a total joint replacement may be needed. The most common surgical techniques are described in the following section.

### 11.3.1. *Autologous chondrocyte implantation*

An autologous chondrocyte implantation (ACI) procedure is mostly performed in younger patients, 15–55 years of age, with small to medium defects of the cartilage ($<10$ cm$^2$). A small amount of healthy cartilage is harvested arthroscopically from a non-load-bearing area of the same patient. In the laboratory the chondrocytes contained in the harvested cartilage are cultured in a solution rich in nutrients over the course of several weeks; thus, the population of cells is expanded from tens of thousands to tens of millions. In a second surgery, the cultured chondrocytes are injected into the affected area of the patient. Theoretically, the implantation of a new large population of chondrocytes should enhance the process of healing by producing new cartilage.

### 11.3.2. *Osteochondral autograft or mosaicplasty*

This is a surgical technique mostly applied to the knee joint where a damaged section of bone and cartilage is removed from the affected area of the joint. A piece of healthy cartilage and underlying bone is removed from an area of the same patient (autograft) where cartilage is less important. If there is not enough cartilage available from the same patient, it can be taken from a donor (homograft or heterograft). The healthy cartilage and bone are transplanted into the site from which damaged tissue was removed. Osteochondral autograft (OATS) and mosaicplasty are based on the same principle, but the difference is that mosaicplasty takes several small plugs of a few millimeters in diameter to cover the entire damaged area (so giving the appearance of a mosaic pattern), whereas OATS utilizes larger plugs.

### 11.3.3. *Microfracture*

Microfracture surgery is the least invasive treatment for cartilage repair and can be performed completely arthroscopically. It consists of creating puncture wounds in the subchondral bone in the affected area in order to expose the inner part of the bone where marrow cells can influx to the injury site. These cells can grow and produce new cartilage in the damaged area.

### 11.3.4. *Paste grafting*

Damaged cartilage and bone are removed, as described above, and the exposed surface is treated with the microfracture technique, creating a bleeding region and activating the healing process. Also, healthy cartilage and underlying bone are harvested from a non-load-bearing zone and are morselized to form a paste in which other elements such as hydroxyapatite and growth factors are usually added. The paste is then impacted into the affected area. The microfracture wounds facilitate the paste to stick to the surface. As in other methods, this procedure requires a long time for recuperation.

### 11.3.5. *Joint replacement*

In the most severe cases of cartilage damage when intolerable pain and loss of functionality are the primary symptoms and other treatments have failed, a total joint replacement is recommended. A common joint replacement is the knee, and Chapter 12, Total Knee Replacements, discusses the options for this technique. This surgery replaces one or both of the load-bearing surfaces of the joint with prosthetic man-made materials. Metal is used for the femoral surface (top), and the tibial component (bottom) is usually made of a metal tray attached to the bone, containing a tough plastic for the articulating surface. The prosthesis can be cemented or screwed to the bone. Currently minimum invasive surgeries assisted by computer navigated instruments aid the surgeon to align the prosthesis precisely.

Surgical approaches represent options when cartilage needs to be replaced and although they are used routinely with success, they are limited by several disadvantages. The transplantation of chondrocytes, as well as entire cartilage, has many problems associated with it. The availability of cells is limited and the differentiated cells do not guarantee the complete regeneration of the tissue. Even if new cartilage is grown, its durability is uncertain. Current treatments need to be assisted by new approaches including tissue engineering, new biomaterials, and use of regenerative medicine. In the subsequent section, hydrogels are analyzed as potential substitutes or promoters for articular cartilage repair or replacement.

## 11.4. Tissue Engineering of Cartilage: Hydrogels as a New Approach

A common strategy in tissue engineering is to develop a scaffold that mimics the matrix of the tissue that is to be repaired. Cartilage consists of three basic components: water, a matrix, and cells. Water occupies up to 60–80% of the total wet weight of cartilage. Even with the apparent simplicity of cartilage (a collagen matrix, few cells, and water), this natural material possesses remarkable mechanical and biochemical properties, which together make it very difficult to mimic.

Hydrogels are produced from natural or synthetic polymers. Natural hydrogels come from proteins or polysaccharides (i.e. alginates, hyaluronic acid, fibrin glue, chitosan, and agarose) while artificial polymers are produced by materials such as poly(lactic) and poly(glycolic) acid, poly(ethylene) oxide, poly(vinyl) alcohols, polydiol citrates, and polyhydroxyethyl methacrylate. Their structure is composed of swollen randomly cross-linked networks of rod-like polymer chains with water filling the interstitial spaces.[6] Water commonly comprises more than 80% of the total volume. The cross-linking density between polymer chains strongly affects the physical properties of the gel.[7] The strength can be improved by the addition of hydrophobic monomers which may create regions of more densely coiled or entangled chains contributing to increased strength.

Hydrogels can be classified based on the type of cross-linking (covalent or ionic) or as physical gels and entangled networks.[8] Though the stability and toughness of covalently cross-linked gels represent an apparent advantage over physical gels, there are other features that make physical gels attractive for certain applications. Physical gels are formed by non-covalent interactions, such as hydrogen bonding and hydrophobic interactions.[9] Physical gels can be re-melted unlimited times and can be highly biodegradable. The following common hydrogels are presented as examples to illustrate the detail of the hydrogel structure.

One of the most commonly used hydrogels is composed of *alginate*. It is a linear polysaccharide obtained from marine algae and various bacteria composed of two acidic monomers: 1,4-linked $\beta$-D-mannuronate (M) and $\alpha$-L-guluronate (G)[10] arranged along a linear chain ranging in size from 3.7–18.2 nm.[11] The interaction among molecules is due to divalent cation bridges such as $Ca_2$, $Mg_2$, $Cu_2$, and $Zn_2$ which facilitates gelation and strongly influences the physical properties. Another common hydrogel is *agarose* which is a neutral polysaccharide extracted from marine red algae (Rhodophyceae). Its idealized structure consists of a double helix with alternating residues of 1,4-linked 3,6-anhydro-$\alpha$-L-galactopyranose and 1,3-linked $\beta$-D-galactopyranose[12] that aggregate, forming long stiff rods organized in a 3D network cross-linked by hydrogen bonds which leaves large voids occupied by solvent. Because an agarose gel is governed by hydrogen bonds, it is classified as a physical gel. Figure 11.1 shows the microstructure of a typical agarose gel. The pore size changes as a function of agarose concentration or other factors. This sample was 1 wt% agarose concentration in water.

The properties of hydrogels are determined by the monomer composition, cross-linking density, and polymerization conditions.[13] Attempts to improve hydrogel properties for load-bearing biomedical applications have included the introduction of composite materials such as rubber or glass, the use of cross-linking agents such as glutaraldehyde, and the use of freeze-thawing procedures to induce partial crystallinity.[14]

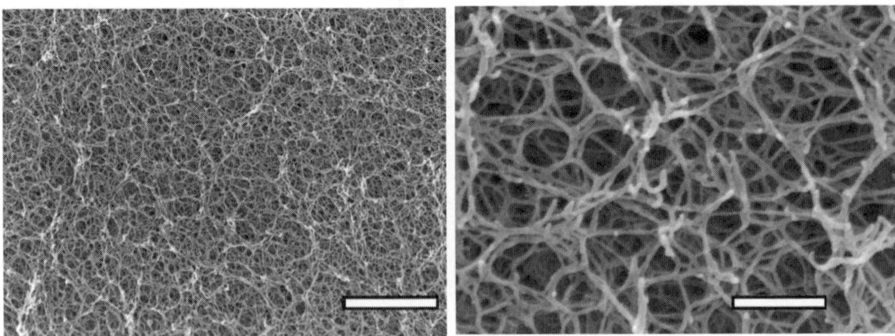

**Figure 11.1.**   FEG-SEM micrographs of the structure of an agarose gel. Left: low magnification (bar = 2 μm). Right: high magnification (bar = 250 nm).

The scientific literature published regarding hydrogels is vast and varied. Several disciplines contribute to this body of work and much of the research relies on multi-disciplinary teams. Tissue engineers use hydrogel scaffolds to support replication of chondrocytes for production of cartilage *in vitro*, whereas regenerative medicine proposes the delivery of chondrocytes or mesenchymal stem cells by injecting a hydrogel-cell suspension to the damaged area to activate the production of new cartilage.[15]

The use of scaffolds in tissue engineering is one of the most promising technologies. Scaffolds are biocompatible and biodegradable materials that support cellular in-growth of a 3D network, provide mechanical strength, allow cells to adhere tightly, promote uniform cell spreading, and conserve the phenotype and functional characteristics of transplanted cells.[16] Many hydrogels are excellent scaffolds for cartilage tissue engineering because they are biocompatible, biodegradable, they mimic the properties of the cartilage extra cellular matrix, and act as a medium for ion diffusion. Several hydrogels can support chondrocyte cultures in 3D structures. Three-dimensionality is critical to retain the phenotype of chondrocytes,[17] providing mechanical integrity to the new tissue, and facilitating the diffusion of nutrients to the cells. Hydrogel systems mimic the *in vivo* conditions essential to create engineered cartilage with proper function.[18]

Temporary scaffolds, which provide shape and serve as conductive templates for initial cell arrangement, can be appropriate cell delivery materials for subsequent *in vivo* implantation. Schagemann and colleagues have proposed a biodegradable and biocompatible hydrogel as a potential temporary scaffold for cartilage regeneration. The hydrogel is composed of alginic acid and gelatin, forming a 3D matrix with high water content (90%), and has proved to be an excellent template for homogeneous chondrocyte growth.[19] Malicev *et al.* reported the use of a collagen-porous fleece used as a biodegradable scaffold for chondrocyte cultures. Chondrocytes were diluted in blood plasma and seeded on the scaffold.[20]

Alternative variants are used to enhance replication of cells. For example, Xu *et al.* have studied the effect of perfusion (flow) over the growth of cartilage in alginate and agarose scaffolds. They compared cultures of chondrocytes in alginate and agarose scaffolds under static and perfusion conditions. Perfusion bioreactors can maintain a stable culture environment by continuously feeding culture medium and removing waste. Bioreactors can also exert forces over cells to provide a mechanical stimulus that may influence the growth of chondrocytes and the generation of extracellular matrix. Xu's team found only a positive effect of perfusion in alginate scaffold, where the cell-alginate perfusion culture produced more collagen and released more collagen into the medium than that under static culture conditions.[21] Another variant was presented by Zehbe *et al.*, in which a combination of a pre-orienting electrochemical process was used with a directional freezing technique to produce a perpendicular pore-channel-structure with structural, mechanical, and biochemical properties similar to native cartilage. They reported that after chondrocyte culture, cells were oriented parallel to the surface, and chondrocytes inside the gelatin-based scaffold were oriented in a columnar fashion, similar to natural cartilage.[22]

Regenerative medicine concepts applied to cartilage could be a promising solution for cartilage diseases. Mesenchymal stem cells (MSCs) are critical in this approach because they can be harvested in a less invasive way and are easily isolated and expanded, including by chondrogenesis. Yokoyama and colleagues achieved the formation of cartilage from composites of synovium-derived MSCs and collagen gel *in vitro* without a scaffold. Collagen gel was mixed with the same volume of cell suspension medium (synovium-derived MSCs) and then incubated for 21 days on culture plates. MSCs-gel composite revealed the differentiation of MSCs towards chondrocytes, demonstrating that synovium-derived MSCs-collagen-gel composites can form cartilage tissues *in vitro*.[23]

Several injectable gels have been used as the carrier of cells for tissue engineering of bone or cartilage. Xu *et al.* studied the formation of bone and cartilage injectable hydrogels containing gene-transduced bone marrow stromal cells (BMSCs) and bone morphogenic proteins (BMPs), a bone growth factor, in a bioreactor. After a period of time, new bone and cartilage-like tissue was formed. They reported that collagen and hyaluronate were the best carriers for bone formation whereas agarose was the best carrier for cartilage formation. They also found that BMPs are excellent promoters of cartilage differentiation and matrix maturation in cartilage.[24] In a similar approach, Stevens *et al.* created a bioreactor *in vivo* by injecting alginate or hyaluronic gels into the subperiosteal space in the tibia of rabbits. The gels provide a scaffold for new bone and enhanced development of osteocytes. They successfully grew a significant amount of bone with excellent mechanical properties. This method, according to the authors, could be used to produce cartilage as well.[25]

Kang *et al.* have developed a novel class of biodegradable polyester elastomers referred to as poly(diol citrates) and have found that poly(1,8-octanediol citrate) (POC) exhibits the characteristics suitable for cartilage tissue engineering.[26] Its degradation and mechanical properties can be modulated by changing the conditions used to cross-link the pre-polymer (i.e. temperature and time). POC-chondrocyte constructs were cultured for several days, and their number increased by almost three-fold. The culture revealed uniform pores. One feature of this gel was its mechanical properties with Young's modulus values in the order of tens of kPa.[27]

Polyvinyl alcohol (PVA) hydrogels can be fabricated with similar water content as cartilage and a low coefficient of friction for load-bearing applications. PVA is soluble but not biodegradable, therefore molecular weights below 16 kDa must be used to ensure the chains can be removed by the kidneys. Biocompatibility studies of PVA have not reported inflammatory or degenerative changes in the cartilage adjacent to PVA hydrogels after several weeks of implantation.[28] Salubria[TM] is a PVA cryogel (PVA-c) which is a PVA hydrogel containing 0.9% saline and thermally crystallized by a repeated process of freeze-thawing.[29] This process has improved the mechanical properties of the PVA-c reporting a failure stress of 5 MPa and strain in tension as high as 226%,[30] values similar to those of articular cartilage. They suggest the incorporation of bioactive agents such as heparin, thrombin inhibitors, growth factors, and glycosaminoglycans can be incorporated into Salubria[TM] cryogel.[13] There is a direct relationship between PVA friction and its stiffness and roughness.[31] Nakashima *et al.* have created a layered film of albumin and gamma-globulin (both present in natural synovial fluid) over the surface of PVA hydrogel; as a protective film it has low friction for reduction of wear, which is essential to prevent failure.[31]

Despite hydrogels having excellent biocompatibility and ability to support cell growth, there are problems to solve in order to achieve an engineered cartilage tissue with characteristics similar to those of real cartilage. The first problem is their insufficient mechanical strength compared to articular cartilage. This severely limits the use of hydrogels as substitutes for natural tissues or for load-bearing applications such as articular cartilage. The second difficulty is related to the regenerative process of natural cartilage. As explained previously, collagen and proteoglycans (produced by the chondrocytes) are two essential components of the cartilage matrix. The collagen fibers resist tensile loads whereas the proteoglycans enhance resistance to compression. Although chondrocytes can synthesize proteoglycans reasonably quickly, reaching natural levels within a few weeks, collagen synthesis is significantly slower. This inability to produce the required levels of collagen in reasonable periods of time has been a major problem in tissue-engineered cartilage due to the crucial function of this matrix constituent.[2] These disadvantages motivate scientists to seek hydrogels with improved properties as well as better techniques for tissue regeneration of cartilage.

## 11.5. Ethical Issues

Allografts always have the risk of pathogen transmission and immunological rejection, and cartilage is no exception. Even though the risk of immuno-rejection is minimal in cartilage, the issue of infection is always present.[32] This is because common aseptic treatments for living tissues, such as cartilage, do not guarantee its complete sterilization.[33] Synthetic grafts offer a safer option for cartilage replacement, for at least the risk of pathogen transmission is eliminated. These materials, however, must be produced under good manufacturing practice conditions to meet FDA regulations. This requires rigorous manufacturing processes under clean room conditions, effective process of sterilization, and high quality control.[34]

Artificial cartilage has been tested for different characteristics such as mechanical strength, frictional behavior, biocompatibility, etc. However, most of these tests do not represent the real conditions of the human joint. For example, the mechanical properties of synthetic cartilage are commonly tested in simulator machines where the material is put under cycling loads simulating the joint movement for a period of time. This is a good approach for testing mechanical properties but it does not assure the same behavior under in vivo conditions where different factors such as the chemical environment, sudden loads application, torsion or shear stresses affect the tissue at the same time. Use of animal trials, specifically primates who possess similar anatomy, physiology, and biomechanics to humans, are becoming more difficult to justify on ethical grounds because of the uncertainty of physiological equivalence. A real need is consistent comparisons of statistical survivability data for current methods of repair to evaluate the risk/reward ratio for new approaches being developed before they are introduced into the clinic.[35]

## 11.6. Economic Issues

Current surgical treatments for cartilage disease in the United States are expensive. Even though the costs varies depending on many factors, a rough estimate follows: costs of an osteochondral autograft, mosaicplasty or a microfracture treatments are in the order of US$15,000–20,000; a total knee replacement or an autologous chondrocytes implantation (ACI) cost between US$20,000 and US$35,000. In all cases, rehabilitation is mandatory which increases the total cost by approximately US$5,000. In the case of hydrogels as potential cartilage replacement, the costs are uncertain. Although hydrogels as raw material are cheap, there are other factors such as manufacturing process, cell culture, and surgical procedures which will increase the prices dramatically. The cost of cartilage replacement with tissue-engineered cartilage will probably be around the same cost of ACI or may be more.

## 11.7. Recent Developments

There is a broad spectrum of materials from which hydrogels can be prepared for use as articular cartilage tissue-engineering constructs. A recent review by Ahmed and Hincke demonstrates the scope of materials and combinations of materials which have been studied for constructing viable tissue-engineering scaffolds for cartilage.[35] The two main types of materials that are employed to engineer an articular cartilage scaffold can be characterized as a natural polymer or an artificial polymer. In either case, the scaffold needs to assume a 3D, highly porous structure which allows for a homogenous distribution of cells. The natural polymer scaffolds can be subdivided further into those derived from carbohydrate-based scaffolds (i.e. agarose, alginate, chitosan, hyaluronic acid derivatives, etc.), protein-based scaffolds (i.e. collagen, fibrin, gelatin, and engineered proteins), and a new class of peptide-based polymer scaffolds.[36,37]

Carbohydrate-based polymers have given rise to a great number of suitable polymeric scaffold materials due to the molecular environment created by the carbohydrate backbone which resembles the glycosaminoglycan- (GAG) rich environment found in native articular cartilage. This distinct characteristic of carbohydrate-based scaffolds provides an inherent advantage in that the GAG-like environment induces chondrogenesis.[38] The use of novel protein-based scaffolds has recently demonstrated their potential as candidates for articular cartilage scaffolds with high tunability of domain size, mechanical properties, and incorporation of subunits for cellular signaling.[39] Recent research of a new class of peptide-based scaffolds investigates self-assembling peptide hydrogels which promote chondrogenesis along with unique cell–cell interactions and cellular signaling properties.[40–42] Considerable progress has also been made in the development of hydrogels for cartilage scaffolds using artificial polymers (e.g. polyethylene glycol polymers) which may also incorporate natural polymers, proteins or other bioactive molecules, and allow for control over hydrogel micro-architectural properties.[43–45]

Additionally, scaffolds engineered on the nanoscale have recently become of particular interest to researchers due to the design control achievable at this level. The development of nanofibers for cartilage tissue-engineering constructs includes manipulation of materials on the nanoscale to create controlled 3D mesh structures for zonal imitation of cartilage and self-assembling polypeptides.[46] The development of nano-structured hydrogel materials promises to become one of the most flourishing areas currently under investigation with the advent of nanotechnology and its many applications throughout scientific research. Hydrogel development for articular cartilage regeneration provides vast future research opportunities due to the possible combinations of materials, processing techniques, cell choice, and incorporated bioactive moieties. Future research also will likely focus on simulating the mechanical forces experienced by developing native articular cartilage tissue *in vitro* and further optimizing the biomechanical properties of hydrogel materials.

## 11.8. Summary

Current treatments for cartilage disease are limited. Thus, new options such as tissue engineering, regenerative medicine, and synthetic materials are being developed. The unique properties of hydrogels, natural or synthetic, make them potential candidates for tissue engineering of cartilage. Despite the fact that hydrogels have many advantages such as excellent biocompatibility, biodegradability, being good scaffolds for chondrocyte growth, availability to be injected, and not representing a risk factor of pathogen transmission, their mechanical properties need to be improved to mimic those of articular cartilage. Many researchers are currently attempting to fabricate hydrogels with improved mechanical properties.

## References

1. NIDCD (National Institute on Deafness and Other Communication Disorders) (2011). *CHID Online Database News*. [Online]. Available at: http://www.chid.nih.gov. [Accessed 30 July 2007].
2. Ateshian, G.A. (2007). Artificial cartilage: weaving in three dimensions, *Nat. Mater.*, **6**, 89–90.
3. He, H., Li, L. and Lee, L.J. (2006). Photopolymerization and structure formation of methacrylic acid based hydrogels in water/ethanol mixture, *Polymer*, **47**, 1612–1619.
4. Tortora, G.J. and Grabowski, S.R. (2003). *Principles of Anatomy and Physiology*, 10th ed., John Wiley & Sons, Hoboken, NJ.
5. Simon, T.M. and Jackson, D.W. (2006). Articular cartilage: injury pathways and treatment options, *Sports Med. Arthrosc.*, **14**, 146–154.
6. Cram, S., Brown, H., Spinks, G., *et al.* (2005). Hydrophobically modified acrylamide-based hydrogels. Smart materials III, *P. SPIE*, **5648**, 153–162.
7. Kong, H.J, Kim, C.J., Huebsch, N., *et al.* (2007). Non-invasive probing of the spatial organization of polymer chains in hydrogels using fluorescence resonance energy transfer (FRET), *J. Am. Chem. Soc.*, **129**, 4518–4519.

8. Kavanagh, G.M. and Ross-Murphy, S.B. (1998). Rheological characterisation of polymer gels, *Prog. Polym. Sci.*, **23**, 533–562.

9. Barrangou, L.M., Daubert, C.R. and Foegeding, E.A. (2006). Textural properties of agarose gels. I. Rheological and fracture properties, *Food Hydrocolloid*, **20**, 184–195.

10. De Boisseson, M.R., Leonard, M., Hubert, P., *et al.* (2004). Physical alginate hydrogels based on hydrophobic or dual hydrophobic/ionic interactions: bead formation, structure, and stability, *J. Colloid Interf. Sci.*, **273**, 131–139.

11. Decho, A.W. (1999). Imaging of alginate polymer gels using atomic-force microscopy, *Carbohyd Res.*, **315**, 330–333.

12. Schafer, S.E. and Stevens, E.S. (1995). A reexamination of the double helix model for agarose gels using optical rotation, *Biopolymers*, **36**, 103–108.

13. Stammen, J.A., Williams, S., Ku, D.N., *et al.* (2001). Mechanical properties of a novel PVA hydrogel in shear and unconfined compression, *Biomaterials*, **22**, 799–806.

14. Zheng-Qiu, G., Jiu-Mei, X. and Xiang-Hong Z. (1998). The development of artificial articular cartilage-PVA-hydrogel, *Biomed. Mater. Eng.*, **8**, 75–81.

15. Peterson, L., Brittberg, M., Kiviranta, I., *et al.* (2002). Autologous chondrocyte transplantation: biomechanics and long-term durability, *Am. J. Sports Med.*, **30**, 2–12.

16. Baek, C.-H. and Ko, Y.-J. (2006). Characteristics of tissue-engineered cartilage on macroporous biodegradable PLGA scaffold, *Laryngoscope*, **116**, 1829–1834.

17. Diaz-Romero, J., Gaillard, J.P., Grogan, S., *et al.* (2005). Immunophenotypic analysis of human articular chondrocytes: changes in surface markers associated with monolayer culture, *J. Cell Physiol.*, **202**, 731–742.

18. Vinatier, C., Guicheux, J., Daculsi, G., *et al.* (2006). Cartilage and bone tissue engineering using hydrogels, *Biomed. Mater. Eng.*, **16**, S107–S113.

19. Schagemann, J.C., Mrosek, E.H., Landers, R., *et al.* (2006). Morphology and function of ovine articular cartilage chondrocytes in 3-D hydrogel culture, *Cells Tissues Organs*, **182**, 89–97.

20. Malicev, E., Radosavljevic, D. and Velikonja, N.K. (2007). Fibrin gel improved the spatial uniformity and phenotype of human chondrocytes seeded on collagen scaffolds, *Biotechnol. Bioeng.*, **96**, 364–370.

21. Xu, X., Urban, J.P.G., Tirlapur, U., *et al.* (2006). Influence of perfusion on metabolism and matrix production by bovine articular chondrocytes in hydrogel scaffolds, *Biotechnol. Bioeng.*, **93**, 1103–1111.

22. Zehbe, R., Liberab, J., Gross, U., *et al.* (2005). Short-term human chondrocyte culturing on oriented collagen coated gelatine scaffolds for cartilage replacement, *Biomed. Mater. Eng.*, **15**, 445–454.

23. Yokoyama, A., Sekiya, I., Miyazaki, K., *et al.* (2005). *In vitro* cartilage formation of composites of synovium-derived mesenchymal stem cells with collagen gel, *Cell Tissue Res.*, **322**, 289–298.

24. Xu, X.L., Lou, J., Tang, T., *et al.* (2005). Evaluation of different scaffolds for BMP-2 genetic orthopedic tissue engineering, *J. Biomed. Mater. Res. B*, **75B**, 289–303.

25. Stevens, M.M., Marini, R.P., Schaefer, D., *et al.* (2005). *In vivo* engineering of organs: the bone bioreactor, *Proc. Natl. Acad. Sci. USA.*, **102**, 11450–11455.

26. Kang, Y., Yang, J., Khan, S., *et al.* (2006). A new biodegradable polyester elastomer for cartilage tissue engineering, *J. Biomed. Mater. Res.*, **77A**, 331–339.

27. Oka, M., Noguchi, T., Kumar, P., *et al.* (1990). Development of an artificial articular cartilage, *Clin. Mater.*, **6**, 361–381.

28. Ku, D.N., Braddon, L.G. and Wootton, D.M. (1999). Poly (vinyl alcohol) cryogel. US patent no. 5981826. [Online]. Available at: http://www.freepatentsonline.com/5981826.html.

29. Swieszkowski, W., Ku, D.N., Bersee, H.E.N., *et al.* (2006). An elastic material for cartilage replacement in an arthritic shoulder joint, *Biomaterials*, **27**, 1534–1541.

30. Covert, R.J., Ott, R.D. and Ku, D.N. (2003). Friction characteristics of a potential articular cartilage biomaterial, *Wear*, **255**, 1064–1068.

31. Nakashima, K., Sawae, Y. and Murakami, T. (2007). Influence of protein conformation on frictional properties of poly (vinyl alcohol) hydrogel for artificial cartilage, *Trib. Lett.*, **26**, 145–151.

32. Schoepf, C. (2006). Allograft safety: efficacy of Tutoplast© process, *Implants*, **1**, 10–15.
33. Persidis, A. (1999). Tissue engineering, *Nat. Biotechnol.*, **17**, 508–510.
34. Hench, L.L. (2001). *Science, Faith and Ethics,* Imperial College Press, London.
35. Ahmed, T.A.E. and Hincke, M.T. (2010). Strategies for articular cartilage lesion repair and functional restoration, *Tissue Eng. B*, **16**, 305–324.
36. Puppi, D., Chiellini, F., Piras, A.M., *et al.* (2010). Polymeric materials for bone and cartilage repairs, *Prog. Polym. Sci.*, **35**, 403–440.
37. Nge, T.T., Nogi, M., Yano, H., *et al.* (2010). Microstructure and mechanical properties of bacterial cellulose/chitosan porous scaffold, *Cellulose*, **17**, 349–363.
38. Segupta, D. and Heilshorn, S.C. (2010). Protein engineered biomaterials: highly tunable tissue engineering scaffolds, *Tissue Eng. B*, **16**, 285–293.
39. Kopesky, P.W., Vanderploeg, E.J., Sandy, J.S., *et al.* (2010). Self-assembling peptide hydrogels modulate *in vitro* chondrogenesis of bovine bone marrow stromal cells, *Tissue Eng. Pt. A*, **16**, 465–477.
40. Kyle, S., Aggeli, A., Ingham, E., *et al.* (2009). Production of self-assembling biomaterials for tissue engineering, *Trends BioTechnol.*, **27**, 423–433.
41. Kisiday, J., Jin, M., Kurz, B., *et al.* (2002). Self-assembling peptide hydrogel fosters chondrocyte extra-cellular matrix production and cell division: implications for cartilage tissue repair, *Proc. Natl. Acad. Sci. USA.*, **99**, 9996–10001.
42. Jukes, J.M., van der Aa, L.J., Hiemstra, C., *et al.* (2010). A newly developed chemically crosslinked dextran poly(ethylene glycol) hydrogel for cartilage tissue engineering, *Tissue Eng. A*, **16**, 565–573.
43. Tan, H., DeFrail, A.J., Rubin, J.P., *et al.* (2009). Novel multiarm PEG-based hydrogels for tissue engineering, *J. Biomed. Mater. Res. A*, **92A**, 979–987.
44. Annabi, N., Nichol, J.W., Zhong, X., *et al.* (2010). Controlling the porosity and microarchitecture of hydrogels for tissue engineering, *Tissue Eng. B*, **16**, 371–383.

45. Shah, R.N., Shah, N.A., Del Rosario Lim, M.M., *et al.* (2010). Supramolecular design of self-assembling nanofibers for cartilage regeneration, *Proc. Natl. Acad. Sci. USA.*, **107**, 3293–3298.
46. Klein, T.J., Rizzi, S.C., Reichert, J.C., *et al.* (2009). Strategies for zonal cartilage repair using hydrogels, *Macromol. Biosci.*, **9**, 1049–1058.

# Total Knee Replacements

CHAPTER **12**

Jason Wertz

## 12.1. Introduction

Total knee replacement, or total knee arthroplasty (TKA), is commonly used when deterioration of the knee occurs from arthritis or by injury, and utilizes prosthetic implants to replace the damaged bearing surfaces of the joint.[1] Arthritis causes deterioration of the articular cartilage that forms a bearing surface between the femur and the tibia, and which also provides a cushioning effect on the joint. Chapters 10 and 11 discuss the structure of cartilage and the damaging effects of arthritis. When cartilage deteriorates, bone comes into direct contact with bone, causing intense pain, and damage to the bone. The improvement of prosthetics and development of implantation techniques has seen an evolution in TKA surgery to the point where it is a routine clinical solution to chronic pain of the knee joint. These changes have helped to improve the quality of life of TKA patients, when no other options are available.

## 12.2. History of Knee Replacement Surgery

In the early 1970s, the biomechanics of the knee were investigated, and this paved the way for replacement of the joint. Tests were performed *ex vivo* to determine the location and the size of the load-bearing area.[2] From these experiments, the first design criteria were established and the concepts of joint replacement could focus on the anatomical portions of the knee relating to load-bearing stresses. One of the first prosthetic knee replacements was designed to create stability, prevent hyperextensions, and have improved fixation without causing toxic reactions from the host tissues. The devices were built with hinges, terminal stops, and rods, in an attempt to create a replacement for the normal knee. Chapter 6 in Kao *et al.*[1] has photographs of the early and later knee prostheses and the survivability of the devices based upon peer-reviewed clinical studies. The hinged elements of the early device were made of a cobalt–chromium alloy and were cemented into the intramedullary canal with poly(methyl)methacrylate (PMMA) bone cement. These devices provided a limited range of mobility since the axial rotation of the knee was not considered, and the device was also bulky and heavy. Most of the problems that occurred were due to the fracture of bone or loosening of parts within the prosthetic.

A second device was then created that utilized linked surface components, which allowed for the extension, gliding, and rotation of the knee. Some limitations from the first device were solved but there were still problems with limited axial rotation.

The third-generation device was an unlinked surface replacement unit, which was unique in that it depended on good quality soft tissue surrounding the prosthesis. Once implanted, the device allowed for the transmission of forces without use of intramedullary stems for fixation.

Many limitations were found with all three types of device, ranging from limitation of movement to not being able to maintain natural loading of the bone. Numerous articles describe the merits and limitations of the devices.[1,3–9] Use of ultra-high-molecular-weight polyethylene (UHMWPE) components improved wear of the mating surfaces in the device. UHMWPE is commonly used as the acetabular cup in total hip replacement procedures and therefore has a long clinical track record. This hard polymer provides a much higher wear resistance compared to other polymers. However, wear debris is still produced from the acetabular cup and this is one of the main reasons for premature failure of hip implants. In the knee, motion still created wear debris, which came into contact with the surrounding tissue creating an immune response and resulting in aseptic bone loss and loosening of the prostheses[7]. Heavier patients who exert more load on the prosthesis also caused an increase in the amount of damage to the polyethylene due to loads as high as 13 MPa, which the polyethylene does not have the ability to withstand. Testing of the polyethylene and metal alloy TKA revealed problems with using the polyethylene parts against the cobalt–chromium alloy femoral surface. Alternative polymers that more closely mimic cartilage properties are needed.[8]

## 12.3. Current Procedures

The knee prosthesis evolved to a new design intended to reduce wear debris from the contact between the polyethylene and alloy components.[10–13] By increasing the thickness of the polymer components, there was an overall decrease in the stresses at the cement–bone interface. The stresses and strains of the prosthesis could also be decreased by inserting the polyethylene component into a metal tray or more simply using a metal tibial plate. The metal plates were still cemented into the tibia, but the stresses at the bone-cement interface were reduced by three to five times.

## 12.4. Future Approaches

New testing and *in situ* monitoring techniques are constantly being developed. One such technique currently being used is the implantation of measuring devices within the implant to measure the stresses that occur within the human knee. In one study, the device was built into the prosthetic to measure the six multi-axial forces within the implant and knee.[10] The testing of these devices could lead to the understanding of the rehabilitation, bracing, and activities that the knee and the implant are subjected to in daily use.

The cementing of the knee replacements continues to be a concern. There is a loosening of the tibial components because of the degradation of the cement–metal alloy interface.[11] One reason is cement shrinkage, which will lead to micromotion of the implant and then damage to the brittle cement, causing wear particles and cracking of the cement. To improve the interface, the metal surface was coated with silicon monoxide particles with a silicate layer protected with a polymer varnish to prevent cracking of the cement.[11] Stress cracking was reduced, thus improving the lifespan of the implant.

Other aspects of the TKA are also being considered to improve the device. One example is to modify the intramedullary plug that connects the prosthesis into the bone. The plug has a sliding mechanism to allow for a snug fit within the canal and also to allow for the extension of the prosthetic component.[12] The concept is to create a tighter fit of the plug that would allow for better cementing of the prosthesis within the canal. The study also discovered that the use of antibiotic-soaked apatite-wollastonite glass ceramics (AW-GC), which is used clinically as a bone substitute under the name Cerabone®, could be used to treat osteomyelitis and infected arthroplasties.[12] See Chapter 8 for a discussion of AW-GC bioactive glass-ceramics.

The effects of wear on the polyethylene component continue to be an issue. Current studies are focusing on the thickness of the insert to determine what effect this variation has on the prosthesis. The use of a 10 mm polyethylene insert showed similar wear properties to those of the thicker inserts.[13] Smaller inserts have been found to be more beneficial, as less bone has to be removed to insert the implant. To test the effect of variations in thickness two tests were performed where the knee prostheses were loaded at two different angles: at 0 degrees, and at 60 degrees. In each case the wear patterns of each insert were observed. Both inserts exhibited similar mechanical properties, but the thinner insert required less bone to be removed, and was therefore selected for future implants.[13]

Computerized programs are being developed that analyze the surgical skills of the surgeon along with the accuracy of the procedure.[14] By considering each individual step, the computer can not only monitor the position of the prosthetic, but also its orientation as it is being implanted into the patient. This computerized program could help ensure the device is accurately aligned with the connecting bones and tissue, resulting in a perfect fit. Although currently the program is being used only as a skills test for surgeons,[14] it may be used in the future for actual surgery. Since the program utilizes an infrared coordinate measuring system, there is a possibility that this type of technology will be used on patients post-operatively to determine the success and accuracy of the surgery.

Robots can assist in the alignment of the prosthetic devices to help eliminate the need for human judgement and accuracy, and to try and determine the correct orientation or position. These robots mount directly to the bone and have specific tasks.[15] For example, one might make cuts of the bone, by first positioning itself to make sure that each cut it makes is at the correct angle and position relative to where the prosthesis is to be placed. The benefit of using the robots is the ability to achieve high consistency of the surgical procedures being performed. The robots reduce the chance of human error and allow for faster recovery of the patient due to the less invasive surgical procedures they are capable of performing. The accuracy of the robot reduces the chance of an early revision being needed due to misalignment of the prosthesis.[15]

## 12.5. Ethics Issues: Survivability of TKAS

The regulatory body, the Food and Drug Administration (FDA) rates a TKA as a Class 3 device, which means that it has a high-risk outcome to the patient and it is an implant that will be placed into the body for a long period of time.[16] While this device is built to be a long-term prosthetic, it carries the risk of failure, as do all prostheses. With revision surgery, the surgical procedures for TKA become more difficult. The age of a patient receiving either a new prosthetic or a revision becomes an ethical issue. There is a risk of the patient not waking up from anaesthesia, dying from a blood clot or heart failure, or getting an infection. Balancing these risks against the reward of walking again is a decision that requires understanding the survivability of the prosthetic. On average, a total of 85% of the

TKA prosthetics will last approximately 13 years.[1,17] From this analysis, there is general agreement that a patient should be over the age of 55 to be a candidate for total knee replacement surgery due to arthritis or injury. TKA implants currently have a 90% survival rate after 18 years of implantation. Younger patients below the age of 55 have not shown such success rates. Revisions are often required because of infection, through wear of the polyethylene components, and owing to loosening of the patellar components.[18]

## 12.6.  Economic Issues

The economic impacts of TKA devices are also important.[19–21] An estimate of the total hospital cost for the total knee replacement surgery was around US$11,500 in 1992,[19] of which the cost of the actual prosthesis was about 24%, or US$2,760. These costs have increased considerably since then. On average, patients stay 6.6 days for total knee replacement surgery.[20] The average stay for a patient for revision surgery is about 7.5 days.[20] This is due to a 41% increase in time spent in the operating room revising the implants.

## 12.7.  New Approaches for Younger Patients: Knee Resurfacing

Improvement of low survivability rates of TKA for patients below 55 years of age is an important goal. Patients in this age group have either damaged their cartilage due to repetitive stress sporting injuries, such as road running; trauma; or they have rheumatoid arthritis, which is a congenital disease. The problem is that a TKA does not adapt to changes in a patient's lifestyle or living environment in the way that natural bone does. There are therefore two options for improving the health-care procedures for younger patients:

1.  Tissue engineering, where new bone and/or cartilage constructs are grown on scaffolds and implanted as discussed in Chapters 10 and 11. However, articular cartilage is difficult to

grow and these procedures are likely to be several years before they enter routine clinical use.

2. A rethinking of the TKA procedure. Clinicians in the United States have modified the hip resurfacing technique (Birmingham Hip), which was developed in the United Kingdom, for use on the knee. The aim of the procedure is to remove very little bone from the tibia so that it is still intact when a full TKA is needed later in life. The tibia is resurfaced with a smooth metal alloy so it is a metal-on-metal-bearing surface. This removes the possibility of polyethylene wear debris. This procedure is only possible for younger patients as the bone must be of high quality.

## 12.8. Recent Developments

Surgical procedures for total knee arthroplasty have experienced a great deal of technological advancement with the advent of robotic assisted surgery, improved imaging, image-guided techniques, and less invasive approaches. Minimally invasive total knee arthroplasty (MIS TKA) has become an increasingly popular technique due greatly to patient preference for faster recovery time and a shorter, more aesthetic scar.[22] MIS TKA uses a 7–10 cm incision, as compared to the 20 cm incision with traditional TKA, and also spares the quadriceps muscle, allowing for reduced recovery time and diminishing the need for rehabilitation. However, a recent study that analyzed surgical complications found that many surgeons have reported significantly higher incidence of severe complications with MIS TKA, as compared to traditional TKA procedures.[22] The most common and potentially serious complications were due to longer tourniquet times (used to cut off the blood supply below the thigh to allow the operation to go ahead), misaligned or improperly placed components, and early implant failure. The study concluded that MIS TKA is associated with unnecessary risk for the patient and that the procedure does not have defined indications or merit for regular clinical use. There is a need to develop improved surgical techniques, by utilizing technology such as robotic assistance and computer navigation, to develop a feasible, less invasive TKA procedure.

Implant failure remains the greatest challenge to overcome in TKA, especially as the demand for extended prosthetic lifetime increases with the increase in patient life expectancy and more demanding active lifestyles. Knee implants will need to be developed which can withstand greater forces over longer periods. Improving implant design by accounting for biomechanical factors and creating advanced materials is essential to prolonging knee prosthesis. Of the more than 150 knee implant models available, none have been found to be optimal in mimicking the natural knee.[23] Failure of knee implants continues to be a result of poor biological fixation, implant surface wear debris, and insufficient mechanical behavior, particularly at the implant interface. Considerable progress has been made in knee implants through *in vitro* testing and the implementation of computational models for examining biomechanics and identifying points of failure during the design process. Finite element analysis (FEA) studies have been employed to improve current models, and to evaluate non-linear material properties of the various knee implant components, and the interactions with the biological tissue present in the knee.[23] Further investigation of the current models of knee implants and the natural knee by finite element analysis (FEA) will be needed, to allow for a smarter, more efficient device design process.

Biocompatibility issues of total knee implants pose some of the most daunting engineering challenges due to the complex inter-related considerations of materials, biomechanics, and *in vivo* response. Three major topics of particular concern include:

1.  initial implant osseointegration;
2.  damage due to wear by-products; and
3.  implant longevity.[24]

Importantly, knee prostheses need to obtain stability within the bone soon after implantation, and the interface of the bone and implant must be capable of tolerating high interfacial shear stress while remaining firmly positioned.[24] Osseointegration can be enhanced by coating the implant with a bioactive material. Use of bioactive ceramic coatings, such as hydroxyapatite, has demonstrated osteoconductive behavior, such as described in Chapter 14 for fixation of total hip prostheses. Development of new bioactive implant coatings with incorporated chemical signals and improved mechanical properties may lead to more rapid osseointegration and therefore greater implant stability. Wear by-products produced at the articulating surfaces of knee implants cause aseptic loosening, periprosthetic osteolysis, and inflammatory response.[24] Much focus has been placed on gaining knowledge of biological interactions with wear debris including particle size, mechanism of formation, debris migrations, and elicited host response. A better understanding of the effects of the implant derived particles at the nano- and molecular-level will be required to develop new materials and improve clinical outcomes in the future.[24]

Recent advances in a wide variety of materials for the construction of knee prosthesis, including metal alloys, polymers, and ceramics, have shown promise for the future of total knee arthroplasty. For example, a review by Geetha *et al.* illustrates the advantages of titanium alloys as superior implantable biomaterials.[25] The study gives a detailed view of the positive aspects of titanium alloys, such as biocompatibility and corrosion resistance, and also discusses the need for further investigation into issues such as wear properties and surface coatings. Recent research has also focused on the improvement of UHMWPE for use as a bearing surface in total joint replacement with the introduction of a second-generation, highly cross-linked UHMWPE that possesses superior mechanical and wear properties.[26,27] The investigation of advanced ceramics such as alumina–zirconia and nano-structured composites has also been an area of significant interest, especially with regard to the potential design of ceramic-on-ceramic implant surfaces.[28] With the number of variables that must be considered, additional research is still required to devise the ideal total knee prosthesis.

## 12.9. Summary

Overall the effectiveness of total knee replacement surgery is increasing, but the improvements are iterative. The main cause of implant failure is wear debris from the polyethylene surface; however, no successful alternative to the metal/polyethylene bearing surface has yet been found. With current advancements in robotic surgery for implanting the device, the increase in accuracy and precision will help to limit the failures. Development of improved procedures for younger patients that delay the need for a total knee replacement will improve their quality of life and reduce the number of revision operations they require.

## References

1. Kao, P., Eggers, S. Graf, N., *et al.* (1996). "A Comparison of Artificial Knee Arthroplasties", in Hench, L.L. and Wilson, J. (eds), *Clinical Performance of Skeletal Prostheses*, Chapman and Hall, London, pp. 71–96.
2. Walker, P.S. and Hajek, J.V. (1972). The load-bearing area in the knee joint, *J. Biomech.*, **5**, 581–589.
3. Haynes, D.W. and McLeod, P. (1984). "The Characteristics of Prosthetic Knees Compared to those of the Normal Knee", in Spilker, R.L. (ed.), *Advances in Bioengineering*, American Society of Mechanical Engineers, New York, NY, p. 101.
4. Hirakawa, K., Bauer, T.W., Stulberg, B.N., *et al.* (1996). Comparison and quantitation of wear debris of failed total hip and total knee arthroplasty, *J. Biomed. Mater. Res.*, **31**, 257–263.
5. Cavendish, M.E. and Wright, J.T.M. (1977). A conservative approach to knee joint replacement, *Eng. Med.*, **6**, 3–11.
6. Endoh, H., Seedhom, B.B. and Takeda, T. (1978). Mobility, stability and stresses in artificial knees, *Eng. Med.*, **7**, 93–100.
7. Wright, T.M. and Bartel, D.L. (1984). "Surface Damage in Polyethylene Total Knee Components", in Spilker, R.L. (ed.), *Advances in Bioengineering*, American Society of Mechanical Engineers, New York, NY, pp. 102–103.

8.  Walker, P.S., Ben-Dov, M. and Askew, M.J. (1981). The deformation and wear of plastic components in artificial knee joints — an experimental study, *Eng. Med.*, **10**, 33–38.

9.  Harris, L.J. and Tarr, R.R. (1979). Implant failures in orthopaedic surgery, *Biomat. Med. Dev.*, **7**, 243–255.

10.  Kirking, B., Krevolin, J., Townsend, C., *et al.* (2006). A multiaxial force-sensing implantable tibial prosthesis, *J. Biomech.*, **39**, 1744–1751.

11.  Mumme T., Marx R., Qunaibi, M., *et al.* (2006). Surface pretreatment for prolonged survival of cemented tibial prosthesis components: full- vs. surface-cementation technique, *Biomed. Technik*, **51**, 95–102.

12.  Fujita, H., Kitaori, T., Iida, H., *et al.* (2005). Novel intramedullary plug with sliding mechanism used in revision total knee arthroplasty, *J. Biomed. Mater. Res.*, **74B**, 419–422.

13.  El Deen, M., García-Fiñana, M. and Jin, Z.-M. (2006). Effect of ultra-high molecular weight polyethylene thickness on contact mechanics in total knee replacement, *Proc. Inst. Mech. Eng. H.-J. Eng. Med.*, **220**, 733–742.

14.  Conditt, M.A., Noble, P.C. and Thompson, M.S. (2007). A computerized bioskills system for surgical skills training in total knee replacement, *Proc. IME H. J. Eng. Med.*, **221**, 61–69.

15.  Davies, B.L., Rodriquez y Baena, F.M., Barrett, A.R., *et al.* (2007). Robotic control in knee joint replacement surgery, *Proc. IME H. J. Eng. Med.*, **221**, 71–80.

16.  Hench, L.L. (2005). "Ethical Issues", in Hench, L.L. and Jones, J.R., (eds), *Biomaterials, Artificial Organs and Tissue Engineering*, Woodhead Publishing Ltd, Cambridge, pp. 265–274.

17.  Morgan, C.G. and Horton, T.C. (2000). Total knee replacement: the joint of the decade, *Br. Med. J.*, **320**, 820.

18.  Diduch, D.R., Insall, J.N., Scott, W.N., *et al.* (1997). Total knee replacement in young, active patients. Long-term follow-up and functional outcome, *J. Bone Joint Surg.*, **79A**, 575–582.

19.  Rissanen, P., Aro, S., Sintonen, H., *et al.* (1997). Costs and cost-effectiveness in hip and knee replacements. A prospective study, *Int. J. Technol. Assess. Health Care*, **13**, 575–588.

20. Ritter, M.A., Carr, K.D., Keating, E.M., *et al.* (1996). Revision total joint arthroplasty: does medicare reimbursement justify time spent?, *Orthopedics*, **19**, 137–139.

21. Mehrota, C., Remington, P.L., Naimi, T.S., *et al.* (2005). Trends in total knee replacement surgeries and implications for public health, 1990–2000, *Public Health Reports*, **120**, 278–282.

22. Gandhi, R., Smith, H.N., Mahomed, N.N., *et al.* (2010). Complications after minimally invasive total knee arthroplasty as compared with traditional incision techniques: a meta-analysis, *J. Arthroplasty*, **26**, 29–35.

23. Carr, B.C. and Goswami, T. (2009). Knee implants — review of models and biomechanics, *Mater. Design*, **30**, 398–413.

24. Goodman, S.B., Gómez Barrena, E., Takagi, M., *et al.* (2009). Biocompatibility of total joint replacement: a review, *J. Biomed. Mater. Res.*, **90A**, 603–618.

25. Geetha, M., Singh, A.K., Asokamani, R., *et al.* (2009). Ti based biomaterials, the ultimate choice for orthopaedic implants — a review, *Prog. Mater. Sci.*, **54**, 397–425.

26. Sobieraj, M.C. and Rimnac, C.M. (2009). Ultra high molecular weight polyethylene: mechanics, morphology, and clinical behavior, *J. Mech. Behavior Biomed. Mater.*, **2**, 433–443.

27. Li, D.S., Garmestani, H., Ahzi, S., *et al.* (2009). Microstructure design to improve wear resistance in bioimplant UHMWPE materials., *J. Eng. Mater. Technol.*, **131**, 041211–041217.

28. Chevalier, J. and Gremillard, L. (2009). Ceramics for medical applications: a picture for the next 20 years, *J. Eur. Ceram. Soc.*, **29**, 1245–1255.

# Total Hip Replacement



## CHAPTER 13

Kim Elshot Alvarez

## 13.1. Introduction

The hip joint is a major source of support in the human body. This is the second largest joint and is primarily used for locomotion. As it is also one of the most flexible free-moving joints and one of the largest weight-bearing joints, it is subject to injury and deterioration over time and often requires replacement. One of the most common reasons is deterioration of the articular cartilage, for example by osteoarthritis. Total hip replacement (THR), also known as total hip arthroplasty (THA), is a surgical procedure in which an arthritic hip joint, or one that has been destroyed, is reconstructed using prostheses. THR is one of the most successful and common surgical procedures in orthopedic surgery. Since 2000, the number of THR surgeries performed each year has been on the rise. In 2005, 235,000 hip joints were replaced in the United States. Worldwide, approximately 800,000 total hip arthroplasties are performed annually. Table 13.1 shows the statistics for primary and revision procedures within the United States.

The number of THRs performed on women is greater than that for men which is primarily a result of female's significantly higher rate of bone density loss with age. It is estimated that by the year 2030 the number of hip replacements will increase to 572,000 annually. In addition, the volume of revision total hip replacements is projected to grow from 48,000 in 2005 to 96,700 in 2030.[2]

**Table 13.1.**   The number of total hip procedures (in thousands) in the US.[1]

| Procedure | Year | # of procedures in women (thousands) | # of procedures in men (thousands) | Total # of procedures (thousands) |
|---|---|---|---|---|
| Total hip | 1998 | 92 | 68 | 160 |
| replacement | 1999 | 95 | 73 | 168 |
| | 2000 | 83 | 69 | 152 |
| | 2001 | 98 | 67 | 165 |
| | 2002 | 112 | 81 | 193 |
| | 2003 | 121 | 99 | 220 |
| | 2004 | 135 | 99 | 234 |
| | 2005 | 126 | 109 | 235 |
| | 2006 | 129 | 102 | 231 |
| Revision of hip | 1998 | 22 | 15 | 37 |
| replacement | 1999 | 19 | 16 | 35 |
| or other | 2000 | 17 | 19 | 37 |
| repair | 2001 | 23 | 17 | 36 |
| | 2002 | 24 | 20 | 44 |
| | 2003 | 25 | 12 | 37 |
| | 2004 | 28 | 20 | 48 |
| | 2005 | 21 | 15 | 36 |
| | 2006 | 22 | 16 | 38 |

Unfortunately revision rates are quite high, with total revision operations approximately 20% of the number of first-time THRs each year. This chapter will discuss the design, material options, success factors and statistics, and alternative solutions for the total hip replacement. Before these topics are addressed, however, it is important to have an historical overview of hip replacement.

## 13.2.  History of Hip Replacement

The replacement of the hip joint has evolved over time. The first successful hip replacement was performed in 1822 by Anthony White at the Westminster Hospital in London. Although his attempt helped the patient's pain and mobility, it did not improve the joint's stability. In 1826, John Rhea Bartonii, an American orthopedic surgeon, performed the first osteotomy, a procedure whereby a bone is cut to shorten, lengthen, or change its alignment. Although the operation appeared to be successful in the short-term, several years after the surgery the patient lost all motion in the joint. Then in 1840, the first artificial joint surface — using wood — was produced by John Murray Carnochan, a surgeon in New York. German Professor Themistocles Glück transformed the use of artificial joints when he created an ivory ball-and-socket joint for the hip. Glück was also first to develop the idea of biocompatibility, a material's inability to elicit an immune response from the body. In 1938, Dr Phillip Wiles from the Middlesex Hospital in the United Kingdom designed and implanted the first THR, substituting for both the socket in the pelvic bone and the head of the femur. Prior to this date, implant replacement surgery involved the replacement of only one surface and the results were substandard. The concept of the modern hip implant was first developed by British orthopedic surgeon Sir John Charnley in 1961. Utilizing a cemented femoral stem component and a high density polyethylene cup, his low-friction prosthetic design is widely regarded to be the "gold standard" in cemented hip arthroplasty.[3]

## 13.3. Design of the Hip Implant

### 13.3.1. *The hip joint*

The human hip consists of two main parts: a femoral head or ball at the top of the thigh bone (femur) that fits into a rounded socket (acetabulum) in the pelvis (Fig. 13.1). Connective tissue and ligaments (the joint capsule) connect the ball to the socket and provide stability to the joint. The bone surfaces of the ball and socket have a smooth tough cover of articular cartilage that cushions the ends of the bones and enables them to move easily. (See Chapter 10 for a discussion of the structure and damage of cartilage.) The rest of the surfaces of the hip joint are covered by a thin, smooth tissue liner called synovial membrane; this secretes a small amount of

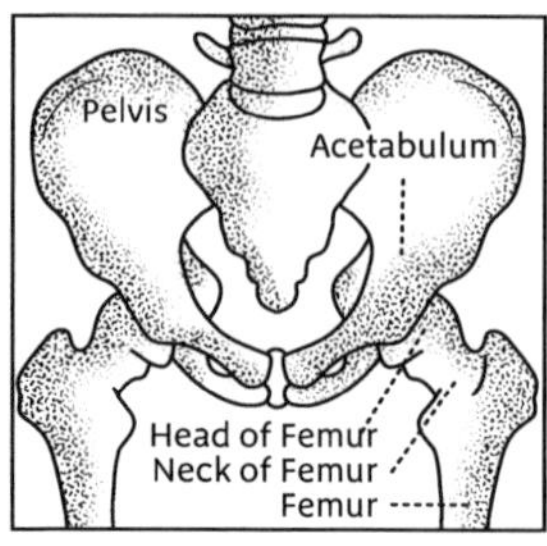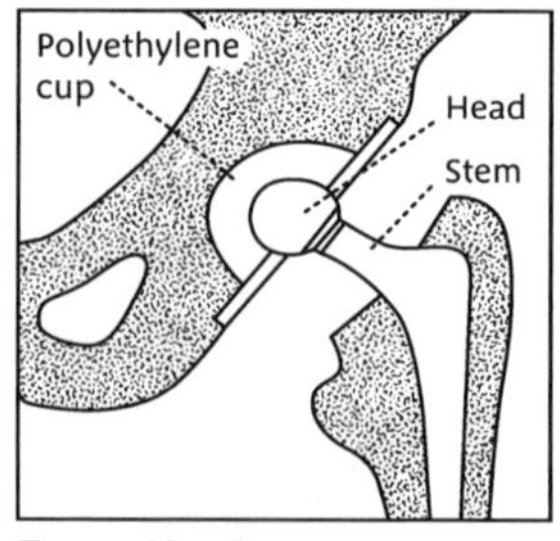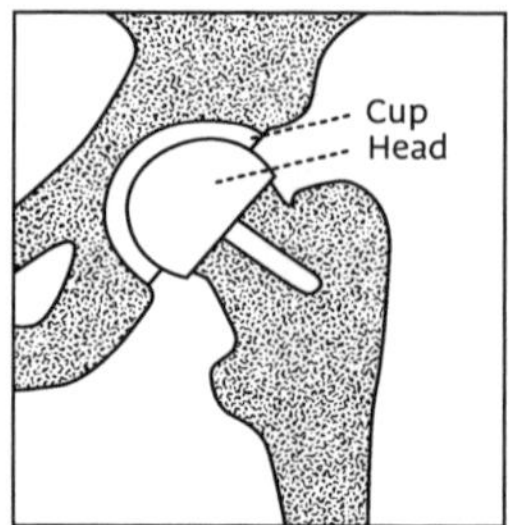



**Figure 13.1.**   Schematic of a hip joint, a total hip replacement, and a hip resurfacing.[4]

synovial fluid which lubricates the joint, further reducing friction and facilitating movement. The goal of hip resurfacing is to restore these characteristics.[4]

## 13.3.2.  *Artificial hip design*

The purpose of the hip prosthesis is to mimic the design of the natural hip.[5] The hip implant consists of four components (Fig. 13.1) that work together to restore the original function of the ball-and-socket joint:

- A hip stem, usually a metal alloy, which is inserted into the top of the thigh bone.
- The femoral head or ball, usually made out of ceramic or metal alloy, which is attached to the hip stem and inserted into the liner to form the ball-and-socket joint.
- A metallic acetabular component or shell fixed into the acetabulum (socket) which holds the cup liner.
- A cup liner which holds the femoral head. The shell is made of metal and the liner can be made of polymer, usually ultra-high-molecular-weight polyethylene (UHMWPE), ceramic, or metal.

Hip implants are designed for one of two methods of fixation: cemented fixation or un-cemented fixation.[5,6] In cemented fixation, the hip implant is designed to be implanted using bone cement, which wedges the implant in place within the intermedullary canal of the bone. The most commonly used bone cement is an acrylic polymer called poly(methylmethacrylate) (PMMA) that hardens *in situ* after placement in the host bone. Bone cement is injected into the prepared femoral intermedullary canal. The implant is positioned within the canal and the cement secures it in the desired position. Current cements are not adhesives. In other words, there is no bond between the implant and the bone.

In un-cemented fixation, hip implants are designed to be placed into the prepared femoral canal without the use of bone cement. Either surface porosity or bioactive fixation can be used to provide the long-term stability of the bone–implant interface. (See Chapter 14 for a discussion of bioactive surface fixation of total hip replacements.) In this method, the femoral canal is prepared so that the implant fits tightly within it. The porous surfaces on the hip implant have a fine mesh of holes on the surface area that are designed to engage with the bone within the canal by encouraging bone to grow into the porous surface. Eventually, this in-growth of bone provides additional fixation to hold the implant in the desired position. Figure 13.2 shows a schematic of the natural hip, a porous implant (no cement), and implant fixed with bone cement. In addition to these fixation methods, there is also a hybrid fixation method which consists of a combination of the cemented and non-cemented techniques.

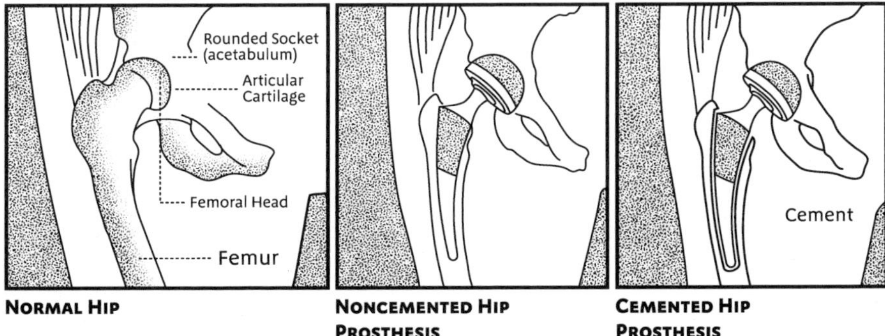


**Figure 13.2.**   Image of a normal hip, a non-cemented hip prosthesis and a cemented hip prosthesis.[7]

### 13.3.3. *Biomaterials for hip implants*

Several factors must be considered when selecting the material for the hip implant prosthetic. These are:[6]

- Biocompatibility: The material's lack of toxicity and its ability to effectively function and interact with the tissues of the human body.
- Strength: The material's ability to withstand these daily stresses without fracture or plastic deformation. Its ability to withstand cyclic loading should also be considered.
- Stiffness: Despite the need for material strength, material stiffness should not be too high to avoid stress shielding. Stress shielding refers to implant's shielding of the underlying bone from the physiological loads resulting in a reduction in bone density, and a high stiffness causes stress shielding. An ideal implant's Young's modulus would match that of the host bone.
- Resistance to wear: This property is particularly significant in maintaining proper joint function and preventing the further destruction of bone caused by generation of undesirable particulates as the implant components move against each other. Wear particles can result in inflammation and progressive deterioration of the host bone, a process termed aseptic loosening.
- Resistance to corrosion: Corrosion can cause implant damage and ultimately lead to implant failure. In addition, it can also result in harmful corrosion particulates in soft tissues and can cause bone damage.

## 13.4. Material Options

THR prostheses can be made of a combination of various materials including metal alloys, polyethylene (plastic), or ceramic.

### 13.4.1. *Metal-on-polyethylene*

This implant is the most commonly used hip replacement implant in the United States due to its durability and performance. In this implant, the femoral head (ball) consists of a metal alloy, usually cobalt-chrome or titanium alloy and a lined acetabular cup made of a ultra-high molecular weight polyethylene (UHMWPE). Advantages of this implant are low cost, low risk for fracture, and it can easily be fit into the bone. In addition, these implants have the longest clinical history and therefore are the most understood. Some of the disadvantages are that the polyethylene may wear over time, producing wear particles that remain in the hip, causing inflammation and aseptic loosening, and requiring a revision procedure.

## 13.4.2. *Ceramic-on-polyethylene*

In this total hip system the ceramic ball component articulates against a polyethylene liner. The ceramic heads are harder than metal and are the most scratch-resistant implant material. Overall, the wear rate for these implants is significantly less than that of metal-on-polyethylene. A disadvantage in early ceramic-on-polyethylene devices was the incidence of fractures. However, improved ceramics with high toughness have resulted in very low rates of failure. In terms of cost, ceramic-on-polyethylene is more expensive than metal-on-polyethylene, but less than ceramic-on-ceramic.

## 13.4.3. *Metal-on-metal*

These metal-on-metal hip joints have both bearing surfaces (the ball and the cup component) made of metal. They offer the potential for greatly reduced wear due to the hard and scratch-resistant joint surfaces, with less inflammation and less bone loss. These devices are generally composed of cobalt-chrome and are used in about one-third of all hip replacement surgeries in the United States each year. The disadvantage of the use of metal on metal hip technology is the generation of metal ions that occur due to the friction of the metal head rotating in the metal socket. This has raised concerns about long-term biocompatibility.

### 13.4.4. *Ceramic-on-ceramic*

Although ceramics have been used in hip replacement applications for about 30 years, they have only recently become prevalent in the United States in the last decade. In these implants, the stem and femoral head components are composed of ceramic, usually alumina oxide or zirconium oxide. They have the lowest friction and wear of any of the bearing surfaces. One problem has been that the zirconia ceramics unexpectedly aged in the body. The body is actually an aggressive environment, and exposure to the conditions (37°C and physiological pH) over long periods (5–10 years) caused phase transformations in the zirconia, which made it more brittle. With the use of stronger ceramics, fracture rates of these implants have been significantly reduced. Two other disadvantages of ceramic-on-ceramic devices are the cost, as they are the most expensive option, and the fact that sometimes these implants can squeak loudly enough to be heard.

## 13.5. Failure and Survivability

Total hip replacement is a very successful procedure. However, as the case with any surgery there is always risk. The most common complications of hip replacement are:[8]

- Blood clots or thrombophlebitis: The most common complication of hip replacement surgery where the blood in the large veins of the leg forms blood clots within the veins.
- Infection: The chance for infection is less than 1% and can result from bacteria invading the bone in the presence of metal and cement.
- Dislocation: The incidence rate is approximately 3%. It can be the result of non-compliance to postoperative restrictions as well as muscle imbalance and tightness around the hip joint.
- Aseptic loosening: The total hip prosthesis and loss of adequate fixation to the bone, not caused by bacterial infection.
- Wear: Caused by local mechanical damage and production of small particles by the bearing surfaces, which trigger an immune

response and bone resorption. Wear will also lead to changes in the geometry of the joint which can result in loosening of the implant.

- Fracture: Can occur in either the device or the host bone as a result of wear, stress shielding of the bone, or overloading of the prosthesis.

The clinical success rate of cemented THA is estimated to be 85–95% at 15 years.[9] Low friction cemented hip implants, which are considered the standard, have demonstrated the highest success rate. The THA clinical success rate can vary and depends on several factors. Some of these include:

- Age: Higher failure reported for patients over 70.[10] In general, reduced THR lifetimes are primarily due to the higher activity levels.
- Gender: Lifetimes for THR in males are lower than those for females.
- Weight: Heavier body weights result in lower than normal success rates.

Other factors that may impact the survivability are the quality of the underlying bone, the skill of the surgeon, and the original cause of the hip degeneration. It is important to note that the success rate of revision implants is lower due to the decreased quantity and quality of remaining natural bone.

## 13.6. Bioactive Fixation

As patients undergoing THR are getting younger and outliving their hip prostheses, the demand for THR revision surgery is increasing considerably and can have a significant economic burden in the future. Consequently, the need for a more robust long-term solution for these load-bearing implants is critical. Alternative methods of implant fixation are being explored and developed as a possible way to address this concern. One such method, called biological fixation, involves the interfacial bonding of the implant to tissue by means of a biologically active hydroxyapatite (HA) layer on the implant surface. The significant benefit of this technique over other fixation methods is the formation of a bioactive bond at the implant–bone interface that has strength comparable, if not better, than bone.[11] As this technology is improved, the use of this coating on implants may extend the device lifetime. Chapter 14 explores these methods.

## 13.7.  Alternatives to THR

As life expectancy is increasing, new methods are needed to delay the need for THR. This is especially pertinent in the hip, where THRs are needed due to damage to the articular cartilage. Quite often, healthy bone is removed to allow a THR to be inserted. While THR involves the replacement of femoral bone, hip resurfacing arthroplasty (HRA) is a less invasive procedure in which only the surface of the hip joint is replaced, thus preserving most of the femoral bone. Although this technique has been in use since the 1960s, improvements in the HRA prosthetic design has resulted in renewed interest in this procedure. The hip resurfacing implant consists of two parts: a metal cap that is placed over the resurfaced femoral head, and a metal cup that is pressed into the socket.[12] Due to the use of the cap in HRA, the implant hip ball size in general is larger than in the THR and therefore reduces risk of dislocation and wear rate. The preservation of natural bone results not only in lower stress shielding caused by the implant, but also improved patient recovery, and retains enough bone for a THR later in life. This means the procedure can only be used on patients with good bone quality (younger patients).

There are a few disadvantages associated with contemporary HRA implants. Revision rates for these implants appear to be higher compared to the conventional THR.[13] This may be because it is generally used on younger, more active patients. But although the femoral head is preserved, HRA does put significant strain on the preserved femoral head and can lead to femoral neck fractures.[14] The bearing surface for THA is metal-on-metal; this is known to increase metal ion levels in the body.[15] Despite these concerns, early results suggest that this relatively new technology is a promising alternative to THR.

## 13.8.  Long-Term Survivabilty

At the present time surgeons implant THR with a wide variety of surface finishes on the femoral stems. A review by Beksac *et al.* of survivability data from numerous clinical reports of cemented THA shows that the surface finish of the femoral stems has a major effect on long-term survivability of the devices.[16] The surface finish of a femoral stem affects the ability of bone cement to adhere to the implant surface. Stems with a rough surface have a stronger interface with bone cement than do smooth stems. However, micro-motion at the metal–cement interface can lead to more fretting, and more cement and metallic abrasion of stems with a rough surface. The comparison of retrieval data of failed prostheses and clinical data of revision surgery shows that survivability of THR with smooth femoral stems is superior to those with rough femoral stems. Table 13.2 summarizes the data from 11 clinical papers reporting on a total of 10,473 total hip replacements with smooth femoral stems of patients ranging from 41 to 69 years at time of surgery.[16] The data from 19 clinical papers presenting results from 2,211 cases of patients ranging in age from 40 to 71 years using rough femoral stems are compared in Table 13.2. The authors conclude that the comparative data and their own clinical experience support the use of cemented femoral stems with a smooth or polished surface finish.[16]

**Table 13.2.**   Survivability of cemented metal/PE hip prostheses with smooth *vs.* rough femoral stems.[16]

| | Survival rates of THRs at years after implantation | | |
|---|---|---|---|
| | 5–10 years | 10–15 years | 15–25 years |
| **Smooth Stems** | 99–100% | 95–100% | 92–97% |
| **Rough Stems** | 75–99% | 93–100% | N/A |

## 13.9. Summary

Hip replacement is a very common orthopedic procedure with one of the highest success rates; however replacements do not last forever, and our life expectancy is increasing. The hip implant has evolved over time and is continually investigated for potential improvements. Currently there are several material options for these prostheses. There are multiple causes for failure and several factors that need to be taken into consideration for the life expectancy of these devices. Alternative methods of fixation are being developed to improve the survivability of the implants. Due to recent advances hip resurfacing has emerged as a viable alternative to THR.

## References


1. Buie, V.C., Owings, M.F., DeFrances, C.J., *et al.* (2010). National hospital discharge survey: 2006 summary. National center for health statistics, *Vital Health Stat.*, **13**, 1–79.
2. Parvizi, J., Eslam Pour, A., Keshavarzi, N.R., *et al.* (2007). Revision total hip arthroplasty in octogenarians. A case-control study, *J. Bone Joint Surg. Am.*, **89A**, 2612–2618.
3. DePuy, Inc. (2010). *Our history*. [Online]. Available at: http://www.depuy.com/corporate-information/about-depuy/history. [Accessed 19 June 2010].
4. Marieb, E.N. and Hoehn, K. (2007). *Human Anatomy and Physiology*, 7th ed., Pearson Education, London.
5. Zimmer, Inc. (2008). *Hip replacement surgery FAQs*. [Online]. Available at: http://www.zimmer.com/z/ctl/op/global/action/1/id/8140/template/PC/navid/10427. [Accessed 19 June 2010].
6. Zimmer, Inc. (2008). *Materials used in orthopaedic implants*. [Online]. Available at: http://www.zimmer.com/ctl?template=PC&op=global&action=1&id=9479. [Accessed 19 June 2010].
7. American academy of orthopaedic surgeons (2009). *Total hip replacement*. [Online]. Available at: http://orthoinfo.aaos.org/topic.cfm?topic=A00377. [Accessed 19 June 2010].

8.  eHealthMD (2010). *Hip replacement: what are the risks and complications?* [Online]. Available at: http://www.ehealthmd.com/library/totalhipreplacement/THR_risks.html. [Accessed 19 June 2010].

9.  Hench, L.L. and Wilson, J. (1996). *Clinical Performance of Skeletal Prostheses*, Chapman and Hall, London.

10. Sanborn, P.M., Cook, S.D., Harding, A.F., *et al.* (1987). Clinical performance of endoprosthetic and total hip replacement systems, *J. Rehabil. Res. Dev.*, **24**, p. 50.

11. Polak, J.M., Hench, L.L. and Kemp, P. (2002). *Future Strategies for Tissue and Organ Replacement*, World Scientific, Singapore, p. 5.

12. Smith and Nephew (2010). *Hip resurfacing.* [Online]. Available at: http://global.smith-nephew.com/us/patients/Hip_resurfacing_11049.htm. [Accessed 14 July 2010].

13. Spierings, P.T.J. (2008). Hip resurfacing: expectations and limitations, *Acta Orthoped.*, **79**, 728.

14. Shimmin, A.J., Bare, J., and Back, D.L. (2005). Complications associated with hip resurfacing arthroplasty, *Orthop. Clin. North Am.*, **36**, 187–193.

15. Cutts, S. and Carter P.B. (2006). Hip resurfacing: a technology reborn, *Post Grad. Med. J.*, **82**, 802–805.

16. Beksac, B., Taveras, N.A., Della Valle, A.G., *et al.* (2006). Surface finish mechanics explain different clinical survivorship of cemented femoral stems for total hip arthroplasty, *J. Long Term Eff. Med. Implants*, **16**, 407–422.

# Bioactive Surfaces for Total Hip Prostheses



## CHAPTER 14

Takao Yamamuro

## 14.1. Introduction

Total hip replacement, or total hip arthroplasty (THA), is commonly used when deterioration of the hip is sustained from arthritis or by injury. Arthritis causes deterioration of the articular cartilage that forms a bearing surface between the head of the femur and the acetabulum. Chapters 10 and 11 discuss the structure of cartilage and the damaging effects of arthritis. The cartilage also provides a cushioning effect of the joint, and when cartilage deteriorates, bone comes into direct contact with bone, which causes intense pain and damage to the bone. THA surgery is a common orthopaedic technique that utilizes prosthetic implants to replace the damaged bearing surfaces of the joint, as discussed in Chapter 13. The objective of this chapter is to discuss the concept and potential of using new concepts of bioactive fixation to improve the long-term survivability of total hip implants.

## 14.2. Bioactive Coatings of Hydroxyapatite

Bioactive fixation of total hips using a hydroxyapatite (HA)-coated hip prosthesis develops a stable fixation with the surrounding bone tissue due to osteoconductive property of the HA-coated layer.[1] In the modern hip prosthesis, the femoral stem made of Ti alloy is coated physically with HA by means of a plasma spray coating over the metal surface.[1] When the hip prosthesis is implanted into the bone, the bone tissue becomes directly bonded with this HA-coated layer by chemical bonding. The interface between Ti alloy and HA layer is physically connected by plasma spray coating, whereas the interface between HA layer and bone tissue is chemically bonded.

However, over time bone remodelling occurs around the implant and osteoclast-mediated resorption of the HA-coated layer can occur. Osteoclasts absorb the HA-coated layer as well as the newly formed bone. In addition to such osteoclastic resorption, disintegration of the HA-coated layer may take place due to combinations of mechanical abrasion, chemical dissolution, and de-lamination, depending upon the composition and crystallinity of the coating.[2] Usually osteoclastic bone resorption is followed by new bone formation by osteoblasts. However, the newly formed bone does not bond directly to the metal substrate. Therefore, in a hip prosthesis which has a 100–200 μm thick HA-coated layer, it is anticipated that the major part of the prosthesis surface may become biologically disconnected from the surrounding bone tissue in about 20–30 years.

## 14.3. Bioactive Coatings of Apatite-Wollastonite Glass-Ceramic

Based on the above analysis, the author and colleagues at Kyoto University designed a cement-less hip prosthesis with two concepts aimed at achieving early bone in-growth by use of an apatite-wollastonite bioactive glass-ceramic (AW-GC) bottom coating.[3] As discussed in Chapter 8, AW-GC has a stronger osteoconductivity than HA. From previous studies, it was confirmed that an AW-GC bottom coating accelerates bone in-growth as well as implant fixation better than an HA-coating or AW-GC full coating.[3]

In 1992, the team developed a cement-less hip prosthesis aimed at obtaining early bone in-growth, minimizing the wear of the poly(ethylene) (PE) socket, and reducing the bone resorption that is due to stress shielding. In this model, a Vanadium-free Ti-alloy femoral stem and a zirconia head with a diameter of 22 mm were used.

A plasma spray coating in combination with AW-GC bottom coating was applied over the whole surface of the socket back and a small circumferential area at the proximal part of the stem. The stem with a minimum coated area at its proximal part is considered advantageous to an extensively coated stem for the prevention of a stress shielding effect on the femur. Therefore, the size of the coated area on the stem was reduced as much as possible, based on the theoretical calculation of the stress to be loaded through this area.

The stem consists of three areas. The coated area is designed not to contact entirely with the bone cortex. Therefore, cancellous bone in the proximal femoral metaphysic is expected to bond with the porous surface of the stem by osseointegration, so that the mechanical stress coming through the neck may be transmitted to the femoral cortex through this zone. The trapezoid area is designed to be press-fitted into the isthmus of the femur for the purpose of initial fixation of the prosthesis, and the round area is only to protect against angulations of the stem.

Using this cement-less hip prosthesis, 106 hip joints underwent total hip replacement at Kyoto University between 1992 and 1998 as a clinical trial to obtain the approval of the Japanese government. There were 87 female and 15 male patients. Their age distribution ranged from 26 to 69 years with an average of 54.1 years. Among them, 86 hips were followed up for five to ten years in 2004. The preoperative hip score evaluated by the JOA hip scoring system was 48.3 points on average. Post-operatively, the average hip score was maintained at around 92 points up to ten years after THR. The survival rate using revision as the endpoint was reasonably high. The revisions were performed due to repeated dislocation in five cases, thigh pain in two, and osteolysis in one case. However, there was no loosening of the components observed for a maximum of ten years postoperatively in 2004.


## 14.4. Zirconia Femoral Heads

There were two reasons why zirconia ($ZrO_2$) femoral heads with a 22 mm diameter were used. First, the mechanical strength of zirconia is significantly greater than that of alumina in terms of bending strength, compressive strength, and fracture toughness, so that the head size can be made smaller without the risk of breakage.[4] Second, the surface roughness of zirconia is identical to that of alumina. The aim was to reduce volumetric wear of the conventional PE socket.

When the linear wear of PE socket was measured on radiograms in 64 cases during a period of 5–10 years, the wear per year ranged from 0.02 to 0.43 mm with an average of 0.16 mm, about the same as that of Charnley's hip prosthesis. (See Chapter 13 for discussion of the Charnley's THA). Thus, wear of the conventional PE socket was not as small as we initially expected for this hip prosthesis.

Retrieved zirconia heads were examined crystallographically, and it was found that a transformation from tetragonal to monoclinic phase had occurred over the surface of all specimens. This was termed "aging" and was due to long-term exposure to the physiological environment. It was established that, due to this crystallographic phase change, the surface roughness of the zirconia head had been increased. Based on these facts, a compositional change was made to minimize the phase transformation of zirconia and, finally, 0.25 wt% of alumina was added. Studies showed that the alumina-added zirconia is very stable against phase transformation.

The wear rate of non-cross-linked and cross-linked PE cups against various ceramic heads was measured by the use of a hip simulator. The least volumetric wear recorded was with a combination of cross-linked PE on alumina-added zirconia. For these reasons, the alumina-added zirconia for the femoral head is now used in combination with cross-linked PE for the socket. This system is named the K-Max series and has widely been used in Japan since 2005.

## 14.5. Bone–Prosthesis Interface

Analysis of the bone prosthesis interface of our hip prosthesis based on the five-year results of the first 300 cases in the K-Max series led to an important conclusion. Special attention was paid to the radiological and histological findings of the coated area. A radiolucent line around the stem was frequently observed in the non-coated area (fibrous encapsulation), but it never appeared in the coated area. In addition, at the coated area of the stem so-called spot welds were frequently observed in the images, which are considered to indicate a sign of active bone in-growth and bone trabecular thickening due to stress transmission through the coated area. Also, along the acetabular component, bone in-growth into the coated surface was observed on X-ray pictures within a few years postoperatively in all cases followed up.

All the retrieved components demonstrated extensive bone in-growth into the AW-GC bottom-coated porous surface. One patient sustained fracture of the operated femur by a car accident only eight months after THR, so the femoral component of hip prosthesis was retrieved. Histological analysis showed bone growth to the bioactive coating within that time. An elemental analysis of the surface of the specimen also demonstrated that bone in-growth had already been completed within eight months of implantation. Thus, in this type of hip prosthesis, a firm bonding between bone and components of the prosthesis is established in the early postoperative stage.

As for the femoral component of this prosthesis, osseointegration is established between the proximal part of the stem and the femur. This type of proximal fixation is preferred to distal fixation for the prevention of stress shielding of the proximal femur. However, it is anticipated that bioactivity of the bottom-coated AW-GC will not last long because of its chemical dissolution and osteoclastic resorption, and therefore the later connection between bone and prosthesis must depend on mechanical micro-anchoring at the porous surface. Thus, it is still unclear if this type of fixation will maintain stability of the prosthesis for a very long period such as 40–50 years, as may be required for very young patients who are only 30–40 years old. This uncertainty leads to the question of what would be the ideal bioactive surface for hip prosthesis?

## 14.6.  The Ideal Bioactive Surface for a Hip Prosthesis

At least three conditions must be satisfied to be the ideal bioactive surface:

1.  early bone in-growth must occur;
2.  persistent firm bone bonding must be ensured by the permanent surface bioactivity; and
3.  bone bonding must be strong enough to connect the prosthesis to bone through a relatively small bioactive area at the proximal part of the stem, to prevent stress shielding of the femur for a very long time.

In 1997, Kokubo and colleagues at Kyoto University developed a new technique to make the Ti metal surface permanently bioactive by an alkali processing method.[5] After degreasing and cleaning, a Ti-alloy device is first alkali treated in a NaOH solution >60°C for 24 hours. By this treatment, a hydrated titania ($TiO_2$) gel layer with a porous structure is formed on the Ti metal surface. After cleaning again, the Ti-alloy is treated at 600°C for one hour. By this heat treatment, the hydrated titania gel layer is condensed and an amorphous layer of alkali titanate is formed firmly over the Ti metal surface.

When an alkali-heat-treated Ti-alloy is soaked in a simulated body fluid, a bone-like apatite layer is formed on the surface in a few days. These *in vitro* experiments confirmed that the surfaces of Ti alloys are made strongly bioactive by the alkali-heat treatment and theoretically it is considered that the bioactivity is permanent. In order to test the strength of the bone–implant interface three kinds of Ti implant with porous surfaces were implanted into canine femurs and push-out tests performed four weeks after implantation. The AW-GC bottom coating significantly increased the failure load of push out as compared to the control, but the alkali-heat treatment further increased the failure load. Thus, the alkali-heat treatment on Ti-alloy seems to produce a nearly ideal bioactive surface to be used for cement-less hip prosthesis.[5]

## 14.7.  Clinical Results

A two-year clinical trial at Kyoto University consisting of 200 cases was very successful and the Japanese Government approved clinical use of the titania-activated coated hip prosthesis in 2007. The design of the new hip prosthesis is a proximal fixation type similar to the K-Max series, but the bioactive surface is alkali-heat treated instead of receiving AW-GC bottom-coating. The longest follow-up time of this new hip prosthesis is nearly ten years now. We are satisfied with the results so far, although much longer clinical observation is necessary to draw any definite conclusion on its longevity.

## References


1.  Lacefield, W.R. (1993). "Hydroxylapatite Coatings", in Hench, L.L. and Wilson, J. (eds), *Introduction to Bioceramics*, World Scientific, Singapore, pp. 223–238.
2.  Klein, C.P.A.T., Wolke, J.G.C. and de Groot, K. (1993). "Stability of Calcium Phosphate Ceramics and Plasma Sprayed Coating", in Hench, L.L. and Wilson, J. (eds), *Introduction to Bioceramics*, World Scientific, Singapore, pp. 199–222.
3.  Yamamuro, T. (1993). "A/W Glass-Ceramic: Clinical Applications", in Hench, L.L. and Wilson, J. (eds), *Introduction to Bioceramics*, World Scientific, Singapore, pp. 75–88.
4.  Hulbert, S.F. (1993). "The Use of Alumina and Zirconia in Surgical Implants", in Hench, L.L. and Wilson, J. (eds), *Introduction to Bioceramics*, World Scientific, Singapore, pp. 25–40.
5.  Kokubo, T., Kim, H.M. and Kawashita, M. (2003). Novel bioactive materials with different mechanical properties, *Biomaterials*, **24**, 2161–2175.

## CHAPTER 15

# Artificial Skin and Wound Dressings

Krystle Placencio, Thomas Frank Jefferson,
Michael Teran and Michael Fenn

## 15.1. Introduction

One of the greatest medical needs is an ideal replacement or means of regenerating damaged skin. It is estimated that there are at least 2.4 million burns reported per year. A sizeable percentage of serious burns, which require professional medical attention, are of such severity that it is necessary to perform skin grafts to replace the dead and damaged skin. The "gold standard" for severe burn treatment involves use of an autograft from the patient's undamaged skin. Split-skin autografting, the graft method of choice by many surgeons, is an operation involving the transplantation of the upper layers of skin from a healthy area to an area with a skin defect. The types of defect which are suitable for closing in this way include ulcers, burns, abrasions, and surgical wounds. However, in many cases there is not enough healthy skin left on the body to harvest an autograft for repair, and the time needed for the regeneration of sufficient skin cells for large scale repair can sometimes be limited. Donor site morbidity associated with large area grafts are also of great concern. Additionally, wound exposure must be limited to prevent infection which can lead to increased morbidity and eventually death. Additionally, the reliance on cadaver skin allografts and porcine xenografts must also be reduced due to various limitations and complications.[1]

Cadaver skin allografts are a common skin graft option. However, there are downsides to use of cadaver skin; in particular, the time-limited viability of cadaver skin which requires transplantation within a short time after harvest. An inherent problem of such transplants, either allograft or xenograft-derived, is the ultimate graft rejection following withdrawal of immuno-suppression therapies.[2–4] The development of synthetic skin substitutes through tissue engineering and regenerative medicine offer potential solutions to these problems.[5,6] The review by Metcalfe and Ferguson[5] is especially important as it covers the basic biological science underlying the use of skin grafts, the limitations of current materials and procedures, and the prospect of creating a new regenerative medicine approach to repair or replace skin.

## 15.2.  Anatomy of the Skin

Skin is the largest organ of the body, and is composed of three main layers: the epidermis, dermis, and subcutaneous adipose tissue (Fig. 15.1). The epidermis is composed mostly of dead cells, keratinocytes, melanocytes, Langerhan cells and Merkels cells. The epidermis is the outer layer which protects the underlying tissue from the environment and infection. Deep beneath the epidermis is the dermis, which is composed of living cells, blood vessels, hair follicles, sweat glands, and nerves; these provide support, structure, and maintain homeostasis. The third layer of the skin is the subcutaneous adipose layer, which is mainly composed of fat cells and is important for insulation, shock absorption, and attachment of skin to underlying muscle and bone. Cells on the surface of the skin are constantly being replaced with new living cells which migrate towards the surface.

Severe burn damage to the skin can be life-threatening. There are at least one million injuries or deaths yearly as a direct result of severe burns.[4] A large majority of deaths from burns are attributed to infections due to exposure of open burn wounds. In the event of a severe burn, the victim's outer layers of skin, the epidermis and dermis, are destroyed. In cases of severe burns when the outer two layers are damaged, extensive immediate emergency care is required to prevent infection and possible death. The graft required

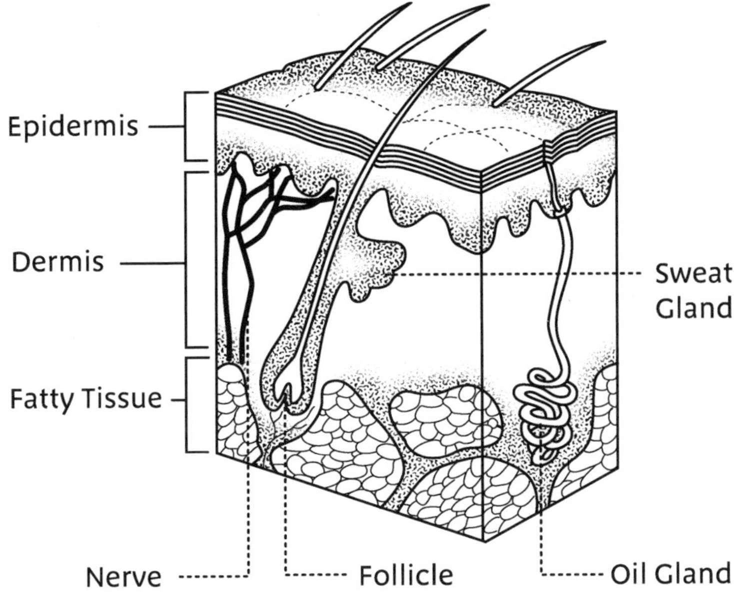



**Figure 15.1.**   Anatomy of the skin.

to repair the damage is directly related to the degree of the burn. There are many precautions but of special concern are rejection and compatibility of the substance used to cover the burn. Many of the common grafts presently used (i.e. cadaver skin) are unable to last for long periods of time, and require patients to take immuno-suppressant drugs to prevent rejection, which is a major risk since their health is already compromised by the trauma. Creating an artificial skin which mimics the structure and function of normal healthy skin could circumvent these difficulties and provide superior treatment of skin wounds.

The primary goal of research in this field is to develop a mate-rial which will act either as a wound dressing or function as a skin substitute, prevent infection, and allow for successful healing. To satisfy this demand the substitute must be flexible, strong, selec-tively permeable, and able to prevent infection and infiltration of pathogens. Several commercial products are available which meet one or more of these requirements, and are summarized below.

Tables 15.1, 15.2, and 15.3 provide an overview of commercial skin substitutes sub-divided into the categories based on:

1. epidermal skin substitutes;
2. dermal skin substitutes; and
3. composite skin substitutes.

Discussions of the characteristics of these materials are presented in Metcalfe and Ferguson,[5] and in Supp and Boyce.[6] Four examples of commercial skin substitutes are presented below. Details of the skin substitutes listed in Tables 15.1, 15.2, and 15.3 are discussed in the references.[5–7]

## 15.3. Examples of Present Biomaterials for Skin Substitutes

### 15.3.1. *Biobrane*

Biobrane, a product of Dow B. Hickman, Inc. (Sugarland, TX) is composed of a bi-laminate material with a nylon film coated with a single layer of silicone rubber, and then coated by porcine peptides derived from type I collagen. The use of the porcine collagen is reported to help promote fibrovascular in-growth, attracting fibroblasts and endothelial cells to the location of the wound site. Biobrane has been used for the treatment of donor sites; noninfected superficial partial thickness burns; as a substitute to cadaver skin prior to grafting;[8] and has also had further development from a wound dressing to a cell carrier application. Experimental results, using neonatal fibroblasts grown on Biobrane and grafted onto athymic mice, showed the seeding provided better adherence to the wound bed, reduced inflammation, and promoted vascularization better than acellular Biobrane.[9] Another treatment based on Biobrane, TransCyte®, is composed of human newborn fibroblasts which are cultured on nylon mesh. TransCyte® provides a thin silicone membrane bound to the mesh which provides a moisture vapor barrier for the wound. This matrix is also used for temporary coverage of partial/full-thickness burns.[10]

**Table 15.1.**   Examples of commercial epidermal skin substitutes (based on[5,6]).

| Product name and manufacturer | Epidermal component | Dermal component | Advantages | Disadvantages |
|---|---|---|---|---|
| Epicel® *Genzyme Tissue Repair Corporation* | Cultured epidermal autograft is grown from biopsy of the patient's skin | None | Provides a large area of permanent wound coverage with low risk of rejection | 3 weeks required before fragile confluent sheets are produced; susceptible to blistering post-grafting |
| Laserskin® *Fidia Advanced Biopolymers* | Keratinocytes taken from biopsy of the patient's skin are cultured and deposited in a perforated hyaluronic acid membrane | None | Provides less fragile delivery system for keratinocytes. Hyaluronic acid and cell interactions improve mechanical stability | Requires a minimum of 3 weeks to expand keratinocyte population |

*(Continued)*

**Table 15.1.**    (*Continued*)

| Product name and manufacturer | Epidermal component | Dermal component | Advantages | Disadvantages |
| --- | --- | --- | --- | --- |
| EpiDex™ *Modex Therapeutics* | Outer root sheath follicle cells taken from the patient are cultured | None | Proliferative capacity of the cells is increased, the product may be cryopreserved. Successful in treatment of chronic ulcers | Product is fragile. It takes up to 6 weeks after harvesting the cells to produce the substitute |
| Myskin™ *CellTran* | Keratinocytes taken from a biopsy of patient's skin is cultured and deposited on PVC polymer coated with a plasma-polymerized surface | Myskin™ with dermal fibroblasts is under development | Keratinocyte attachment and proliferation is encouraged by PVC polymer, provides stable delivery platform. For repeated application keratinocytes can be thawed for repeated application | Cell expansion can take up to 14 days and repeated application is needed to provide a good clinical outcome |

**Table 15.2.**  Examples of commercial dermal skin substitutes (based on[5,6]).

| Product name and manufacturer | Epidermal component | Dermal component | Advantages | Disadvantages |
|---|---|---|---|---|
| Alloderm® *Life Cell Corporation* | None | Allograft skin that has been processed from a cadaver | Antigenic components reduced due to processing. Successful in resurfacing full-thickness burns | There are issues with graft rejection and disease transfer |
| Dermagraft® *Advanced Biohealing, Inc.* | None | 3D bioabsorbable scaffold on which is deposited allogeneic neonatal fibroblasts | Proliferation of neonatal fibroblasts produces collagen, GAGs, and growth factors that aid wound healing | Potential risks of both rejection and disease from the fibroblasts; however none have been reported |
| Integra® *Johnson & Johnson* | Synthetic polysiloxane polymer | GAGs and bovine type I collagen | Ingrowth of fibroblasts and epithelial cells is encouraged. After 14 days epidermal equivalent is replaced with an autograft | Risk of antigenicity and disease due to use of bovine collagen. Takes 3 weeks to expand the dermal autograft |

*(Continued)*

**Table 15.2.**  (*Continued*)

| Product name and manufacturer | Epidermal component | Dermal component | Advantages | Disadvantages |
| --- | --- | --- | --- | --- |
| TransCyte® *Advanced Biohealing, Inc.* | Thin silicone layer (Biobrane®) | Neonatal allogeneic fibroblasts are seeded onto a collagen-coated nylon mesh | Used to treat 2nd- and 3rd-degree burns. Collagen, GAGs, and growth factors that aid wound healing are secreted by dermal fibroblasts | Fibroblasts present rejection and disease risk; nylon mesh not biodegradable |
| Permacol™ *Tissue Sciences Laboratories* | None | Acellular porcine-derived matrix | Removal of non-collagenous and cellular material makes it non-immunogenic. Supports fibroblast infiltration and revascularization | Revascularization is sometimes insufficient to support an overlying epidermal graft |

**Table 15.3.**    Examples of commercial composite skin substitutes (based on[5,6]).

| Product name and manufacturer | Epidermal component | Dermal component | Advantages | Disadvantages |
|---|---|---|---|---|
| Apligraf® *Organogenesis* | Human allogeneic neonatal keratinocytes | Human allogeneic neonatal foreskin fibroblasts in bovine type I collagen, ECM proteins and cytokines | Graft take is comparable to autografts; good cosmetic results. Granulation tissue deposition improved; no signs of rejection have been observed | Risk of chronic graft rejection and disease from allogeneic keratinocytes and fibroblasts. Requires repeated applications |
| OrCel® *Ortec International* | Human allogeneic neonatal keratinocytes | Human allogeneic neonatal foreskin fibroblasts in a bovine collagen sponge | Provides a favorable environment for host cell migration and a source of cytokines and growth factors | Not intended as a permanent skin replacement, but as a biological dressing. Risk of rejection and disease from bovine collagen and other cells |

### 15.3.2. *Integra*

Integra Artificial Skin is a cell-seeded wound-dressing biomaterial product designed using the principles of molecular biology in 1984 by Drs Yannas and Burke at MIT and Massachusetts General Hospital.[11] This Artificial Skin comprises a lower layer made up of a matrix of cross-linked bovine collagen and chondroitin-6-sulfate. With pore sizes between 70 and 200 microns, the matrix structure has similar layer characteristics of the dermal layer of skin. A temporary poly(siloxane) layer simulates a barrier function normally provided by a skin epidermal layer present during wound healing. Application of the product in clinical practice, once the wound site is stabilized, involves peeling the Silastic layer to leave a vascularized neodermis suitable for epidermal grafting by using a widely meshed autologous split-thickness skin graft.[11] This type of wound-dressing technique is a relatively complex procedure and depends on following surgical protocol carefully for high-quality results.[12] Recent studies of seeding fibroblasts and epidermal cells into collagen matrix of Integra are promising.[13]

### 15.3.3. *Dermagraft*

Dermagraft is a product of Advanced Tissue Sciences (La Jolla, CA) and is a bioactive bi-layered skin substitute used for partial/full-thickness burns. Production involves placing human fibroblasts taken from neonatal foreskins onto the inner layer of Biobrane in a culture medium environment. Fibroblasts then proliferate and secrete the natural proteins fibronectin, type I human collagen, glycosaminoglycans, and other extracellular matrix proteins.[14] The product is then cryopreserved at the end of the 17-day growth cycle. The time period for growth is chosen to achieve peak fibroblast-derived protein content in the matrix. One complexity with cryopreservation is that it destroys the fibroblasts but preserves the bioactivity of the fibroblast-derived proteins. Dermagraft is designed for use with mid-dermal to deep, partial-thickness skin grafts. The wounds are at high risk for conversion to deep injury from surface inflammation. The potential advantage of this product over regular synthetic substitutes is the presence of wound-healing stimulants on the membrane.[14,15]

### 15.3.4. *Alloderm (dermal replacement)*

Alloderm is a product of LifeCell Corp. (The Woodlands, TX) and is used as an epithelial replacement for the treatment of permanent full-thickness wounds. The initial application of this treatment was for a full-thickness allograft using tissue-banked cadaver skin with dermabrasion of the epidermal layer which allows for graft vascularization and attachment to the wound bed.[16] The current product is used together with an ultrathin (0.003–0.006 inch) meshed split-thickness autograft. Clinical study with the current product yielded an equivalent 14-day take rate of dermal matrix when comparison was done on split-thickness autografts.[16] A follow-up study demonstrated that at six months there were equivalent clinical assessments for the Alloderm ultra-thin autografts combination compared with traditional grafts. The material has been shown to act as a durable skin replacement in human subjects with the concomitant use of allogenic dermal replacement in combination with cultured epidermal autografts (CEA).[17]

### 15.4. Limitations of Commercially Available Skin Substitutes

All presently available skin substitutes have serious long-term limitations, as reviewed by Metcalfe and Ferguson.[5] A synopsis of their conclusions follows.

### 15.4.1. *Reduced vascularization*

Long-term survivability requires the skin graft to acquire a viable blood supply and integrate with the host tissue. Enhanced angiogenesis is needed for most artificial skin substitutes.

### 15.4.2. *Scarring*

Scar tissue forms at the margins of skin grafts and limits the function and appearance of the graft. Future skin substitutes should incorporate anti-scarring technology to attend to the issue of scar formation which can be debilitating and can cause an undesirable cosmetic appearance.[5]

### 15.4.3. *Absence of differentiated structures*

A multiplicity of differentiated cells and 3D organization of the cells into normal skin tissues are absent in artificial skin substitutes. Consequently many of the biochemical and biophysical characteristics of normal skin are absent, such as Langerhans cells which provide immune regulation, nerves, and temperature control by sweat glands, and lubrication by sebaceous glands. Research into organogenesis to create this level of complexity in skin substitutes is in progress.[5]

### 15.4.4. *Delay in cell expansion in culture*

The need for cell culture and proliferation time upwards of two or three weeks is a severe limitation for present skin-substitute technologies. Solutions include having immuno-tolerant cells available in ready-to-use packaging, or the development of a new generation of bioreactors for rapid cell expansion.

### 15.4.5. *Persistence of cells in heterologous grafts*

Cell age and other factors may affect viability and persistence of donor cells (allogenic) used *in vivo*. Understanding these variables is essential to develop ready-to-use large-scale skin substitutes.

### 15.4.6. *Biocompatibility, biomechanical behavior and handling properties*

The next generation of skin substitutes needs to possess biomechanical characteristics that more closely mimic the durability of normal skin and also possess a handling behavior that enables clinicians to manipulate the graft as required within the surgical setting.

### 15.4.7. *Cell source*

Additional studies are required to establish the most effective cell sources to produce skin substitutes at a large scale. As discussed in Chapter 2, stem cells are the ideal starting point for organogenesis of skin, but much progress is needed before a commercially viable skin substitute can be produced by differentiation of stem cells.

### 15.4.8. *Safety and cost*

Development of the next generation of skin substitutes is limited by the cost of development, pathway for regulatory approval, requirements for packaging, delivery and maintenance of sterility, and shelf life.

### 15.5. **Alternative Approaches**

Hydrogels are polymers with properties suitable for use as a wound dressing for burn victims, as well as other wound abnormalities. A hydrogel is a network of polymer chains that are water insoluble (Fig. 15.2) and that can be formed as a colloidal gel with water as the dispersion medium. These polymers can be either natural or synthetic, depending on specific needs and uses. Hydrogels may be constructed from a wide variety of polymers such as polyvinyl alcohol, polyacrylates, polyethylene glycol, polyesters, polyurethanes, and many naturally occurring polymers.[18] See Chapter 11 for additional discussions of use of hydrogels in tissue engineering.

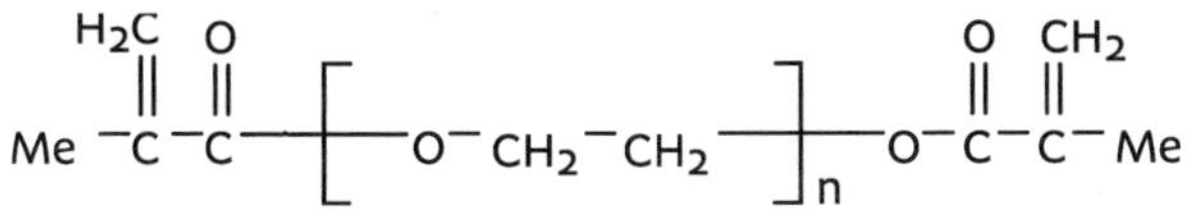

$$\mathrm{Me-C(=CH_2)-C(=O)-[O-CH_2-CH_2]_n-O-C(=O)-C(=CH_2)-Me}$$

**Figure 15.2.**   Hydrogel chemical structure.

In November 1999, the US Food and Drug Administration (FDA) proposed five main dressing categories:

1.  Non-reabsorbable gauze/sponge dressings for external use;
2.  Hydrophilic wound dressings;
3.  Occlusive wound dressings;
4.  Hydrogel wound and burn dressings; and
5.  Interactive wound and burn dressings.

The purpose of establishing these dressing categories is to define boundaries for safety and effectiveness of their uses, as well as for future development of products within this area. In the case of treatment for burn victims, category (4) takes precedence. According to the FDA, these dressings are intended for use to cover a wound, control bleeding or fluid loss, and to protect against abrasion, friction, desiccation, and contamination. The principal advantage of hydrogel polymers is a range of physical and chemical properties that are similar to that of natural tissues, the ability to tailor their water content, which can be as high as 99%, and high surface area, such as illustrated in Fig. 15.3. Another important aspect of hydrogels is that they have the ability to respond to changes in pH and temperature.

Hydrogels are also pliable and can be sterilized or created in a sterile environment.[7,19] Hydrogel dressings can be produced in

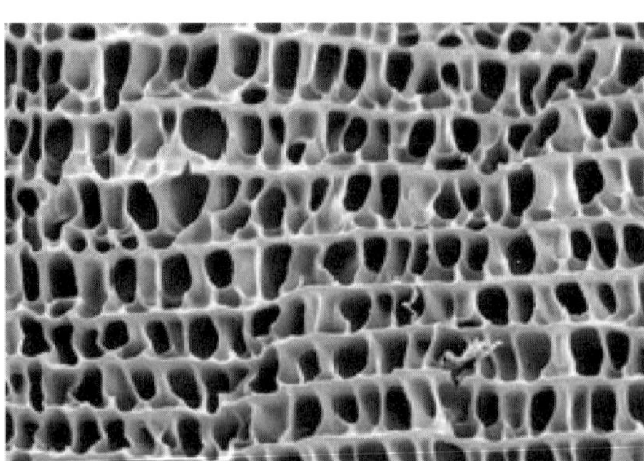

**Figure 15.3.**   SEM of a swollen biodegradable hybrid hydrogel.

many shapes or forms, such as sheets which can conform to the wound bed.

In 2007, Lin and colleagues sought to develop a new type of skin that could function both as a wound dressing as well as a type of artificial skin. Their approach was to develop a membrane composed of three layers for extensive skin burns.[1] The bottom layer, contacting the wound bed, consists of a 3D tri-copolymer sponge of gelatin/hyaluronan/chodroitin-6-sulfate with pore sizes ranging from 20–100 $\mu$m. This dermal analogous layer serves to stimulate capillary penetration, promote dermal fibroblast migration, and induce the secretion of extra-cellular matrix, providing an improved physiological environment for wound healing. The second layer is an "auto-stripped" layer composed of poly-N-isopropyacrylamide (PNIPAAm). This "auto-strip" layer is removed from the tri-copolymer layer once the wound site has recovered sufficiently. The third layer is made up of a polypropylene (PP) non-woven fabric, which provides an open structure for drainage, reducing the risk of second infection.[1] The 3D layering of this material was determined to be a successful candidate for a new type of skin replacement as it allowed for the regeneration and growth of skin and within the scaffold, and upon completion of this growth the membrane splits off without adversely affecting the wound site.

Another new skin substitute material, Suprathel, is a synthetic co-polymer mostly consisting of D,L-lactide and minor constituents of caprolactone and trimethylenecarbonate.[20] It is has been used as a dressing or wrap to keep wounds moist, a requirement for re-epithelialization. Patients treated with cortisone gauze improved epithelialization, but used together with Suprathel the treatment was even more effective.[20,21] Adding Corticoltulle, and older material used in treating skin burn wounds, to Suprathel did not lead to further improvements.[20,21] Suprathel dressings do not need to be changed as often which makes the recovery less painful for the patient. It received CE certification and entered European markets classified as a class III medical device May 2004.[20] Suprathel works as an epidermal skin substitute until the damaged skin layers are able to regenerate and had been investigated for six years in clinical studies prior to CE approval.[20]

Genetically engineered and regenerated tissues are the most exciting new approach to repair of skin damage. Details of the understanding required to produce regenerated skin are discussed by Metcalfe and Ferguson.[5] For example, the outer epidermal layers of skin containing large amounts of keratin proteins can be used to "over express" cytokines involved into the wound healing process, particularly those shown to improve healing upon topical application.[22–24] In such a case, a vector may be targeted to a specific DNA sequence causing an alteration of the genetic code and preventing or up-regulating the expression of certain phenotypic traits. Thus, modification of the genes could allow for improved wound healing.

## 15.6. Economic and Personal Issues

According to Boyce, skin loss injuries resulting from burns, traumatic injury, congenital or acquired disease caused by acute and chronic disabilities exceed US$1 billion dollars/year in the United States.[7] Skin loss from burn injuries constitutes a large portion of this cost and it has been determined that earlier wound closure, with full restoration of function, could reduce this socio-economic cost.[7]

In an interview conducted by one of the authors (KP), a burn victim shared her feelings and opinions on the process of skin grafting. This patient was involved in a severe car accident resulting in massive skin trauma which left the patient with graft scars covering 60% of her body. For well over a year after her initial graft, she had to undergo numerous surgeries to replace the cadaver skin used, until the skin from her feet could be harvested for replacement of the cadaver allograft. Through the many painful surgeries, recovery periods, and physical therapy, she has grown to be thankful of the physicians and scientists who have developed the process of grafting. The grafted skin has continued to remain competent, although it does not possess the same properties or appearance of natural skin. Continued technical advances in this field will be welcomed by her and other burns victims.

## 15.7.  Recent Developments

Continued development and advancement of artificial skin technology has dramatically increased survival rates of burn patients and recently such materials have also shown promise in applications for wound healing. Of the more than 20 commercial products currently available for skin substitutes, many are acellular and all have varying degrees of bioactive and regenerative capacity.[25] Autologous skin grafts remain the primary standard of care for burn patients in cases where severe or extensive area burns have occurred, although many patients may not possess the necessary quantity of healthy autologous skin needed for transplantation. In such cases allografts or skin substitutes must be utilized to provide critical care and help repair and regenerate the damaged skin. Wound dressings for protecting and promoting wound healing should be considered, in tandem with the development of artificial skin, as both materials must possess similar capabilities. It is important to recognize that wound dressings are for temporary treatment, whereas skin substitutes need to be designed for permanent transplantation and incorporation into native tissue. The optimal wound dressing must be able to provide protection from infection, allow for a moist healing environment, and remain biocompatible.[26] On the other hand, viable artificial skin has greater complexity due to the anatomy and physiology of skin which it must attempt to imitate. There is a tremendous need to develop a skin substitute which can be transplanted in large quantities, with the least amount of surgical intervention, and which also limits scarring and restores skin competence in the long term.[27]

The multi-layered nature of skin makes it a challenging tissue to mimic and most severe trauma requires regeneration of both epidermal and dermal layers. Although an epidermal layer can save the life of a severely burned patient, a dermal layer is needed to allow the skin substitute to permanently integrate into the adjacent tissue, as well as provide better healing and cosmesis.[28] A suitable skin substitute for such applications should consist of a derma-epidermal bi-layered structure constructed on a biodegradable scaffold which can be vascularized rapidly, promote cell proliferation, provide optimal mechanical properties, and be easily handled during surgical intervention.[27] The skin substitute must also act as a barrier immediately after transplantation to prevent bacterial infection, as infection of burn injuries are a leading cause of mortality in burn patients. Bacteria have also been found to have an adverse effect on the growth and proliferation of transplanted keratinocytes which have been cultured *in vitro*.[28] Due to these many complexities few successful engineered derma-epidermal substitutes have been developed.

Advances in scaffold design for tissue engineering of a skin substitute has been an area of great focus and recently scaffolds for bi-layered substitutes have received much attention. Huang *et al.* recently developed a bi-layered skin substitute combining a tissue-engineered extracellular matrix (TEECM) with a gelatin hydrogel (GG) containing epidermal-growth factor (EGF) microspheres.[29] The TEECM was constructed by culturing human fibroblast on a type I collagen matrix followed by removal of the cells *via* a freeze-thawing regimen and subsequent lyophilization. The TEECM was then coated with a layer of a gelatin hydrogel containing microspheres of epidermal growth factor which was to serve as the epidermal layer. Figure 15.4 shows a schematic of

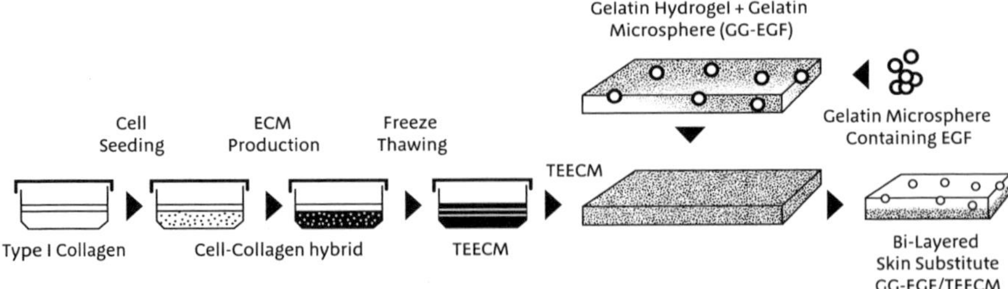


**Figure 15.4.**   Schematic outlining the construction of GG-EGF/TEECM. Following preparation of the TEECM for the lower dermal layer, the upper GG-EGF epidermal layer was bound, to form the bi-layered composite (GG-EGF/TEECM) (adapted from[29]).

the process used in this study to create the GG-EFG/TEECM bi-layered scaffold. *In vivo* studies of the GG-EFG/TEECM scaffold demonstrated that the construct was biocompatible, promoted cell migration and proliferation, efficiently released EFG, and improved overall wound healing compared to control constructs.

Lee and colleagues developed a method for creating a multi-layered skin tissue-engineered composite by using a 3D freeform bio-printing technique which dispensed a collagen hydrogel gel incorporated with human skin cells.[30] The process allowed for the construction of two distinct layers in which a fibroblast-collagen inner layer mimicked the dermis and the keratinocyte–collagen layer mimicked the epidermis. The cell-seeded hydrogel was then printed on a non-planar contoured surface which demonstrated the feasibility of developing an on-demand artificial skin grafting technique having a high degree of control and reproducibility of the construct morphology. This study provides an example of the recently developed novel techniques for constructing a scaffold-based tissue-engineering skin substitute. Other methods such as electro-spinning have also been employed for constructing promising scaffolds for tissue engineering artificial skin.[31]

Hydrogels have been found to be excellent candidates for developing wound dressings as they have the ability to conform to the wound bed and keep the wound sufficiently moist.[32] Hydrogels also possess a barrier function, allowing for protection from bacteria yet remaining permeable to oxygen.[26,32] Lu *et al.* recently created a chitosan-derived hydrogel which is capable of being formed *in situ* by photo-polymerization.[26] It was shown that the hydrogel membrane was biocompatible and exhibited good barrier function. This hydrogel material is novel in that it can be formed *in situ* and therefore has the potential to be used for creating patient injury-specific wound dressings. In another study by Mishra and Chaudhary, hydrogel-based composites were constructed from polyvinyl alcohol, carboxymethylcellulose, gelatin, and cross-linked polyacrylamide, and incorporated with the anti-microbial agent povidone-iodine.[33] The antimicrobial hydrogels were shown to be biocompatible and promoted wound healing in a mouse model, while tunability of the hydrogel properties were made possible by varying the amounts of polymer constituents. This hydrogel material displays the potential for developing an anti-microbial wound dressing in which the flexibility, strength, and other physical properties are highly controllable.

Creating an optimal tissue-engineered skin substitute which mimics natural skin includes further challenges in addition to creating a semi-permeable, biocompatible, protective barrier. One of the greatest concerns for developing a superior tissue-engineered skin substitute is achieving early vascularization of the artificial dermal layer. Preparation of a suitable wound by debridement and other techniques is critical but the dermal substitute must also promote angiogenesis to allow the graft to succeed. A major limitation of the neovascularization process is the thickness of the dermal layer, which can delay vascularization and lead to failure of the skin substitute.[34] Berthod *et al.* attempted to overcome this by developing a collagen sponge, co-cultured with human fibroblast and endothelial cells.[37] The collagen sponge was transplanted onto nude mice and within four days a reconstructed endothelialized epidermis was observed, demonstrating successful vascularization throughout the entire thickness of the graft construct. This study shows the promising potential for creating a skin substitute which can rapidly form a capillary-like structure and integrate into the native vasculature of a wound site. Along with vascularization, a robust skin substitute must also be able to maintain homeostasis by the same mechanisms of native tissue as well as regulate temperature. In a recent study, sweat glands were cultured on EGF-gelatin microspheres and incorporated into a bi-layered keratinocyte–fibroblast collagen matrix.[35] When the artificial skin constructs were transplanted into full-thickness cutaneous wounds in mice, it was observed that sweat glands had formed in a natural skin-like pattern and the implants had excellent *in vivo* performance with regard to wound healing. To achieve the goal of developing a skin construct which has all analogous functions to natural skin, the manifestation of additional features such as pigmentation, neural regeneration, and the formation of sebaceous glands and hair follicles are also being studied.

Stem cells have recently become of great interest for use in skin regeneration, wound healing, and tissue-engineered skin substitutes. Tissue-engineered skin could potentially benefit by utilizing stem cells of various lineages for directed differentiation into a variety of tissues, initiating complex reconstruction processes in which epidermal, dermal, endothelial, neural, and other tissues and skin structures form concomitantly.[36] For example, the recruitment of endothelial progenitor cells (EPCs) has been thought to play an important role in the healing of epidermal skin and recently it was shown that circulating EPCs can greatly improve neovascularization and enhance engraftment of artificial skin grafts.[37] Other groups have used stem cells in novel approaches such as bio-printing skin substitutes and developing nanofiber-based skin tissue-engineering scaffolds.[28,38] Much more research will be needed in the field of stem cells to realize their full potential for developing tissue engineering skin and wound-healing constructs.

## 15.8. Summary

Skin substitutes and improved wound dressings are critical to treating and saving lives of skin-afflicted trauma patients and also reducing morbidity of patients suffering chronic wounds. Tremendous advances have been made in the past thirty years in the development of commercially available artificial skin. Trauma patients presenting skin damage, especially those suffering severe burns, have been able to receive life-saving acute care and also the benefit from long-term preservation and integration of skin substitutes. Recent developments have shown promising potential for improving artificial skin and wound dressing. Continued research and development in the field will prove invaluable to the treatment of traumatic and chronic skin damage.

# References

1. Lin, F.-H., Tsai, J.-C., Chen, T.-M., *et al.* (2007). Fabrication and evaluation of auto-stripped tri-layer wound dressing for extensive burn injury, *Mater.Chem. Phys.*, **102**, 152–158.

2. LaRossa, D.F., Rahman, A.H., and Turka, A.T. (2007). The innate immune system in allograft rejection and tolerance, *J. Immunology*, **178**, 7503–7509.

3. Ratner, B.D., Hoffman, A.S., Schoen, F.J., *et al.* (eds) (1996). *Biomaterial Sciences: An Introduction to Materials in Medicine*, Academic Press, San Diego, CA, p. 362.

4. Lorenz, C., Petracic, A., Hohl, H.P., *et al.* (1997). Early wound closure and early reconstruction. Experience with a dermal substitute in a child with 60 per cent surface area burn, *Burns*, **23**, 505–508.

5. Metcalfe, A.D. and Ferguson, M.W.J. (2007). Tissue engineering of replacement skin: the crossroads of biomaterials, wound healing, embryonic development, stem cells and regeneration., *J.R. Soc. Interface*, **4**, 413–437.

6. Supp, D.M. and Boyce, S.T. (2005). Engineered skin substitutes: practices and potentials, *Clin. Dermatol.*, **23**, 403–412.

7. Boyce, S.T. (1996). "Skin Repair with Cultured Cells and Biopolymers", in Wise, D.L., Trantolo, D.J., Altobelli, D.E., *et al.* (eds), *Human Biomaterials Applications*, Humana Press, Totowa, NJ, pp. 347–377.

8. Zapata-Sirvent, R., Hansbrough, J.F., Carroll, W., *et al.* (1985). Comparison of biobrane and scarlet red dressings for treatment of donor site wounds, *Arch. Surg.*, **120**, 743–745.

9. Hutmacher, D.W., Ng, K.W. and Khor, H.L. (2005). "Skin Tissue Engineering Part I – Review", in Reis, R.L. and San Román, J. (eds), *Biodegradable Systems in Tissue Engineering and Regenerative Medicine*, CRC Press, Danvers, MA, p. 521.

10. Noordenbos, J., Doré, C., and Hansbrough, J.F. (1999). Safety and efficacy of transCyte for the treatment of partial-thickness burns, *J. Burn Care Rehab.*, **20**, 275–281.

11. Hutmacher, D.W., Ng, K.W. and Khor, H.L. (2005). "Skin Tissue Engineering Part I – Review", in Reis, R.L. and San Román, J. (eds),

*Biodegradable Systems in Tissue Engineering and Regenerative Medicine*, CRC Press, Danvers, MA, p. 513.

12. Moiemen, N.S., Staiano, J.J., Ojeh, N.O., *et al.* (2001). Reconstructive surgery with a dermal regeneration template: clinical and histologic study, *Plastic Reconstructive Surgery*, **108**, 93–103.

13. Kremer, M., Lang, E. and Berger, A. (2001). Organotypical engineering of differentiated composite-skin equivalents of human keratinocytes in a collagen-gag matrix (integra artificial skin) in a perfusion culture system, *Langenbecks Arch. Surg.*, **386**, 357–363.

14. Orgill, D.P., Park, C. and Demling, R. (2000). "Clinical Use of Skin Substitutes", in Garg, H.G. and Longaker, M.T. (eds), *Scarless Wound Healing*, Marcel Dekker, New York, NY, pp. 279–306.

15. Hutmacher, D.W., Ng, K.W. and Khor, H.L. (2005). "Skin Tissue Engineering Part I – Review", in Reis, R.L. and San Román, J. (eds), *Biodegradable Systems in Tissue Engineering and Regenerative Medicine*, CRC Press, Danvers, MA, p. 515.

16. Orgill, D.P., Park, C. and Demling, R. (2000). "Clinical Use of Skin Substitutes", in Garg, H.G. and Longaker, M.T. (eds), *Scarless Wound Healing*, Marcel Dekker, New York, NY, p. 293.

17. Pangarkar, N. and Hutmacher, D.W. (2003). Invention and business performance in the tissue engineering industry, *Tissue Eng.*, **9**, 1313–1322.

18. Sperling, L.H. (2006). *Introduction to Physical Polymer Science*, 4th ed., John Wiley & Sons, Hoboken, NJ, pp. 477–478.

19. Yoshida, R., Uchida, K., Kaneko, Y., *et al.* (1994). Comb-type grafted hydrogels with rapid de-swelling response to temperature-changes, *Nature*, **374**, 240–242.

20. Uhlig, C., Rapp, M., Hartmann, B., *et al.* (2007). Suprathel® — an innovative, resorbable skin substitute for the treatment of burn victims, *Burns*, **33**, 221–229.

21. Ulubayram, K., Aksu, E., Gurhan, S., *et al.* (2002). Cytotoxicity evaluation of gelatin sponges prepared with different cross-linking agents, *J. Biomater. Sci. Polym. Ed.*, **13**, 1203–1219.

22. Greenhalgh, D.G. and Warden, G.D. (1992). Transcutaneous oxygen and carbon dioxide measurements for determination of skin graft "take", *J. Burn Care Rehab.*, **13**, 334–339.

23. Tagami H., Ohi, M., Iwatsuki, K., *et al.* (1980). Evaluation of the skin surface hydration *in vivo* by electrical measurement, *J. Invest. Dermatol.*, **75**, 500–507.
24. Boyce, S.T. and Williams, M.L. (1993). Lipid supplemented medium induces lamellar bodies and precursors of barrier lipids in cultures analogues of human skin, *J. Invest. Dermatol.*, **101**, 180–184.
25. Auger, F.A., Lacroix, D. and Germain, L. (2009). Skin substitutes and wound healing, *Skin Pharmacol. Physiol.*, **22**, 94–102.
26. Lu, G., Ling, K., Zhao, P., *et al.* (2010). A novel *in situ*-formed hydrogel wound dressing by the photocross-linking of a chitosan derivative, *Wound Repair Regen.*, **18**, 70–79.
27. Bottecher-Haberzeth, S., Biedermann, T. and Reichmann, E. (2010). Tissue engineering of skin, *Burns*, **36**, 450–460.
28. van Blitterswijk, C., Thomsen, P., Hubbell, J., *et al.* (2008). *Tissue Engineering*, Elsevier Academic Press, San Diego, CA.
29. Huang, S., Zhang, Y., Tang, L., *et al.* (2009). Functional bilayered skin substitute constructed by tissue-engineered extracellular matrix and microsphere-incorporated gelatin hydrogel for wound repair, *Tissue Eng.*, **15**, 2617–2624.
30. Lee, W., Debasitis, J.C., Lee, V.K., *et al.* (2009). Multi-layered culture of human skin fibroblast and keratinocytes through three-dimensional freeform fabrication, *Biomaterials*, **30**, 1587–1595.
31. Cui, W., Zhu, X., Yang, Y., *et al.* (2009). Evaluation of electrospun fibrous scaffolds of poly(Dl-lactide) and poly(ethylene glycol) for skin tissue engineering, *Mater. Sci. Eng. C.*, **29**, 1869–1876.
32. MacNeil, S. (2008). Biomaterials for tissue engineering of skin, *Mater. Today*, **11**, 26–35
33. Mishra, A. and Chaudhary, N. (2010). Study of povidone iodine loaded hydrogels as a wound dressing material, *Trends Biomater. Artificial Organs*, **23**, 122–128.
34. Berthod, F., Germain, L., Pouliot, R., *et al.* (2010). How to achieve early vascularization of tissue engineering skin substitutes, *Adv. Wound Care*, **1**, 445–450.
35. Huang, S., Xu, Y., Wu, C., *et al.* (2010). *In vitro* constitution and *in vivo* implantation of engineered skin constructs with sweat glands, *Biomaterials*, **31**, 5520–5525.

36. Charruyer, A. and Ghadially, R. (2009). Stem cells and tissue-engineered skin, *Skin Pharmacol. Physiol.*, **22**, 55–62.

37. Morris, L.M. and Cromblehome, T.M. (2010). The role of bone marrow-derived endothelial progenitor cells in neovascularization of engineered skin substitute, *Adv. Wound Care*, **1**, 508–513.

38. Drago, H., Marin, G.H., Sturla, F., *et al.* (2010). The next generation of burns treatment: intelligent films and matrix, controlled enzymatic debridement, and adult stem cells, *Transplantation Proc.*, **42**, 345–349.

# Degenerative Disc Disease in the Lumbar Spine

CHAPTER **16**

Russell Beal

## 16.1. Introduction to the Spine

Functioning as the axial support system to the trunk, the spine extends from the base of the skull to the pelvis allowing for weight to be transmitted vertically to the lower limbs. The spine is critical for posture, flexibility, and stability. It consists of 26 irregular bones, or vertebrae, with numerous articulations held in place by scores of ligaments and strips of muscle. In addition the spine accommodates and protects the spinal cord and its millions of nerve fibers which extend outward, innervating all parts of the body. Thus, this is one of the most complex and essential structures in the human body and along with such complexity, the spine is very delicate as well. With all of the functions that it performs it can often become the focal point of many problems. These problems can be neural-related, dealing with the spinal cord nerve roots and peripheral nerves, or they can be structural, involving the system of bones and ligaments. One of the most common ailments of the spine is degenerative disc disease (DDD) which originates from degeneration of intervertebral discs between the vertebral bones of the spine.

The spine is composed of four regions; cervical in the neck, thoracic in the trunk, lumbar in the lower back, and sacral in the pelvic. The sacral region has five fused vertebrae and is superior to the coccyx, the inferior tip of the spinal column, which consist of four fused vertebrae. The uppermost part of the spine is the cervical region (Fig. 16.1) and consists of the first seven vertebrae of the spine. The cervical section functions primarily as a medium for

## NORMAL SPINE

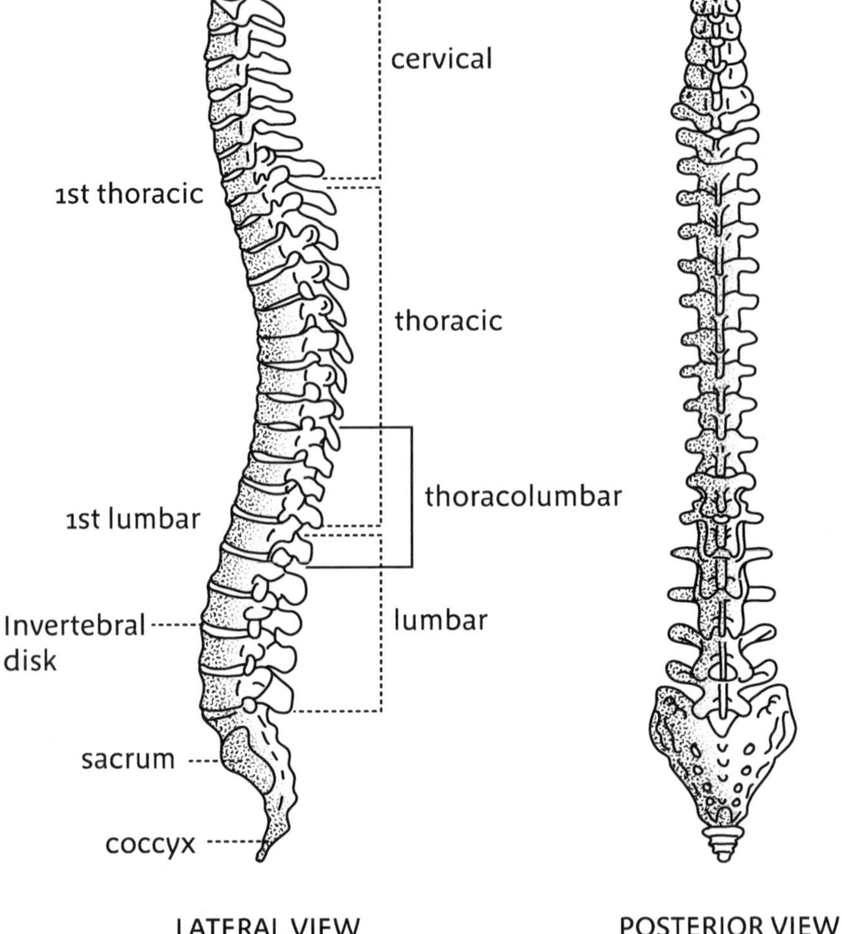



**Figure 16.1.**   Anatomy of the spine.

rotation. The thoracic section encompasses the largest group of vertebrae, 12 in total. The thoracic spine connects with the ribs forming a cage which serves to protect the internal organs, most importantly the heart and lungs. Degenerative disc disease is most commonly presented in the lumbar region, and will therefore be the focus of this chapter.

Surgical repair of the cervical spine is reviewed in *Clinical Performance of Skeletal Prostheses.*[1] Grooms, Bianchi, and Chan present a comprehensive discussion of the etiology of spinal injuries; the biology and biomechanics of the vertebral bones; statistics on the need for surgical repair; complications and use of autograft *versus* allograft in repair; and the biology and biomechanics of instrumentation used to stabilize the spine. The reader may refer to this text for detailed discussion of cervical spine injury and repair. The authors discuss and analyze the findings of 19 published clinical results that used the Smith–Robinson anterior cervical spinal fusion technique to relieve pain from chronic cervical damage. They also discuss seven clinical studies for outcomes of the internal plating technique. Clinical success of the techniques ranged from 84% to 91%. The comparison of these alternative methods of treatment led to the conclusion that to achieve long-term stability fusion is necessary. Since plating systems have high fusion rates plates should be used for multiple level surgeries, whereas the standard Smith–Robinson technique without a plate is sufficient for single level procedures.[1]

However, the most vulnerable part of the spine to DDD is the lumbar region. This section of the spine has the fewest vertebrae — only five — but they are the largest in diameter and bear the highest load. The lumbar vertebrae do not have added support from other parts of the body, unlike the upper regions of the spine. For these reasons, problems associated with DDD often present in this range and may contribute to the onset of DDD.

The spinal cord runs through a corridor in the interior of the spinal column called the vertebral canal. This is made up of successive vertebral foramen through which the spinal cord passes. The spinal cord is protected by the vertebral arch, which forms the posterior aspect of the vertebral foramen. The vertebral body forms the anterior aspect of the vertebral foramen. The intervertebral discs are sandwiched between the vertebral bodies of the spine. The vertebrae structure consists of the body and the pedicles (Fig. 16.2). The pedicles are protrusions on the sides of the vertebrae

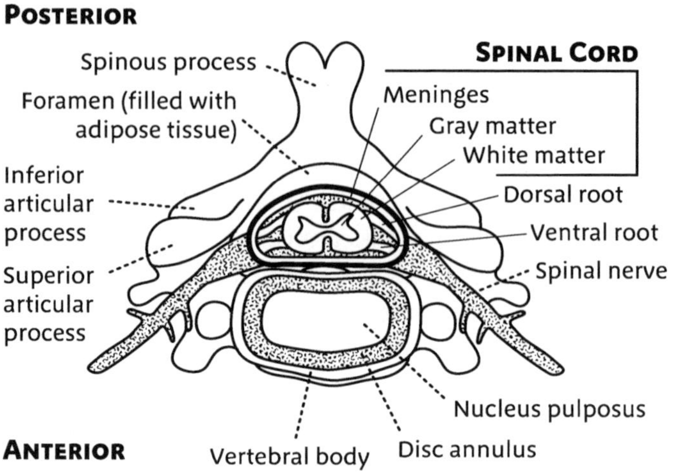

**Figure 16.2.**   Anatomy of the vertebra.

and are connection points for the many ligaments. These are often harvested for use as autograft by surgeons. The muscles of the spine connect to the transverse processes, which lend support and strength to the spine by providing tension in bending and rotation.[1,2] The body is the largest bony mass of the vertebrae. It consists of both hard, dense, cortical bone and soft, spongy, trabecular bone. The trabecular bone is the site of the bone marrow and therefore great care must be taken to lessen any impact on this region. If the disc degradation is severe the vertebrae can start to be damaged. This will usually cause considerable pain and also inhibit the functioning of the bone marrow. In addition, a decrease in stability is experienced. A major concern and cause of pain for patients suffering from DDD is the impingement and impairment of neural pathways, such as the spinal cord or nerve roots. When any surgery is performed on the spine great care must be taken to not damage the spinal cord.

The intervertebral discs are composed of two main components: the nucleus pulposus and the annulus fibrosis. The nucleus pulposus is the center of the disc and the annulus is made up of collagen and serves to contain the nucleus pulposus from over-expansion.

Composed of cartilage tissue, the disc acts as a cushion and when a stress is applied to the spine it spreads the force evenly throughout the vertebrae. The nucleus pulposus has high water content and a gel-like consistency. The annulus fibrosis connects to an end plate and is made of fibrocartilage. The end plate provides the junction to the trabecular bone. These components, working in unison, carry the majority of the stress on the spine as well as providing proper spacing between the vertebrae. Therefore when the discs are weakened by DDD the ramifications can be severe.

As the core of body movement, the spine is exposed to multiple forces, which are passed through the intervertebral discs.[1,2] There is a compression force from normal body weight with an added force whenever a load is carried. Shear forces are also commonplace from normal movements such as walking or running, and are applied when bending occurs, often magnified by picking up a load at the same time. Rotational movements from twisting the trunk of the body are another source of stress on the spine, although the forces involved are of a lesser degree. Shear stresses and rotational forces are shared by the vertebral arch to a small extent.[1] DDD affects all aspects of the load-bearing capabilities of the intervertebral discs.

DDD is one of the most common indications for spinal surgery. The name of the condition is a source of confusion as it is a misnomer. Degenerative disc disease, while involving the discs of the spine, is not a disease. Although it can be thought of as a degenerative process, DDD does not necessarily worsen with time. The term has been used by physicians to describe almost any problem associated with back pain. For the purposes of this chapter the term will refer to the weakening of the intervertebral discs that cause instability in the spine. The weakening of the discs can lead to spondylolisthesis, a slipping of the discs, or rupture through the annulus fibrosis. Even this definition is unclear through the use of the term "instability." To date there is no clear quantitative definition of instability. Generally speaking instability is an excess of movement, either translational or rotational, but these measurements are difficult to assess for a patient. Also, hyper-mobility of the spine is

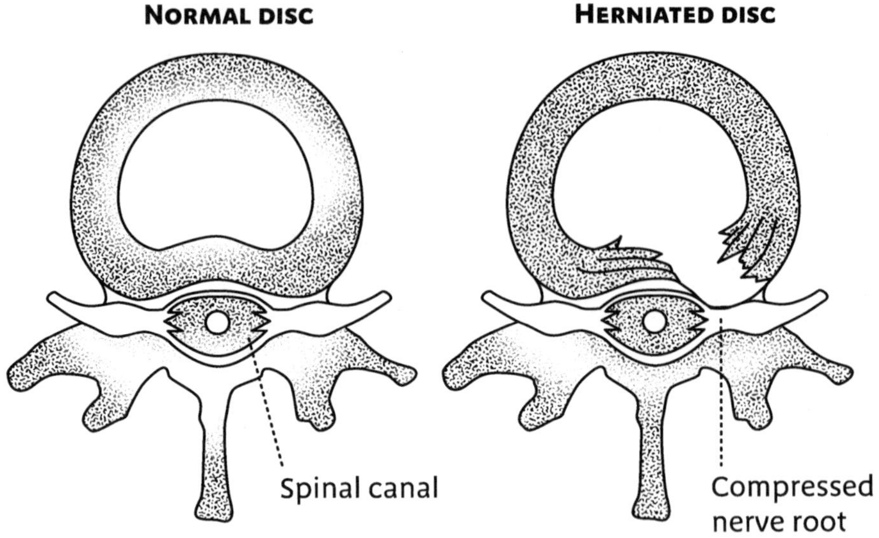



**Figure 16.3.**   Schematic of normal disc and herniated disc.

not necessarily conducive to DDD.[2] Although a person may not have a ruptured or slipped disc, some symptoms may be present, mostly pain and sensitivity, that are associated with it.

DDD causes the disc to become more fibrous and may progress so there becomes no clear distinction between the nucleus pulposus and the annulus fibrosis. Also, the disc can decrease in volume by more than 90% due to dehydration.[3] As the nucleus pulposus begins to dry out it does not transfer the load evenly to the rest of the spine. This causes the annulus fibrosis to weaken and can cause it to rupture or separate from the cartilage end plate as shown in Figure 16.3. In turn compression of the spine can cause pain and decrease body movement. In essence all humans are likely to suffer from some degree of DDD, but not to the point where medical attention is required. Some cases may progress and require further care, but the majority of people do not experience any complications. If left untreated DDD can cause excessive pain and also degrade the vertebral body. This can lead to further degradation of adjacent discs creating a domino effect;

in other words, poorly functioning discs put more pressure on healthy discs and they quickly begin to degrade.

The causes of DDD are generally unknown and many factors appear to influence its onset and progression. DDD seems to be part of the aging process, although why it becomes more severe for some people and not others is not well understood. Studies investigating a possible genetic link to DDD[4] have been on-going, although no conclusive relationship has been established. Strenuous activity has the largest correlation to onset of disc problems. Strenuous activity is highly correlative to the onset of DDD, which includes such activities involving heavy and/or improper lifting. Both sexes are equally affected by DDD and it can affect individuals of all ages, although DDD is more common among those of >50 years of age.

One of the reasons that DDD is so elusive is the lack of predictive diagnostic techniques. Pain is the most common symptom but is not exclusive to DDD. When someone has demonstrated back problems, diagnostic imaging is usually in order. Magnetic resonance imaging (MRI) has shown to be the most effective non-invasive technique. MRI images can clearly show an intervertebral disc, but cannot always distinguish between healthy and weakened discs.[5] Magnetic resonance imaging is not a routine part of annual medical exams because of cost, which often leaves cases of DDD undetected until the damage has reached a more advanced stage. MRI images are becoming clearer with improved technology but a correct diagnosis still requires an expert. This is because it is difficult to distinguish between soft tissue and the nucleus pulposus, so even when a disc does rupture the images may not be conclusive. Also, MRI images do not show the strength of the annulus fibrosis nor tell if there is an increased load on a vertebra.

After it has been determined that a patient is suffering from DDD there are several treatments available. Some have suggested that chiropractic alignment may relieve pain in mild cases of DDD by helping to maintain the proper spacing between vertebrae.[6] Pain associated with spondylolisthesis may also be reduced by regular adjustments of the spine. These treatments may help ease the pain but do not address its source. Other methods have tried to regenerate the disc itself, using such methods as electrical impulses or injections of growth factors.[7] In severe cases, the only solution available is surgery, although the outcomes and success rates are variable and many limitations exist with current surgical procedures.

## 16.2.  Repair by Fusion

A number of techniques have evolved for the repair of degenerative discs and relief of lower back and radicular pain; of which spinal fusion, a procedure in which the vertebrae are physically fused together, has demonstrated the greatest success. The theory supporting the fusion procedure is based upon the concept that if no forces are applied to the disc, other than those of normal loading, then they cannot be a source of pain or instability.[2]

Spinal fusion has become increasingly popular as a remedy for back pain. In 2003 Medicare spending on spinal fusions was about US$482 million, about a 500% increase from 1992, and accounted for roughly 47% of its total expenditure on back pain.[8]

Fusion of the vertebrae can be achieved through various methods. The goal is arthrodesis, or solid fusion. Plates and rods are the traditional means to stabilize the spine; their purpose is to prevent movement of the vertebrae while they fuse. Generally the disc is removed so fusion can be accomplished. Although several variations of spinal fusion are applied routinely there is no rating system for comparison of the techniques.[1] Fusion techniques are often elaborate, highly invasive procedures, which may also run the future risk of weakening of the vertebrae through stress shielding, yielding unfavorable long-term result.[9] In some cases fusion devices such as plate and rods are temporarily implanted for several months until arthrodesis is achieved, after which they can be removed to allow the fused bone to carry the stress.[10] Studies done on spinal implants have helped to show the effectiveness of the various methods on stability.[11]

Success rates for these types of implants are difficult to gauge. The ultimate goal is to eliminate pain and improve stability, but that can be met with varying degrees of success. Most patients feel that their surgery was a success, but many say that if they had to do it again they would choose another option if one was available.[10,11] Significant recovery time is also common, taking six or more months, and often requiring a regimen of physical therapy.[12] Examination of several patients after surgery leads to the conclusion that arthrodesis was sometimes never fully achieved, yet a reduction in pain was experienced.[2] These mixed outcomes help to demonstrate the complexity of diagnosis and treatment of spinal problems as well.

## 16.3.  Cages

Cages are another type of implantable device used in spinal fusion (Fig. 16.3).[13] Unlike rods and plates, cages are designed to remain in place after bone growth has occurred. This means that later revision surgery will not have to be performed to remove the devices, and for that reason they have become more popular. Cages are designed to be placed in the space of an intervetebral disc after the disc has been removed. But, just as rods and plates, cages expose the spine to the risk of infection. A substantial amount of research has focused on improving cage design and implantation techniques as well as understanding the biomechanical forces incurred by these devices. For instance, several cage devices have undergone modifications so as to reduce the cage's impact on bone during placement by decreasing surface area of the cage.[13] However, this has been shown to increase loading forces experienced by the spine and therefore makes arthrodesis more difficult. The cage is designed with holes so it can be filled with autograph bone, or a mixture of autograft bone and synthetic bone grafting material, as discussed below, to help the healing process.

Some spinal fusion procedures include the use of autogeneous bone to strengthen the site and promote fusion.[14] Autograph bone is the gold standard; however, as discussed in Chapter 7 there are limitations, such as a limited supply and donor site morbidity.

Allograft material has improved greatly and is now commonly used in fusion procedures but concerns and limitations remain with this material.[15] These limitations have made innovation of materials that can bond to bone a necessity. These biocompatible materials must promote bone growth and resorb in a controlled manner leaving new bone with mechanical properties similar to natural bone.[14,16,17] Use of synthetic bone graft material that stimulates osteogenesis is reviewed in Chapter 3 and a description of its use in maxillofacial reconstruction appears in Chapter 7.

Some implant devices have been made from polymeric biomaterials. Such materials could potentially minimize further degradation through contact between the bone and the implant, but leaves the area with decreased strength.[18] Other procedures use a mesh cage to surround the spine to provide proper confinement while filling the construct with graft material. This method may leave the bone intact, but plates are still needed to stabilize the vertebrae.[19] While these applications have helped to improve spinal fusion, they fall short of an ideal situation by inhibiting natural movement and having a high associated cost. An ideal procedure would be relatively inexpensive, reduce surgical complexity, and retain full functionality of the joint.

Artificial discs are still in being investigated.[20] Unlike fusion procedures which require plates, rods and graft material to stabilize the spine, disc implant devices will serve to replace the defective disc and allow the joint to remain functional. Disc implants would be positioned between the cartilage end plates with the hope of avoiding damage to the vertebral body. Unfortunately they do not bond with the cartilage and can weaken the end plate which could cause loosening. With people having spine fusion procedures at an increasingly younger age, survivability of the implant is crucial. The notion of a transplanted disc is unlikely because of the need to remove the intact disc entirely, which creates additional issues and limitations. While resorbable biomaterials are used for many implantable devices, most are not suitable for a replacement disc.[21] Hydrogels have been considered for applications in disc replacement, since they exhibit soft tissue bonding and do not reabsorb, while still remaining somewhat flexible.[22] However, current hydrogels lack sufficient strength to be used directly as disc replacements. See Chapter 10 for discussion on cartilage repair and Chapter 11 for a discussion of the strengths and limitations of the use of hydrogels in cartilage repair.

## 16.4.  Stem Cells for the Future?

An ideal solution to DDD appears to require a combination of several techniques. Growth factors and stem cells are promising, if they can be seeded onto an appropriate scaffold. The scaffold would need to have a 3D interconnected porous structure which is able to bond to the surrounding tissue and be absorbable while maintaining strength.[23] Stem cells could be used with the scaffold to regenerate the worn disc with the proper spacing.[24] One difficulty will be to control the extent of stem cell differentiation into nucleus pulposus tissue or the annulus fibrosis tissue. As a disc ages the nucleus pulposus becomes more random and fibrous, thus added formation of the annulus fibrosis would only magnify the problem and not solve it. Other approaches currently under investigation involve the use of growth factors to stimulate repair and regeneration of the damaged disc.[25]

## 16.5.  Recent Developments

Degenerative disc disease continues to persist as one of the most prevalent spinal disorders. A poor understanding of its etiology and the complex biomechanical nature of the intervertebral disc make repair and replacement a challenging task. Advances in surgical techniques, fusion devices, and bone graft alternatives have greatly improved the outcomes of patients undergoing spine surgery for DDD relief. Spine fusion has been the gold standard for treating patients suffering from DDD, with a majority of patients undergoing posterior lumbar interbody fusion (PLIF) procedures. Other procedures such as anterior lumbar interbody fusion (ALIF) and transforaminal lumbar interbody fusion (TLIF) are also commonly indicated and have experienced progressive advances. In all cases some form of instrumentation, such as pedicle screws or interbody cages; or graft materials, such as autograft or allograft, are used to fuse vertebral levels and stabilize the spine.[14,26,27] Much research has focused on development of a superior synthetic bone graft substitutes with bioactive materials, reducing the morbidity and tissue-availability issues associated with harvesting autograft.[14]

Metallic interbody cages have been increasing in popularity as these devices are biomechanically adequate and have displayed high success rates for fusion. Nevertheless, the use of cages has several limitations including increased morbidity, possible neural agitation during PLIF, the need for supplemental fixation in ALIF, and post-operative cage migration.[27] Recently, fusion procedures have seen the introduction of a combination cage and synthetic bioactive bone graft material incorporated with biological molecules such as growth factors to improve fusion. For example, Medtronic's Infuse Bonegraft®/LT-Cage® Lumbar Tapered Fusion Device uses an interbody cage to restore disc space height which is then filled with a synthetic bone graft material containing recombinant human bone morphogenic protein.[28] Such combinatorial devices are leading the way in technological advances for improving spinal fusion procedures and long-term outcomes for DDD patients.

Disc arthroplasty has also seen great advances and is viewed by some as a promising alternative to spinal fusion procedures for patients suffering from DDD. Disc arthroplasty is the partial or complete removal of the defective disc followed by replacement or repair with a prosthetic disc and no need for fusion, which allows the patient to maintain flexibility and motion of the joint.[29] Disc replacement devices must be able to transmit load, allow joint flexibility, and maintain the proper disc height, although as yet none of the current implants have the ability to reproduce the biomechanics and kinematics of intervertebral discs.[29] Only a limited amount of data has been collected from clinical trials using disc replacement devices, and the results are not yet conclusive.

Ongoing research has been focusing on developing a functional disc implant from a variety of materials. Since intervertebral discs are composed of cartilage much of the focus has been placed on the investigation of hydrogels. Recent research in the use of hydrogels for lumbar disc replacement and regeneration have seen promising developments in the use of hyaluronan matrices and protein-incorporated polymeric hydrogel scaffolds.[30,31] Chapters 10 and 11 give overviews of the advances in cartilage repair and the use of hydrogels for tissue engineering of cartilage tissue.

Stem cells have also been studied for the repair and regeneration of cartilage tissue (Chapter 10) and likewise have demonstrated successful outcomes in various animal models.[32,33] Furthermore, gene therapy has been investigated for treatment of DDD. Sobajima *et al.* review the possible applications of gene therapeutic approaches and outline the body of work that remains to be completed before human trials can commence.[34] Continued research into such regenerative medicine and tissue-engineering approaches will be instrumental in creating an unparalleled treatment for DDD, which would provide a long-term solution while allowing the patient to regain full range of motion and flexibility.

## 16.6. Conclusion

Degenerative disc disease is the most common indication for lumbar fusion procedures, which have become increasingly popular for its treatment. Although DDD is one of the most commonly diagnosed ailments of the spine, the cause of the disorder is still unknown. Great advances have been made in the surgical procedures and devices used to treat DDD; however a number of limitations and concerns remain. A better understanding of the etiology and the cause of failures related to current surgical procedures will be key to improving the treatment of DDD. Hydrogel materials and stem cell technology offer promising potential for treating and repairing defective discs and allowing DDD patients to retain full functionality in the long term. Regenerative medicine, improved biomaterials, and tissue-engineering strategies have shown potential for developing a successful treatment for DDD in the future.

## References


1. Grooms, J.M., Bianchi, J. and Chan, J. (1996). "Success of Surgery on the Anterior Cervical Spine: Smith-Robinson Technique *vs.* Internal Plates", in Hench, L.L. and Wilson, J. (eds), *Clinical Performance of the Skeletal Prostheses*, Chapman and Hall, London.

2.  Muggleton, J.M., Kondracki, M. and Allen, R. (2000). Spinal fusion for lumbar instability: does it have a scientific basis?, *J. Spinal Disord.*, **13**, 200–204.

3.  Modic, M.T. (1989). "Degenerative Disorders of the Spine", in Modic, M.T., Masaryk, T.J. and Ross, J.S. (eds), *Magnetic Resonance Imaging of the Spine*, Year Book Medical Publishers, Chicago, IL.

4.  Dickson, R.A. and Butt, W.P. (2003). *The Medico-Legal Back: An Illustrated Guide*, Greenwich Medical Media Ltd, London.

5.  Modic, M.T. (1989). "Normal Anatomy", in Modic, M.T., Masaryk, T.J. and Ross, J.S. (eds), *Magnetic Resonance Imaging of the Spine*, Year Book Medical Publishers, Chicago, IL.

6.  Madigan, L., Vaccaro, A.R., Spector, L.R., *et al.* (2009). Management of symptomatic lumbar degenerative disk disease, *J. Am. Acad. Orthoped. Surg.*, **17**, 102–111.

7.  Filler, A.G. (2004). *Do You Really Need Back Surgery? A Surgeon's Guide to Neck and Back Pain and How to Choose Your Treatment*, Oxford University Press, New York, NY.

8.  Weinstein, J.N., Lurie, J.D., Olson, P.R., *et al.* (2006). United States trends and regional variations in lumbar spine surgery: 1992–2003, *Spine*, **31**, 2707–2714.

9.  Hench, L.L. (2005). "Repair of Skeletal Tissues", in Hench, L.L. and Jones, J.R. (eds), *Biomaterials Artificial Organs and Tissue Engineering*, Woodhead Publishing Ltd, Cambridge, UK.

10. Glassman, S.D., Bazzi, J., Puno, R.M., *et al.* (2000). The durability of small-diameter rods in lumbar spinal fusion, *J. Spinal Disord.*, **13**, 165–167.

11. Jacobs, R.R., Nordwall, A. and Nachemson, A. (1982). Reduction, stability, and strength provided by internal fixation systems for thoracolumbar spinal injuries, *Clin. Orthoped. Rel. Res.*, **171**, 300–308.

12. Bossons, C.R., Levy, J., Sutterlin, C.E. III (1996). Reconstructive spinal surgery: assessment of outcome, *South Med. J.*, **89**, 1045–1052.

13. Murakami, H., Boden, S.D. and Hutton, W.C. (2001). Anterior lumbar interbody fusion using a barbell-shaped cage: a biomechanical comparison, *J. Spinal Disord.*, **14**, 385–392.

14. Hench, L.L., Day, D.E., Höland, W., *et al.* (2010). Glass and medicine, *Int. J. Appl. Glass Sci.*, **1**, 104–117.
15. Schultheiss, M., Sarkar, M., Arand, M., *et al.* (2005). Solvent-preserved, bovine cancellous bone blocks used for reconstruction of thoracolumbar fractures in minimally invasive spinal surgery-first clinical results, *Eur. Spine J.*, **14**, 192–196.
16. Hench, L.L., Xynos, I.D., Edgar, A.J., *et al.* (2001). Gene-expression profiling of human osteoblasts following treatment with the ionic products of bioglass® 45S5 dissolution, *J. Biomed. Mater. Res.*, **55**, 151–157.
17. Thompson, I.D. (2005). "Biocomposites", in Hench, L.L. and Jones, J.R. (eds), *Biomaterials Artificial Organs and Tissue Engineering*, Woodhead Publishing Ltd, Cambridge, UK.
18. Fujibayashi, S., Shikata, J., Tanaka, C., *et al.* (2001). Lumbar posterolateral fusion with biphasic calcium phosphate ceramic, *J. Spinal Disord.*, **14**, 214–221.
19. Chuang, H.-C., Cho, D.-Y., Chang, C.-S., *et al.* (2005). Efficacy and safety of the use of titanium mesh cages and anterior cervical plates for interbody fusion after anterior cervical corpectomy, *Surg. Neurol.*, **65**, 464–471.
20. Hench, L.L. (2005). "Joint Replacement", in Hench, L.L. and Jones, J.R. (eds), *Biomaterials Artificial Organs and Tissue Engineering*, Woodhead Publishing Ltd, Cambridge, UK.
21. Hill, R.G. (2005). "Biomedical Polymers", in Hench, L.L. and Jones, J.R. (eds), *Biomaterials Artificial Organs and Tissue Engineering*, Woodhead Publishing Ltd, Cambridge, UK.
22. Burdick, J.A. and Stevens, M.M. (2005). "Biomedical Hydrogels", in Hench, L.L. and Jones, J.R. (eds), *Biomaterials Artificial Organs and Tissue Engineering*, Woodhead Publishing Ltd, Cambridge, UK.
23. Jones, J.R. (2005). "Scaffolds for Tissue Engineering", in Hench, L.L. and Jones, J.R. (eds), *Biomaterials Artificial Organs and Tissue Engineering*, Woodhead Publishing Ltd, Cambridge, UK.
24. Polak, P. and Hench, L.L. (2005). Gene therapy progress and prospects: in tissue engineering, *Gene Therapy*, **12**, 1725–1733.

25. Haid, R.W., Subach, B.R. and Rodts, G.E. (eds) (2003). *Advances in Spinal Stabilization*, Karger Publishers, Basel, Switzerland.

26. Rodriguez-Gonzalez, F.A. (2009). *Biomaterials in Orthopaedic Surgery*, ASM International, Materials Park, OH, pp.113–143 and 171–175.

27. Arnold, P.M., Robbins, S., Paullus, W., *et al.* (2009). Clinical outcomes of lumbar degenerative disc disease treated with posterior lumbar interbody fusion allograft spacer: a prospective, multicenter trial with 2-year follow-up, *Am. J. Orthoped.*, **38**, E115–E122.

28. Medtronic, Inc. (2010). *Medtronic, the world leader in medical technology and pioneering therapies.* [Online]. Available at: http://www.medtronic.com. [Accessed 22 July 2010].

29. Fekete, T.F. and Porchet, F. (2010). Overview of disc arthroplasty — past, present and future, *Acta Neurochir.*, **152**, 393–404.

30. Berlemann, U. and Schwarzenbach, O. (2009). An injectable nucleus replacement as an adjunct to microdiscetomy: 2 year follow-up in a pilot clinical study, *Eur. Spine J.*, **18**, 1706–1712.

31. Shen, B., Wei, A., Bhargav, D., *et al.* (2010). Hyaluronan: its potential application in intervertebral disc regeneration, *Orthoped. Res. and Rev.*, **10**, 217–226.

32. Ciacci, J., Ho, A., Ames, C.P., *et al.* (2008). Stem cell horizons in intervertebral disc degeneration, *Stem Cells Cloning: Adv. Appl. 2008*, **1**, 31–39.

33. Yan, F., Leung, V.Y.L., Luk, K.D.K., *et al.* (2009). Mesenchymal stem cells arrest intervertebral disc degeneration through chodrocytic differentiation and stimulation of endogenous cells, *Mol. Ther.*, **17**, 1959–1966.

34. Sobajima, S., Kim, J.S., Gilbertson, L.G., *et al.* (2004). Gene therapy for degenerative disc disease, *Gene Therapy*, **11**, 390–401.

# Shoulder Repair



## CHAPTER 17

Shane D. Smith

## 17.1. Introduction

The shoulder is a unique and important joint. When a person's shoulder is damaged, that entire arm is compromised. Because hands are our primary means of interacting with the world around us, people who have lost the use of even a single arm suffer from an extreme handicap, reducing or completely eliminating their ability to perform such activities as personal hygiene, driving, writing, typing, opening jars, or even taking money from their wallets to make payments. The shoulder's importance in everyday life warrants extra attention to keeping it functional. However, the unique construction of the shoulder makes this task more difficult than for other joints, and also means that the shoulder is more easily injured. This chapter will classify common problems with the shoulder joint, and present modern solutions to those problems. It will also describe solutions which may not be practical or even possible in today's hospital environment, but which may become the standard for shoulder repair in the near future.

## 17.2. Anatomical Considerations

The human shoulder is the critical point of attachment for the arm, arguably the most important limb for most humans of any time period, prehistoric or modern. The loss of function in any limb can be extremely debilitating, but the arms are critical for independent living. As such, maintaining the health of the shoulder is paramount

**THE SHOULDER JOINT**

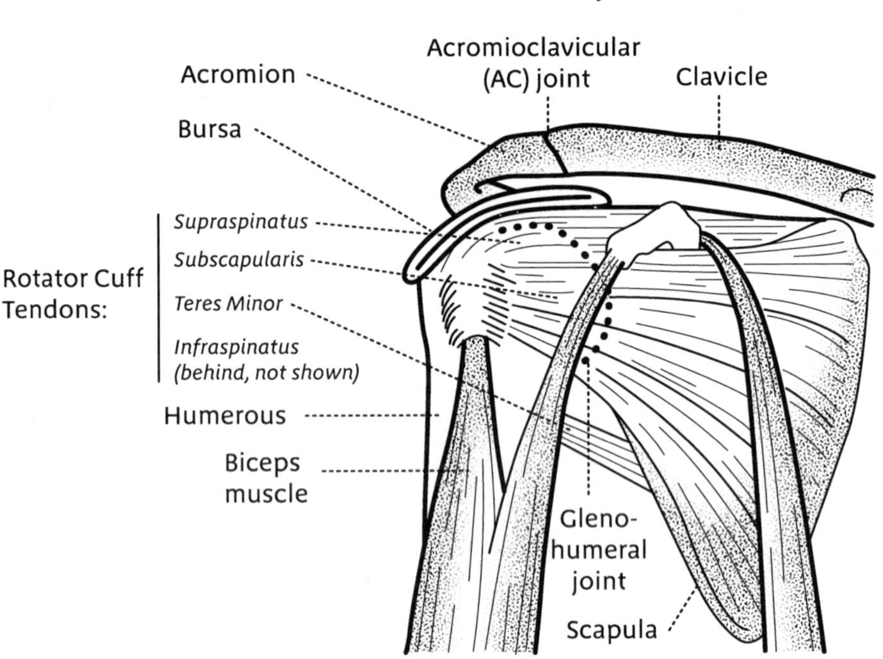


**Figure. 17.1.**   Anatomy of the shoulder joint.

to a healthy life, but the shoulder's structure makes it both delicate and vulnerable. The shoulder is the most flexible joint in the body, allowing for more than 180 degrees of motion in some directions. It must be capable of elevation; depression; lateral and medial rotation; flexion; extension; abduction; and adduction. In order to achieve this level of mobility, the shoulder has also evolved to be an extremely shallow joint, precariously held in place with muscles, tendons and ligaments, as illustrated in Fig. 17.1.[1–3]

While this arrangement allows for the shoulder to have an incredibly high range of motion, it also results in a high probability of dislocation. According to the National Institutes of Health, the shoulder joint is the most frequently dislocated joint in the body.[4] In the year 2003, the Center for Disease Control and Prevention reported 13.7 million cases of medical care for shoulder-related problems in the US, much higher than the 8.7 million reported between 1977 and 1978.[5] While many of these injuries are sports-related, some people seem to have a higher propensity for dislocation than others and this increasing trend may also point towards inadequate safety guidelines in the workplace, where seemingly routine actions may lead to chronic shoulder problems.[6] Strenuous activity can easily cause separation of the ball and cup, which stretches the stabilizing rotator cuff. With repeated abuse, these tendons can become permanently distorted, making the shoulder even easier to dislocate again. Eventually, surgical repair is required to stabilize the joint.

## 17.3.  Present Solutions to Shoulder Problems

When a shoulder is dislocated, it stretches the supporting ligaments holding the joint together. Usually, the best way to treat a dislocated shoulder is to manipulate it back into place and keep it immobile for several days, using ice to keep down the swelling. This treatment gives the rotator cuff a chance to heal and tighten up again, so the shoulder will restabilize. It is then advisable to begin light exercises to further tighten the muscles, eventually moving onto light weightlifting to complete the recovery program. As with all medical practices, surgery is to be avoided until all other alternatives have been exhausted. However, it is sometimes the case that no other method will provide adequate relief. When the shoulder has been dislocated many times, the tendons holding it in place may be torn beyond the body's ability to repair them, such as when they are torn completely from the bone. When that happens, there are generally two approaches to reattaching and repairing the torn tendons. Detachment of the ligament and labrum from the glenoid is often termed a Bankart lesion. Either a traditional open surgery can be used, or the surgeon can operate less invasively using a small scope. In either case the tendons are reattached using sutures after cleaning the joint of any damaged tissue debris.

Open surgery provides the surgeon with a better view of the damage, and therefore is often more successful. However, open surgery damages a lot of tissue in exposing the shoulder joint, and requires a long recovery time. Using the small scope is less reliable, but less damaging to the surrounding tissues, and so patients may begin moving their shoulder again in about six weeks, after which they should begin physical therapy on their newly healed but stiff shoulder to regain joint mobility.

Sometimes the problem does not lie in the rotator cuff, but is instead a problem of the joint itself, as with advanced arthritis of the shoulder or fracture from injury. A total shoulder replacement, called a total shoulder arthroplasty (TSA), should be considered when the cartilage in the shoulder joint is sufficiently damaged so that the joint becomes debilitating and dysfunctional. A shoulder prosthesis resembles a prosthetic hip joint in many ways, with a shaft, neck, and a head that fits into a socket.

There is little significant difference between the failure rates of cemented and un-cemented shoulder prostheses.[7] Thus, a patient should typically rely on the method of fixation that their surgeon recommends. Because of the musculature involved, a total shoulder replacement procedure is considered very complex when compared to hip or knee surgery.[8] Shoulder replacement also requires that the patient must follow a rigorous rehabilitation program involving extensive physical therapy in order to achieve full recovery.

Fortunately, if a total shoulder arthroplasty is performed properly and the patient is compliant with the recovery program, the procedure has a high probability of success. According to the Maryland Health Services Cost Review Commission database of patients undergoing shoulder, hip or knee arthroplasty between 1994 and 2001, no deaths resulted from shoulder surgery, whereas 0.18% of hip and 0.16% of knee surgery patients died after their operations. Also, patients of shoulder arthroplasty tended to spend less time in the hospital, and had fewer complications.[9] While shoulder surgery complications are not unheard of, these data should give a high degree of confidence to patients considering the procedure.

For special cases, there are other options for shoulder surgery. Osteocapsular arthroplasty is designed to smooth the joint for patients with abnormal bone spurs and growths in their shoulder. In a hemi-arthroplasty, only the head of the shoulder joint is replaced with a prosthesis, and patients are left with their original natural socket. This is best for people who still have a healthy joint socket, because the cup in a traditional total shoulder arthroplasty is subject to wear. Arthrodesis is a process in which the shoulder joint is fused, removing its mobility but saving the arm of a patient whose musculature can no longer support any sort of mobile joint. Choosing the best surgical procedure is a decision which must be made with much care by the patient, family, and surgeon, as the patient's ability to live independently is at stake.

Repetitive dislocation can also lead to bone defects in the humerus. As the shoulder joint dislocates, the humerus can move forward and rotate, impacting on the socket of the joint. The bone on the back of the ball of the joint is not as hard as in other places and is easily damaged. This defect is called the Hill–Sachs defect and is often the shape of an orange segment. The defect makes it very difficult for individuals to raise their arms above the horizontal, as the defect comes into contact with the socket. For this reason the defect is difficult to heal as it must have a smooth exterior surface that will contact with the cartilage of the socket. Current treatment relies on the use of autograft. If the defect is large, bone graft extender material can be mixed with the autograft, otherwise a metal prosthesis with the shape of an incandescent light bulb is used. Alternative solutions are needed.

## 17.4. Alternative Shoulder Repair Systems

From 1970 up to the present day, a number of advances in shoulder arthroplasty procedures have been made which provide alternatives to reconstruction or replacement of the shoulder. The approaches fall into the six categories listed in Table 17.1, and are reviewed in Chapter 7 of *Clinical Performance of Skeletal Prostheses*.[7]

A brief summary of each type is presented here.

**Table 17.1.**   Types of shoulder replacement.[7]

| Type of shoulder replacement | Complications | Failures |
| --- | --- | --- |
| Unconstrained total arthroplasty | 24% | 4% |
| Semi-constrained total arthroplasty | 5% | 3% |
| Constrained total arthroplasty | 21% | 0% |
| Unconstrained hemi-arthroplasty | 24% | 0% |
| Bipolar shoulder replacement | 25% | 0% |
| Humeral head cup hemi-arthroplasty | 4% | 0% |

### 17.4.1. *Unconstrained total arthroplasty*

This class of implant system consists of a metal humeral stem/ball component (316L stainless steel or Co-Cr alloy) that glides on an ultrahigh molecular-weight polyethylene (UHMWPE) surface that has been surgically attached to the scapula.[9] Several references discuss these biomaterials.[10,11] The glenoid component may consist of UHMWPE alone or have a metal backing. The bearing surface is spherical with a large radius; the ball end of the humeral surface is spherical or ellipsoid with a radius smaller than the glenoid surface. See Neer[12] for illustrations and details. Clinical results are discussed in several papers.[13–17]

### 17.4.2. *Semi-constrained total arthroplasty*

Rotator cuff damage was recognized as a major limitation early in clinical evaluation of unconstrained shoulder replacements. Laurence developed a system in which the humeral end is loosely constrained by a glenoid component which envelops the humeral head but is not a socket.[18] The range of motion is not fully constrained because the ball can "pull" in an outward-in direction and still be constrained in the lateral directions. Thus, subluxation is reduced. The materials used for this system are stainless steel for the humeral component, and UHMWPE. Both components are cemented with PMMA, with reinforcement from screws in the glenoid component. Unlike the Neer and Dana systems, the glenoid component is screwed into the scapula, and is fixed to all three contact points of the scapula, the acromion, the coracod, and the lower lip of the glenoid. A typical, semi-constrained prosthesis is illustrated by Laurence.[18]

### 17.4.3. *Constrained total arthroplasty*

Initial attempts at shoulder replacements focused on the development of the constrained or fixed-fulcrum type, in which the "ball" of the humeral component articulates within a socketed glenoid component, which is also fastened to the scapula.[19] They are viable alternatives for replacement surgeries with a need for joint stability due to soft tissue deficiency. However, these implants have very limited resultant range of motion (ROM), and kinematic compromise, in which large loads can cause loosening and failure. Generally, a metal, humeral stem/ball component is attached to a metal-encased UHMWPE socket, which is in turn cemented or screwed to the scapula. The attachment of the stem to the humerus is achieved through cementing, bone in-growth, press-fit or morphological fixation. Clinical results with supporting evidence have been described.[20]

### 17.4.4. *Unconstrained hemi-arthroplasty*

In hemi-arthroplasties, no glenoid component is used.[21] The existing glenoid bone and associated tissue must be adequate to withstand the bearing of the loads of the smooth humeral implant surface. The complications involving this type of system obviously do not involve failure of a glenoid component, thus reducing the potential for implant failure. However, if the existing glenoid surface and the supporting musculature is inadequate, shoulder instability and tissue trauma can result. Clinical results with supporting evidence are presented in references.[21–23]

### 17.4.5. *Bipolar shoulder replacement*

This system was designed primarily as a salvage procedure for arthritic shoulders with torn rotator cuffs. It also has an application when the Neer-type implant cannot be placed without glenoid bone-grafting. The bipolar implant configuration (see Swanson *et al.* for an illustration[24]) is, in a sense, a hemi-arthroplasty implant. The implant is inserted within the humerus in a fashion similar to the other implant configurations. However, no glenoid component is attached to the scapula. Instead, a ball with a socket within it is placed in the natural socket of the shoulder and may articulate to some extent in the socket. Within this ball is the socket that constrains the proximal ball-end of the humeral component. This pair of nested ball-and-socket joints provides an excellent range of motion when properly positioned in a relatively healthy shoulder. When subluxation or minor dislocation occur the joint still should maintain a limited range of movement through motion between the inner-cup and the head of the prosthesis. Clinical results have been described.[24–25]

### 17.4.6. *Humeral head cup hemi-arthroplasty*

A unique variation on hemi-arthroplasty shoulder implants is re-surfacing with a metal cup cemented over the proximal end of the humerus, providing a new, smooth articulating surface. This is the least invasive of all available arthroplasties, preserving the medullar cavity and reducing the overall trauma experienced by the shoulder. Further discussion of this approach to shoulder repair can be found in Chapter 7 of *Clinical Performance of Skeletal Prostheses.*[7]

### 17.5. Clinical Results

Chapter 7 of *Clinical Performance of Skeletal Prostheses*[7] compares the relative clinical success of all types of shoulder prostheses up to 1995 based upon published clinical papers in peer-reviewed journals. Thirteen clinical studies are reviewed. The chapter also describes alternative evaluation methods used by the surgical teams. Since there are no standard evaluations it is difficult to make direct comparisons, and only relevant trends are meaningful. In addition to different evaluation criteria, the following factors are among those recognized as having a major effect in the success, failure or complications arising from the shoulder repair: a) age of the patient; b) general state of health; c) lifestyle, such as smoking, drug or alcohol abuse; d) gender; e) weight, exercise; f) compliance with rehabilitation requirements; and g) importantly, the condition of damage to the shoulder prior to surgery. The studies indicated that the use of cemented prostheses was preferred for both humeral and glenoid fixation. Screws should be used in the glenoid component of constrained implants and should be of the maximum strength suitable for the condition of the bone.

The general findings of percentage of complications and failures[7] are summarized above in Table 17.1. In the absence of contradictions, the order of preferences for the selection of a shoulder prosthesis is:

1. None.
2. Unconstrained hemi-arthroplasty.
3. Unconstrained total shoulder arthroplasty.
4. Semi-unconstrained total shoulder arthroplasty.
5. Constrained total shoulder arthroplasty.
6. Humeral head cup hemi-arthroplasty.

The unconstrained hemi-arthroplasty yields a greater postoperative range of motion and function, as well as satisfactory pain reduction, given proper operative procedures and postoperative therapy. The absence of the glenoid component gives fewer parts to fail and a less complicated surgical procedure, which reduces the chance of surgical error.

In the case of severe rotator cuff deficiency, the semi-constrained total shoulder arthroplasty appears to have performed the best. This trial included a sizeable population and up to ten years follow-up. The failure rate was unexpectedly low, considering the condition of many of the shoulders. The bipolar implants seemed to have unpredictable results, depending on the patient and the geometry of the sub-acromional arch. This makes it difficult to define appropriate contraindications, making success of this system very dependent on the ability and judgement of the surgeon.

Of all of the systems examined, the most promising system appears to be the humeral cup hemi-arthroplasty. Its success rates were as high as any other, with fewer complications.


## 17.6.  New Developments

A relatively recent development is the reverse shoulder prosthesis. This seemingly backwards design switches the ball and cup, placing the ball of the joint in the patient's torso, while the cup is attached to the stem in the patient's arm. This unique arrangement is suitable for older patients who might be unable to receive the full benefits of a traditional arthroplasty, notably those with certain forms of arthritis or osteoporosis and severe problems with the rotator cuff. In such patients, a standard shoulder implant is likely to fail or provide inadequate mobility due to the placement and orientation of stress on the shoulder. By shifting the shoulder's centre of rotation, sheer forces on the interface between musculature and bone are reduced, while taking stress off of the deltoid.[21] When properly implemented, this can greatly increase the range of motion for some patients. The reverse shoulder prosthesis technique is currently only used by a few surgeons, but is likely to see a gain in acceptance and popularity as patients and doctors get a better sense of its specialized purpose, and when such a technique is appropriate.

An indirect way to treat shoulder injury is to prevent it from ever happening, or to at least take precautionary measures against known risks. Research shows that any sort of industrial work which requires repetitive reaching upwards causes stress and fatigue in the shoulder joints.[16] This conclusion can likely be applied to any job which requires such activities, from mail sorting to restocking grocery items in a food store. One study[15] shows that a dynamic arm support may reduce stress on the shoulder for certain tasks, and that mechanical improvement of the device may increase its usefulness for more applications. In the future, if employees at risk for shoulder fatigue were provided with some kind of similar dynamic support, the incidence of shoulder-related injuries may be reduced.

Rotator cuff tears are one of the most problematic areas of shoulder repair and are also the one of the most common causes of disability of the shoulder.[25] Surgery still remains the standard of care although recently it has been reported that rotator cuff repair failure or re-tear rates are between 20% and 95% depending on the patient, size of the injury, and the surgical procedure used for repair.[26-30] Although great advances have been made in pathology and surgical techniques, failure rates have persisted for rotator cuff repair. Even with the rapid evolution of arthroscopic techniques and the introduction of reverse shoulder arthroplasty, an optimal shoulder repair technique has not been conceived as surgical outcomes remain similar to classical open repair procedures.[31,32] Moreover, chronic repeated rotator cuff tears may not be repairable without major reconstructive surgery leaving the patient permanently disabled.[27] Due to the recurring limitations and difficulties observed in current rotator cuff repair procedures there is an overwhelming need for new repair methods that will enhance mechanical strength and promote long-term injury healing. Tissue engineering and regenerative medicine approaches may offer a variety of techniques for improving the repair of rotator cuffs and other tendons.

The rotator cuff consists of a group of muscles, ligaments, and tendons which form the glenohumeral joint and permits most shoulder movements. Injury of the rotator cuff is almost always synonymous with a deficit or tear of the tendons. Tendon tissue, like cartilage, is a unique connective tissue which has a poor capability for self-repair, therefore many of the same considerations must be made when investigating tendon repair and regeneration as were made for cartilage repair. Two of the most significant considerations are mimicking the mechanical properties of tendon and providing a template or mechanism for healing, including the reattachment of tendon to bone. Currently xenografted and synthetic materials such as those derived from porcine small intestine submucosa (SIS) are beginning to be implemented clinically for repairing or providing strength to soft-tissue tendon repair. The data (albeit limited) for these procedures so far has been discouraging, and the mechanical properties of these materials have been shown to be significantly inferior to that of the natural tendon of the rotator cuff.[26,29,30]

The use of stem cells for tendon repair has recently been under investigation. Bone marrow mesenchymal stem cells (MSCs) have been successfully induced to differentiate into tendon and also into bone *in vitro*.[25,33,34] The ability of the MSCs to differentiate under controlled conditions into either tendon or bone yields a unique approach to rotator cuff repair, as some groups of researchers have focused on the tendon–bone interface.[30] It has been observed that the modulation of matrix metalloproteinase (MMP) and its inhibitors plays an important role in the healing of rotator cuff tendon–bone interface after repair.[35] In a recent study, it was shown that MSCs can be induced by genetic modification to over express membrane type 1(MT1) MTPs to aid healing at the tendon-bone insertion of the rotator cuff.[33] Mazzocca *et al.* recently demonstrated a method by which viable MSCs could be harvested from a patient intra-operatively during arthroscopic rotator cuff repair, providing insight into developing a clinical procedure which combines both surgical and biological repair.[25] Current uses of stem cells in rotator cuff repair give rise to many potential avenues to further advance such a modality by use of genetic modification and the biochemical induction mechanisms of MSCs.

Tissue-engineering scaffolds have also become a topic of interest for repair and regeneration of tendon tissue for rotator cuff repair. Numerous materials have been used to develop scaffolds for tendon tissue growth, including poly-L-lactic acid (PLLA), poly-glycolic acid (PGA), polytetrafluoroethylene, polyurethane, chitin, and collagen.[26,27,35,37] The success of the various materials was greatly dependent on the material mechanical properties and the level, if any, of bioactivity.[37] Developing a scaffold possessing mechanical properties which mimic those of tendon *in vivo* has been attracting great attention.[38] Tendon tissue must have a high tensile strength and also be highly elastic, so an appropriate scaffold material must also have these mechanical properties. Santoni *et al.* have recently shown the design and optimization of polyurethane scaffolds, taking into account the mechanical properties, which were demonstrated to conduct tissue in-growth and promote healing at the tendon–bone interface.[35] An optimal scaffold material and configuration has yet to be established for rotator cuff repair, although many possibilities abound, especially when the possible combinations of scaffold matrices, cells, and other bioactive molecules are considered. The conception of an alternative method to conventional surgical techniques may involve scaffolds, biological augmentation or a combination of the two, but it is clear that a great deal of work in the field will be needed to accomplish this goal.

## 17.7. Summary

Shoulder prostheses are a successful clinical means to repair severely damaged shoulder joints. Survivability of shoulder joint prostheses is similar to other total joint prostheses. New developments offer the potential to provide repair by tissue-engineered constructs but clinical use is still in the future.

## References


1. Gray, H. (1858). *Gray's Anatomy: The Anatomical Basis of Clinical Practice*, 39th ed., 2004, C.V. Mosby, New York, NY.
2. Martini, F., Timmons, M. and McKinnley, M. (2000). *Human Anatomy*, 3rd ed., Prentice Hall, Upper Saddle River, NJ.
3. Calais-Germain, B. (1993). *Anatomy of Movement*, Eastland Press Seattle, WA.
4. National Institutes of Health, Department of health and human services (2010). *Arthritis, musculoskeletal and skin diseases*. [Online]. Available at: http://www.niams.nih.gov. [Accessed 8 November 2010].
5. Cypress, B.K. (1981). Patients' reasons for visiting physicians: national ambulatory medical care survey, United States, 1977–1978, *Vital Health Stat.*, **13**, 1–134.

6.  Sood, D., Nussbaum, M.A. and Hager, K. (2007). Fatigue during prolonged intermittent overhead work: reliability of measures and effects of working height, *Ergonomics*, **50**, 497–513.

7.  Hench, L.L. and Wilson, J. (eds) (1997). *Clinical Performance of Skeletal Prostheses*, Chapman and Hall, London.

8.  University of Maryland Medical Center (2010). *Orthopaedics*. [Online]. Available at: http://www.umm.edu/orthopaedic. [Accessed 8 November 2010].

9.  Farmer, K. (2007). Shoulder arthroplasty *versus* hip and knee arthroplasties: a comparison of outcomes, *Clin. Orthopaed. Rel. Res.*, **455**, 183–189.

10. Ratner, B.D., Hoffman, A.S., Schoen, F.J., *et al.* (2004). *Biomaterials Science: An Introduction to Materials in Medicine*, Academic Press, Philadelphia, PA.

11. Hench, L.L. (2005). "Joint Replacement", in Hench, L.L. and Jones, J.R. (eds), *Biomaterials, Artificial Organs and Tissue Engineering*, Woodhead Publishing Ltd, Cambridge, UK.

12. Neer, C.S. II (1985). Unconstrained shoulder arthroplasty, *Instructional Course Lectures*, **34**, 278–286.

13. Neer, C.S. II and Morrison, D.S. (1988). Glenoid bone-grafting in total shoulder arthroplasty, *J. Bone Joint Surg.*, **70A**, 1154–1162.

14. Amstutz, H.C., Sew Hoy, A.L. and Clarke, I.C. (1981). UCLA anatomic total shoulder arthroplasty, *Clin. Orthopaed. Rel. Res.*, **173**, 109–116.

15. Amstutz, H.C., Thomas, B.J., Kabo, J.M., *et al.* (1988). The dana total shoulder arthroplasty, *J. Bone Joint Surg.*, **70**, 1174–1182.

16. Barret, W.P., Franklin, J.L., Jackins, S.E., *et al.* (1987) Total shoulder arthroplasty, *J. Bone Joint Surg.*, **69A**, 865–872.

17. Cofield, R.H. (1979). Total joint arthroplasty, the shoulder, *Mayo Clinic Proc.*, **54**, 500–506.

18. Laurence, M. (1991). Replacement arthroplasty of the rotator cuff deficient shoulder, *J. Bone Joint Surg.*, **73B**, 916–919.

19. Lugli, T. (1978). Artificial shoulder joint by Pean (1893): the facts of an exceptional intervention and the prosthetic method, *Clin. Orthopaed. Rel. Res.*, **133**, 215–218.

20. Post, M. and Jablon, M. (1983). Constrained total shoulder arthroplasty, *Clin. Orthopaed. Rel. Res.*, **146**, 161–174.

21. Kay, S.P. and Amstutz, H.C. (1988). Shoulder hemi-arthroplasty at UCLA, *Clin. Orthopaed. Rel. Res.*, **228**, 42–48.

22. Jonsson, E., Egund, N., Kelly, I., *et al.* (1986). Cup arthroplasty of the rheumatoid shoulder, *Acta Orthoped. Scand.*, **57**, 542–546.

23. Sim, F.H., Chao, E.Y., Pritchard, D.J., *et al.* (1980). Replacement of the proximal humerus with ceramic prosthesis: a preliminary report, *Clin. Orthopaed. Rel. Res.*, **146**, 161–174.

24. Swanson, A.B., de Groot, K., Swanson, G., *et al.* (1989). Bipolar implant shoulder arthroplasty, *Clin. Orthopaed. Rel. Res.*, **249**, 227–247.

25. Mazzoca, A.D., McCarthy, M.B.R., Chowaniec, D.M., *et al.* (2010). Rapid isolation of human stem cells (connective tissue progenitor cells) from the proximal humerus during arthroscopic rotator cuff surgery, *Am. J. Sports Med.*, **20**, 1–10.

26. Working, B.S. and West, W.V. (2010). Use of matrices as a tissue substitute in shoulder surgery, *Op. Techniques Orthopaed.*, **20**, 154–160.

27. Derwin, K.A., Codsi, M.J., Milks, R.A., *et al.* (2009). Rotator cuff repair augmentation in a canine model with use of a woven poly-L-lactide device, *J. Bone Joint Surg.*, **91**, 1159–1171.

28. Derwin, K.A., Baker, A.R., Iannotti, J.P., *et al.* (2010). Preclinical models for translating regenerative medicine therapies for rotator cuff repair, *Tissue Eng. B*, **16**, 21–30.

29. Gumina, S., Patti, A.M., Vulcano, A., *et al.* (2009). Culture of human rotator cuff cells on orthobiological support (porcine small intestinal submucosa), *Musculoskeletal Surg.*, **93**, 65–70.

30. Nho, S.J., Delos, D., Yadav, H., *et al.* (2010). Biomechanical and biologic augmentation for the treatment of massive rotator cuff tears, *Am. J. Sports Med.*, **38**, 619–629.

31. Abboud, J.A. (2010). Current concepts in rotator cuff disease and treatment, *Clin. Orthopaed. Rel. Res.*, **468**, 1467–1468.

32. Piasecki, D.P., Verma, N.N., Nho, S.J., *et al.* (2010). Outcomes after arthroscopic revision rotator cuff repair, *Am. J. Sports Med.*, **38**, 40–46.

33. Gulotta, L.V., Kovacevic. D., Montgomery, S., *et al.* (2010). Stem cells genetically modified with the developmental gene Mt1-Mmp improve

regeneration of the supraspinatus tendon-to-bone insertion site, *Am. J. Sports Med.*, **20**, 1–8.

34.  Violini, S., Ramelli, P., Pisani, L.F., *et al.* (2009). Horse bone marrow mesenchymal stem cells express embryo stem cell markers and show the ability for tenogenic differentiation by *in vitro* exposure to BMP-12, *BMC Cell Biol.*, **10**, 1–10.

35.  Bedi, A., Fox, A.J., Kovacevic, D., *et al.* (2010). Doxycycline-mediated inhibition of matrix metalloproteinase improves healing after rotator cuff repair, *Am. J. Sports Med.*, **38**, 308–317.

36.  Santoni, B.G., McGilvray, K.C., Lyons, A.S., *et al.* (2010). Biomechanical analysis of an ovine rotator cuff repair via porous patch augmentation in a chronic rupture model, *Am. J. Sports Med.* **38**, 679–686.

37.  Chen, X., Zou, X.H., Yin, G.L., *et al.* (2009). Tendon tissue engineering with mesenchymal stem cells and biografts: an option for large tendon defects? *Frontiers in Bioscience*, **1**, 23–32.

38.  Saber, S., Zhang, A.Y., Ki, S.H., *et al.* (2010). Flexor tendon tissue engineering: bioreactor cyclic strain increases construct strain, *Tissue Eng. A*, **16**, 2085–2090.

# Drug-Eluting Stents: Worth the Risk?



## CHAPTER 18

Adam Estelle

## 18.1. Introduction

Coronary heart disease is one of the leading causes of death in the United States. With 6,363,000 cardiovascular procedures and operations performed in 2004, there is continuing need for improvement in cardiac care.[1] Clinical use of bare metal stents in 1994 provided an alternative to the standard coronary bypass procedure. Hospital stays were reduced and a minimally invasive procedure for treating occluded arteries was made available. However, this breakthrough procedure was accompanied by several pitfalls that are still being investigated.

Foremost among the problems surrounding stents is formation of a potentially fatal blood clot, known as a thrombosis. Recipients of bare metal stents (BMS) have an increased risk of forming blood clots that can lead to a stroke or myocardial infarction. To combat this effect, patients are placed on blood-thinning drugs indefinitely, immediately following implantation of a stent. Another concern is the issue of in-stent restenosis. This phenomenon is caused by a condition known as neointimal hyperplasia in which the body overcompensates to heal the arterial wall. Endothelium and smooth muscle cells from the arterial wall grow and surround the stent, thereby restricting blood flow. Scar tissue can form, requiring additional angioplasties or even a bypass procedure.

Advances in materials processing led to the development, testing and US Food and Drug Administration (FDA) approval of drug-eluting heart stents (DESs) in 2003. These devices are similar in structure to BMS but once inserted they elute a specific drug at a controlled rate that prevents the formation of clots and restenosis. Drug-eluting stents were widely accepted upon approval and have the potential to eliminate the need for blood-thinning drugs and additional procedures. However, the advancements are shadowed by several issues discussed in this chapter. The increased cost of DES raises the question of whether or not the benefits balance the risks. Insurance coverage of DES is becoming increasingly selective. Along with economic issues are ethical concerns, including post-operative drug treatment, the use of unapproved devices, and patient reimbursement. Although DESs are widely used, they are so new that long-term (>10 years) data for analysis are not yet available to make statistically valid comparisons of costs, risks, and rewards. To understand the potential concerns of these devices, the current technology must be fully understood.

## 18.2. Current Technology and Limits

Stents are essentially a metallic scaffold that holds open a blocked portion of an artery (see Fig. 18.1). Structure and performance have a delicate relationship in the success of stents. Composition, length, pore size, and thickness of the stent wall play a critical role in the long-term health of recipients. Full color images and schematics of various stent designs can be found in references 2, 3, and 4.

A study involving 611 patients showed that stents 50 microns thick had a 17.9% chance of restenosis, while stents 140 microns thick showed a 31.4% chance.[4] Stents must possess significant radial strength to prevent collapse from the elastic stress of the artery. Designs are typically tubular and have ranged from slotted columns to spiraling helices with supporting ring structures. Traditional BMS stents are composed of metals, and alloys including stainless steel alloy 316L Nitinol (a shape memory alloy composed of Ni and Ti).[5] Newer models have employed a Teflon® coating to minimize restenosis. Although structure and composition

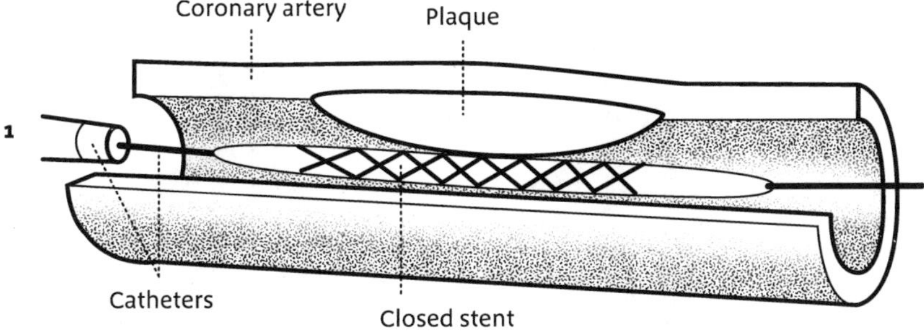


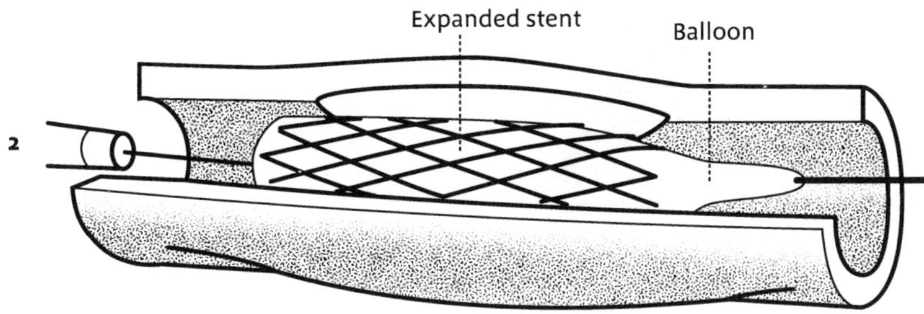


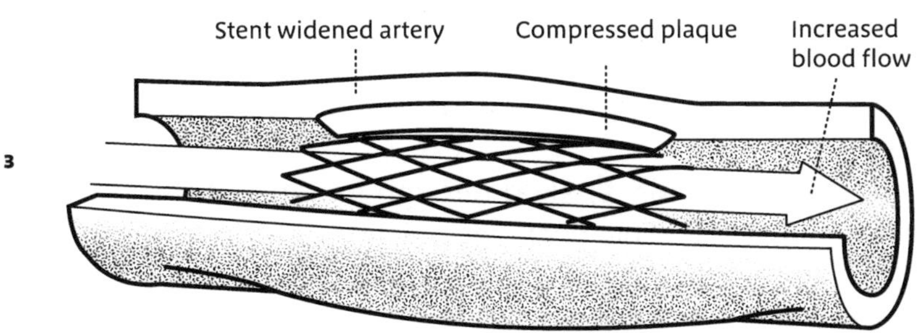


**Figure 18.1.** Schematic of stent repair of a blocked artery.

have a great impact on the effectiveness of the stent, recent research has focused on special coatings that interact with the body directly on a cellular level. These efforts resulted in the invention of the drug-eluting stent. Before these new devices are

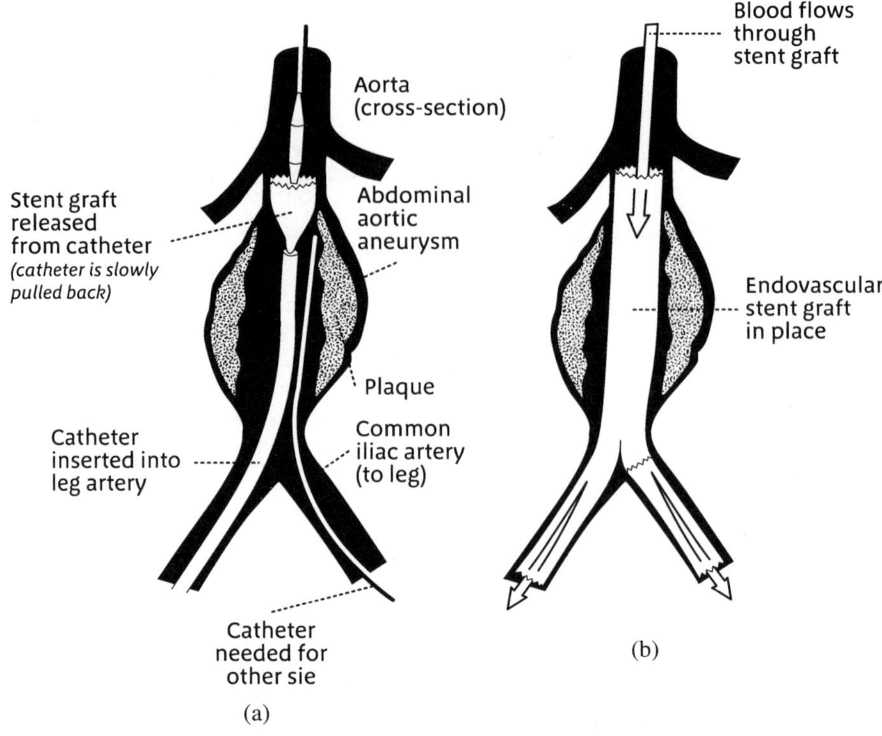



**Figure 18.2.**    Application of a stent.

discussed, it is important to understand how the devices are implanted.

The stent is usually snaked through the femoral artery using a catheter (Fig. 18.2). The occluded section of artery is cleared by inflating a balloon which compresses the buildup of plaque in a procedure known as an angioplasty. Stents are inserted in a collapsed form and expanded either mechanically or thermally. Current procedures place the stent simultaneously with the angioplasty using the balloon to expand the stent. Several risks are involved with the procedure, including damage to the arterial wall, and localized inflammation or infection. Also, the balloon may dislodge pieces of the plaque build-up which leads to additional inflammation. Improvements such as nitrate infusion, intracoronary beta-blockers, adenosine, clopidogrel, and IIb/IIIa inhibitors during implantation have led to reduced risks associated with implantation of stents.[6] The latest goal has been to create a stent that could release drugs over time and prevents restenosis, the most prominent problem associated with stents.

The approach taken was to find a drug that prevented the body's own cells from overtaking the stented portion of the artery. The first successful drugs investigated were sirolimus, tacrolimus, and paclitaxel.[7] Sirolimus is a next generation immunosuppressant that slows T-cell production caused by the presence of IL2 cytokines. Tacrolimus prevents cytokine gene expression.[4] Paclitaxel acts by disrupting the microtubule growth of cells during mitosis. The drugs used are cell-cycle or cell-proliferation inhibitors that prevent cell division, the direct cause of in-stent restenosis. Sirolimus stops the cell from dividing in the G1 phase of the cell cycle (Fig. 18.3). After potential drugs were selected, a method for coating the stents had to be developed, as well as a way to make sure the right dosage was released over the right amount of time.

Drug delivery is a very sensitive factor in DES. The potent cell-cycle inhibitors are delivered directly at the site and must be

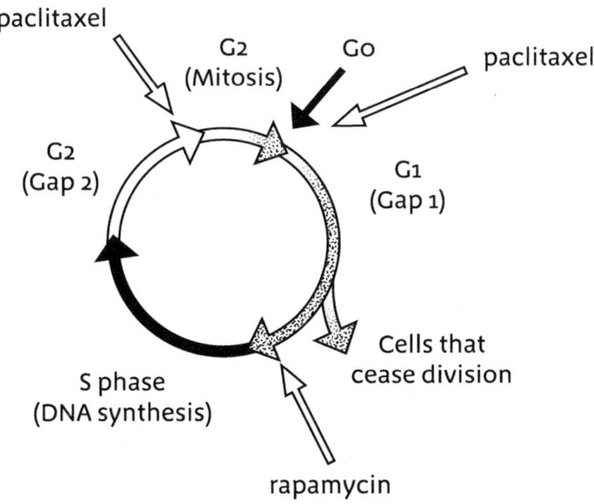


**Figure 18.3.**   The cell cycle.

controlled accordingly. Patients are most susceptible to restenosis during the first few weeks following surgery. Therefore, it is important that the drugs are in ample supply over those first critical weeks. A typical drug-delivery profile consists of a heavy dosage for the first few days followed by a slowed release for about two weeks. The majority of the drug is diffused out after about a month. Methods for drug release include diffusion through a polymer film, degradation of a polymer matrix, osmotic pressure, and ion exchange. Each method of controlled release has specific kinetic mechanism that can be tailored by adjusting properties such as polymer type, thickness, and membrane size.[8] Methods for depositing the drugs on metal stents include dipping, spraying, and sputtering. All have relative advantages and disadvantages. After clinical testing and data analysis of two alternative devices, the FDA approved the Cypher® and Taxus® drug-eluting stents in early 2003.

Careful control of the diffusion of drugs into the bloodstream is essential since there are dangers involved in overdosing. Drugs used such as paclitaxel and sirolimus can be toxic in large doses and can damage the vessel wall if not carefully controlled. Kinetics of drug delivery must be correctly tailored or coronary aneurisms may form. In one such case, a 43-year-old man developed a coronary aneurysm after receiving a paclitaxel-eluting stent six months earlier. Investigation revealed poor contact between the stent and the majority of the vessel wall and an aneurysm 6 mm in diameter.[9] Two similar cases have been documented concerning the sirolimus-eluting Cypher® stent. Excess amounts of these cell-cycle inhibitors heighten the risk for vessel rupture and vessel dilation as well.

While the delivery of the proper dosage of drugs poses one problem, the effectiveness of the drugs poses another. A Korean study compared the effectiveness of sirolimus- and paclitaxel-eluting stents in 440 patients. The study showed that the paclitaxel-eluting stents had a larger occurrence of binary restenosis (17.9%) *versus* the sirolimus stents (4.5%).[10] Additional studies involving larger populations of patients comparing the various drugs used in DES and traditional BMS devices are needed in order to determine the relative effectiveness of preventing thrombosis and restenosis.

## 18.3. New Approaches

The latest stent technology is a new generation of biodegradable, drug-eluting stents (BDES). The concept of BDES devices is to have them resorb (dissolve) into harmless compounds that are metabolized by the body after about a year. Before resorption, BDES devices will provide adequate vessel support and also supply a controlled dosage of the drugs already used in regular DES during the first critical stages after implantation. Thus, the objective of these devices is to prevent most long-term problems associated with permanent DES devices including thrombogenicity, physical irritation, chronic inflammatory local reactions, long-term endothelial dysfunction, delayed re-endothelialization, and mismatches in mechanical behavior.[11]

One goal of the BDES is to restore the original function of the vessel before the occlusion and before the stent was implanted. After the vessel has healed, the stent is no longer necessary, in most cases. The stent is a foreign body and ideally would be removed after its function was performed, but this is only possible through additional surgery. Thus, there are numerous advantages of a stent that completely dissolves. Two alternative biomaterials approaches are being investigated: self-corroding metals and resorbable polymers. BDES made entirely of polymers provide a larger drug reservoir. This allows drugs to be released over a longer period of time, and it is hoped this will eliminate the problem of late in-stent restenosis. The polymer approach to design of a BDES gives increased control over the drug release kinetics. The stent can consist of several differing polymer layers, some for structure and others that contain different drugs that are released at specified rates.[11] Through this method, less toxic drugs can be released over controlled times which may prevent damage due to overdosing.

One of the first BDES to be investigated was constructed from a magnesium metal alloy. The aim was to utilize the natural antithrombotic, anti-proliferative and anti-arrhythmic properties of the alloy as it corroded in the vessel.[12] Several studies were conducted that analyzed the effectiveness of the device when implanted in animals. In addition to the natural properties of the alloy, monitoring the stent *in vivo* using magnetic resonance imaging is easier. The structural properties gained by using a bio-corrodible metal alloy rather than a biodegradable polymer are also worth noting. Studies show promise for this BDES device and suggest the design should be used in conjunction with drug-eluting polymers.

A second metallic BDES being studied is constructed from an iron alloy. In theory, the stent breaks down into ferrous ions which stimulate surrounding cells on a genetic level, preventing restenosis. One study exposed human umbilical venous smooth muscle cells to ferrous ions and studied the reaction on a molecular and phenotypic basis.[13] This showed a reduction in the mRNA of the cells responsible for cell proliferation, which ultimately could slow restenosis caused by smooth muscle cell in-growth. Exploring the possibilities of bioactive materials that stimulate cells on a genetic level for use in DES could potentially eliminate the problem of restenosis.

A concern for all BDES devices is obtaining predictable structural behavior while they undergo degradation in the vessel. There continues to be debate over whether or not BDES can provide the luminal support necessary to allow the damaged vessel to heal. A recent study implanted sirolimus-eluting stents in the common carotid arteries of pigs. This showed that the polymeric poly l-lactic acid (PLLA) BDES used demonstrated sufficient mechanical stability without thrombotic occlusion of the arteries,[14] and demonstrated that a polymer-based BDES can provide adequate, temporary support while eluting drugs that prevent restenosis and inflammation at a controlled rate.

A similar study was conducted that investigated the effects of materials properties on the collapse pressure of a poly l-lactic acid BDES. Variables tested included molecular weight, design, and drug dispersion. The study found that collapse pressure declined with increasing drug concentration. It also determined that collapse pressure could be maximized by increasing the load-bearing surface area.[15] Therefore, in order to create the most stable BDES, drug concentration must be closely monitored, and the load-bearing surface area must be as large as possible.

Accompanying the structural problems of BDES is failure by elastic recoil. The elastic nature of the polymers used can cause the stents to collapse and block the vessel again after inflation through angioplasty. When BMS are implanted in a vessel, they experience plastic deformation after balloon inflation. This effect is undesirable, but the stents still retain their structural integrity. Polymer-based stents will undergo elastic deformation which greatly increases the chances of a potentially fatal collapse. Through temperature conditioning, a two-layer BDES was created using a combination of PGLA and PLLA that possesses elastic memory. This processing method allows the stents to be expanded thermally *in vivo* as opposed to balloon inflation. Thermal expansion is more subtle and the elastic memory helps prevent collapse due to elastic recoil.[16]

A self-expanding BDES is implanted using a sheath method. A sheath composed of the naturally occurring polymer chitosan is used to contain the expansion of the stent during catheterization and placement. The sheath is designed to degrade on exposure to moisture. This raises the problem of early expansion caused by the sheath dissolving too soon during implantation,[16] and variations in composition and cross-linking are being investigated to rectify this problem. However, the heat provided by the body is not enough to anchor the stent in the vessel. In order to secure the BDES in the artery, data from the study show that the stent must be heated to 70°C.[16] This presents the problem of cellular damage caused by exposure to excessive heat. Ideally, the stent would fully expand and anchor at body temperature. The study concluded that the glass transition temperature of the outer layer greatly impacts the self-expandability of the stent. More research is necessary, but a self-expandable stent may soon be ready for clinical trials. If improvements continue to be made, this technology may potentially solve the problem of elastic recoil associated with BDES.

BDES are slowly reaching approval status and are currently being researched in clinical settings. Several problems have occurred, mostly involving inflammation at the implant site, and different polymers are being investigated. These include PGA/PGLA, PEA, PCL, PBT, and PLLA.[11] Structural properties are also slowing the progress of BDES because polymers possess different elastic moduli and strengths than the standard stainless-steel models. Even after the engineering problems are solved, there will still be numerous economical hurdles that must be overcome before BDES are ready for the market.

Also under development are multi-layered stents. One such stent utilizes a hydrophobic, heparinized top layer on a BMS to improve biocompatibility and drug release kinetics. This additional layer acts as an interface between the body and the polymerized drug layer. This barrier provides added control over drug release while also preventing platelet adhesion which reduces restenosis and thrombosis *in vitro* and *in vivo*.[17] By experimenting with additional layers, greater control over drug kinetics and restenosis/thrombosis rates can be achieved.

Drug-eluting stents are being advanced on several fronts. As research continues, it is hoped that advancements will be combined to create a device that will completely eradicate the potential for in-stent restenosis and thrombosis. Many of these new improvements are close to clinical trial and may soon be available to the public at a reasonable price. The new features will drive up prices originally, but if the effectiveness is proven, new DES could become the standard treatment for occluded vessels in the near future.

## 18.4. Ethics

Although DESs have been hailed as revolutionary devices that are changing the face of cardiac care, they have also been the subject of scrutiny and investigation. A prime example of the controversy surrounding DES took place in India in June 2005. It was reported that unapproved Axxion® DESs were implanted in nearly 60 patients at the JJ Hospital in Mumbai.[18] The stents were manufactured by a company in the Netherlands, but were never approved. An Indian-based company obtained the stents and made them available to the market where patients could buy the devices directly, and patients brought the stents to the hospital where they were implanted. Hospital records dating back to 2004 were discovered, documenting these illegal purchases and procedures.

Issues arose when government bodies in India disagreed on the classification and legality of DES. Since DES devices are coated with a drug, it was unclear whether or not they should be classified as a drug or as a device. Litigation proceeded until it was decided that all DES devices used for treatment be classified as drugs and regulated accordingly. DES devices are now controlled in India under the Drugs and Cosmetics Act.

Numerous cardiologists have heralded DES to be the greatest advancement in minimally invasive treatment of heart disorders. This advocacy affects the decisions of those in need of care and leads to some of the concerns discussed in Chapter 1. In the United States the FDA has the authority to monitor and assure compliance of usage of these devices. However, patients such as those in Mumbai, who take illegal measures to obtain these devices, place themselves in danger. In all instances it is the physician's responsibility to inform the patient of all the potential hazards involved and the relative risk/reward ratio of the use of DES *versus* the standard BMS. The absence of long-term statistical survivability differences need to be addressed by the field.

Another ethical concern surrounds use of a post-operative anticoagulant treatment.[19,20] There is the potential problem of late in-stent restenosis developing after completion of the release of the drugs from DES devices. BMS are covered by endothelial cells within several weeks and intense anticoagulant therapy is often slowed after this period. However, DES devices require longer to develop an endothelial lining, which can result in late in-stent restenosis which requires longer anti-platelet treatment than BMS.[19] Since this phenomenon was rarely encountered with BMS, there is a population of doctors implanting these devices that may not be aware of the need for prolonged antiplatelet therapy.

## 18.5. Economics

Economic issues are an important factor influencing use of DES, as the devices cost approximately four times as much as BMS. Thus, it is debatable whether or not DES devices are worth the added cost, which insurance companies may not reimburse. An economic study conducted in Canada analyzed the cost-effectiveness of DES devices (sirolimus- and paclitaxel-eluting) compared to BMS. This was determined by comparing the added cost of DES to the amount saved due to prevented additional procedures (i.e. re-stenting, by-pass surgery or angioplasty). The study showed that significant money was saved by avoiding repeat procedures, proving that DES are indeed worth the added cost.[21] However this study was only conducted over a period of one year and does not estimate long-term cost-effectiveness (see Table 18.1). Long-term effectiveness could be reduced by complications such as late in-stent restenosis. The study also indicates that the purchasing of DESs from stent manufacturers by hospitals is often confidential and poorly documented. In order to maximize the cost-effectiveness of DES, this economic information should be made public so that

**Table 18.1.**    Budget impact analysis: total cost projections for DES *versus* BMS in Canada.[22]

| | |
|---|---|
| Population | 31.3 million |
| # annual angioplasty procedures | 39,136 |
| # patients receiving stents (90% of angioplasties) | 35,222 |
| # stents (1.5 per angioplasty) | 52,833 |
| Total cost with 100% BMS implants (C\$608 per BMS) | C\$32.1 million |
| Total cost using 40% DES implants in high-risk patients | C\$50.7M DES + C\$19.3M BMS = C\$70M |
| Total cost with 100% DES implants | C\$126.8 million |

healthcare and insurance companies can assess coverage and costs, and tailor plans accordingly.

Insurance coverage of DES is partially controlled by the patients' willingness to pay extra to prevent additional procedures. Even if patients are willing to pay extra, the increased number of these expensive procedures drives up health-care costs. Sometimes a standard heart bypass procedure or a BMS implantation may have the same outcome depending on the patient's condition. Doctors are increasingly implanting DESs regardless of condition, which ultimately drives up insurance costs. Another study in Canada has shown that using sirolimus-eluting stents to treat a lesion in a native vessel actually increases the net health-care cost in the Canadian healthcare system.[22] To reduce health-care costs, doctors must assess each patient individually and determine the best treatment option that accounts for patient safety, economic consequences, and long term outcome.

## 18.6. Recent Developments

Much of the focus of emerging stent technology has recently been directed towards developing BDESs. The majority of the research has centered on using biodegradable polymers, especially the use of poly-L-lactic acid (PLLA), poly-glycolic acid (PGA), and their copolymers (PLA/PGA). To date, no universal bioresorbable polymer has been established and even the most promising materials have demonstrated some limitations including potential biocompatibility issues. For example, it is known that PLLA and PGA biodegrade completely and the by-products are either eliminated or bio-assimilated, but some studies have shown that PLLA can elicit an immune inflammatory response.[23,24] Issues observed with the bioresorbable polymers are related to concerns of toxicity and elimination of the degradation products, as well as inflammatory response. Some concerns relate to formation of crystalline residues or particles from polymer degradation causing documented inflammation, and also the toxicity from the release of entrapped polymerization initiators.[24] It should also be noted that not enough data are available at the present to determine if a permanent stent construct is unnecessary, and that a stent with bioresorbable capability would be of greater benefit in the long term.[23] Further research is needed if an optimized polymeric stent is to be developed.

It is of the utmost importance that investigations into the interplay of biology and materials are examined in order to develop devices which promote healing and integrate seamlessly into the patient. To promote the investigation of biological interactions with materials, some research has focused on creating *in vitro* methods which better mimic the *in vivo* conditions that a device will encounter. In a recent study by Yazdani *et al.*, a porcine artery was used to model vascular smooth muscle growth on various stents in a bioreactor; this allowed for native explanted tissue to proliferate on the stent constructs in simulated *in vivo* conditions.[25] The development of a feasible system to simulate conditions *in vivo* could be used to assess stent design with respect to biological interactions *in vitro* with more rigorous experimental parameters.

Although BDES have become the main focus in current research, some groups have begun to explore other routes for improving drug-eluting stents. Recently, nano-fabrication of potential stents, and stent coatings with drug-eluting properties have come under investigation, especially the application of PLGA nanoparticles for clinical use.[26] Mutsuga *et al.* recently developed a tacrolimus-eluting biodegradable nanofiber (TEBN) with a "cotton-wool" structure of nanofibers, which demonstrated the potential to prevent pulmonary venous obstructions post-operatively.[27] It was found that the TEBN reduced neointimal hyperplasia in a rat anastomotic model and therefore has the potential to prevent recurrent intimal hyperplasia and pulmonary venous obstruction observed after vascular surgery such as stent placement.

Nonporous inorganic coatings for sustained drug delivery have been investigated as substitutes for stent coatings, with results similar to those of current technologies. A wide variety of nanoporous materials are currently being studied, including aluminum oxide, titanium oxide, and porous silicon.[28] Peng *et al.* examined the elution of albumin, sirolimus, and paclitaxel from titanium oxide nanotubes demonstrating controlled release rates based on the eluting molecule's size in relation to the nanotube dimensions.[29] This study showed a means by which nanostructures can be tuned for controlling drug release, and their potential for use as improved BDES coatings. Advances in stent coatings have also been an area of increased interest for improving drug-eluting stents. Incorporation of bioactive surfaces by coating stents with a wide variety of moieties including antibodies, peptides, nucleotides, and other genetic modifying molecules are innovative concepts to promote vascular healing and stent adhesion.[30]

Drug-eluting stent technology has come a long way, but an optimized vascular stent has yet to be developed. Future research will be needed to improve BDESs and allow for translation of the technology to clinical application. Progress in nano-structured materials and bioactive surfaces will most likely play an important role in fueling innovations.

## 18.7. Summary

Stenting occluded vessels changed the face of cardiac care. The technology and the minimally invasive implantation techniques have improved the length and quality of life for millions of patients. Innovative DES devices take this approach to the next level by combining drugs and devices. As the technology advances, there may be a time when occluded arteries can be fixed without the need for additional procedures and a lifetime prescription of antithrombogenic drugs. New prototypes provide a temporary device that dissolves once the vessel has healed, and current studies are perfecting this device which ideally will leave the body as it was before the occlusion. Ethical issues are still under discussion, including regulation policies and post-op drug therapy. This breakthrough technology carries a lofty price tag, and cost efficiency must be analyzed in great detail with analyses of cost/risk/reward ratios. Practicing cardiologists must be informed of all risks involved with DES, and treatment of patients should be based on specific needs. Drug-eluting stents have the potential to revolutionize cardiac care and drive down the cost of health-care by reducing the number of repeat procedures.

## References


1. Rosmand, W., Flegal, K., Friday, G., *et al.* (2007). *Heart disease and stroke statistics — 2007 Update.* [Online]. Available at: http://www. americanheart.org. [Accessed 8 November 2010].

2. The Cleveland Clinic. (1995). *Stent.* [Online]. Available at: http://www.clevelandclinic.org [Accessed 8 November 2010].

3. Cordis Corporation (2004). *Cypher® drug-eluting stent for coronary heart disease.* [Online]. Available at: http://www.cypherstent.com/ Pages/index.aspx. [Accessed 16 August 2010].

4. VA pharmacy benefits management strategic healthcare group & medical advisory panel (2003). *National PBM drug monograph: Aprepitant (Emend®).* [Online]. Available at: http://www.pbm.va.gov. [Accessed 8 November 2010].

5. Hara, H., Nakamura, M., Palmaz, J.C., *et al.* (2006). Role of stent design and coatings on restenosis and thrombosis, *Adv. Drug Deliver. Rev.*, **58**, 377–386.

6. Pasceri, V., Patti, G., Sciascio, G., *et al.* (2006). Prevention of myocardial damage during coronary intervention, *Cardiovasc. Hematol. Disord. Drug Targets*, **6**, 75–83.

7. Wang, J. (2003). *Overview of the cell cycle.* [Online]. Available at: http://www.scq.ubc.ca/wp-content/cellcycle.gif. [Accessed 30 November 2007].

8. Acharya, G. and Park, K. (2006). Mechanisms of controlled drug release from drug-eluting stents, *Adv. Drug. Deliver. Rev.*, **58**, 387–401.

9. Vik-Mo, H., Wiseth, R. and Hegbom, K. (2004). Coronary aneurysm after implantation of a paclitaxel-eluting stent, *Scand. Cardiovasc. J.*, **38**, 349–352.

10. Suh, J.-W., Park, J.-S., Cho, H.-J., *et al.* (2007). Sirolimus-eluting stent showed better one-year outcomes than paclitaxel-eluting stent

in a real life setting of coronary intervention in Koreans, *Int. J. Cardiol.*, **117**, 31–36.

11. Commandeur, S., van Beusekom, H.M.M., van der Giessen, W.J., *et al.* (2006). Polymers, drug release, and drug-eluting stents, *J. Interv. Cardiol.*, **19**, 500–506.

12. Di Mario, C., Griffiths, H., Goktekin, O., *et al.* (2004). Drug-eluting bioabsorbable magnesium stent, *J. Interv. Cardiol.*, **17**, 391–395.

13. Mueller, P.P., May, T., Perz, A., *et al.* (2006). Control of smooth muscle cell proliferation by ferrous iron. *Biomaterials*, **27**, 2193–2200.

14. Bunger, C.M., Grabow, N., Sternberg, K., *et al.* (2007). Sirolimus-eluting biodegradable poly-L-lactide stent for peripheral vascular application: a preliminary study in porcine carotid arteries, *J. Surg. Res.*, **139**, 77–82.

15. Venkatraman, S., Poh, T.L., Vinalia, T., *et al.* (2003). Collapse pressures of biodegradable stents, *Biomaterials*, **24**, 2105–2110.

16. Venkatraman, S.S., Tan, L.P. Joso, J.F., *et al.* (2006). Biodegradable stents with elastic memory, *Biomaterials*, **27**, 1573–1578.

17. Lee, Y.-K., Park, J.H., Moon, H.T., *et al.* (2007). The short-term effects on restenosis and thrombosis of echinomycin-eluting stents topcoated with a hydrophobic heparin-containing polymer, *Biomaterials*, **28**, 1523–1530.

18. Magotra, R. (2006). The controversy of drug-eluting cardiac stents, *Indian J. Med. Ethics*, **3**, 25–26.

19. Hirshfeld, J.W. Jr and Wilensky, R.L. (2004). Drug-eluting stents are here — now what? Implications for clinical practice and health care costs, *Cleveland Clin. J. Med.*, **71**, 825–828.

20. McFadden, E.P., Stabile, E., Regar, E., *et al.* (2004). Late thrombosis in drug-eluting coronary stents after discontinuation of antiplatelet therapy, *Lancet*, **364**, 1519–1521.

21. Mittmann, N., Brown, A., Seung, S.J., *et al.* (2005). *Drug Eluting Stents: An Economic Evaluation [Technology Overview no 15]*, Canadian coordinating office for health technology assessment, Ottawa, ON.

22. Rinfret, S., Cohen, D.J., Tahami, M.A.A., *et al.* (2006). Cost effectiveness of the sirolimus-eluting stent in high-risk patients in Canada: an analysis from the C-sirius trial, *Am. J. Cardiovasc. Drugs*, **6**, 159–168.

23. Doyle, B. and Holmes, D.R. Jr (2009). Next generation drug eluting stents: focus on bioabsorbable platforms and polymers, *Med. Device: Evid. Res.*, **2**, 47–55.

24. Vert, M. (2009). Degradable and bioresorbable polymers in surgery and in pharmacology: beliefs and facts, *J. Mater. Sci.-Mater. Med.*, **20**, 437–446.

25. Yazdani, S.K. and Berry, J.L. (2009). Development of an *in-vitro* system to assess stent-induced smooth muscle cell proliferation: a feasibility study, *J. Vasc. Interv. Radiol.*, **20**, 100–106.

26. Lü, J.-M., Wang, X., Marin-Muller, C., *et al.* (2009). Current advances in research and clinical applications of PLGA-based nanotechnology, *Expert Rev. Mol. Diagn.*, **9**, 325–341.

27. Mutsuga, M., Narita, Y., Yamawaki, A., *et al.* (2009). Development of novel drug-eluting biodegradable nano-fiber for prevention of post-operative pulmonary venous obstruction, *Interac. Cardiovasc. Thorac. Surg.*, **8**, 402–407.

28. Gultepe, E., Nagesha, D., Sridhar, S., *et al.* (2010). Nanoporous inorganic membranes or coating for sustained drug delivery in implantable devices, *Adv. Drug Deliv. Rev.*, **62**, 305–315.

29. Peng, L., Mendelsohn, A.D., LaTempa, T.J., *et al.* (2009). Long-term small molecule and protein elution from $TiO_2$ nanotubes, *Nano Lett.*, **9**, 1932–1936.

30. Wessely, R. (2010). New drug-eluting stent concepts, *Nat. Rev. Cardiol.*, **7**, 194–203.

# Treating Aortic Aneurysms



## CHAPTER 19

Rochellee Manygoats

## 19.1. Introduction

In 2006, an estimated 2 million people in the USA were diagnosed with aneurysms of the aorta. Aneurysm is the 13th leading cause of deaths in the United States[1] and causes the death of 15,000 people there annually. Thus, there is need for an intervention that can minimize or eliminate the chances of further dilation or rupture. There are four types of aneurysm, named according to where they occur: abdominal, thoracic, cerebral, and peripheral. Abdominal aortic aneurysms (AAA) occur most often, whereas thoracic aortic aneurysms (TAA) are the most difficult to detect.

In order to understand treatment of aortic aneurysms, a brief overview of the anatomy and physiology of the aorta is needed. The heart is the critical organ of the circulatory system. The pulmonary artery connects the lungs to the heart, and the vena cava and inferior vena cava bring in the oxygen-depleted blood (Figs. 19.1a and 19.1b). The oxygenated blood moves through the chambers of the heart and out through the aortic artery. From the aortic artery the oxygenated blood flows throughout the body.

The muscles of the heart and arteries are made from connective tissue, primarily elastin and collagen. It is the elasticity of the heart and arterial muscles that allow for proper function, so as a disease degrades the elasticity of the heart or the arterial tissues, major complications and eventually death results.

There are three ways aneurysms may occur: trauma, loss of elasticity due to age, and genetic disorders. Individuals with the

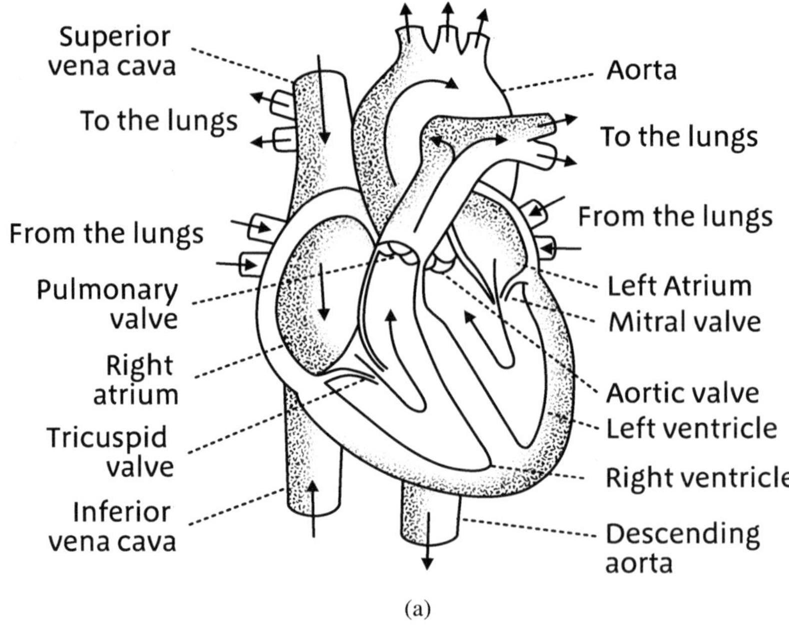


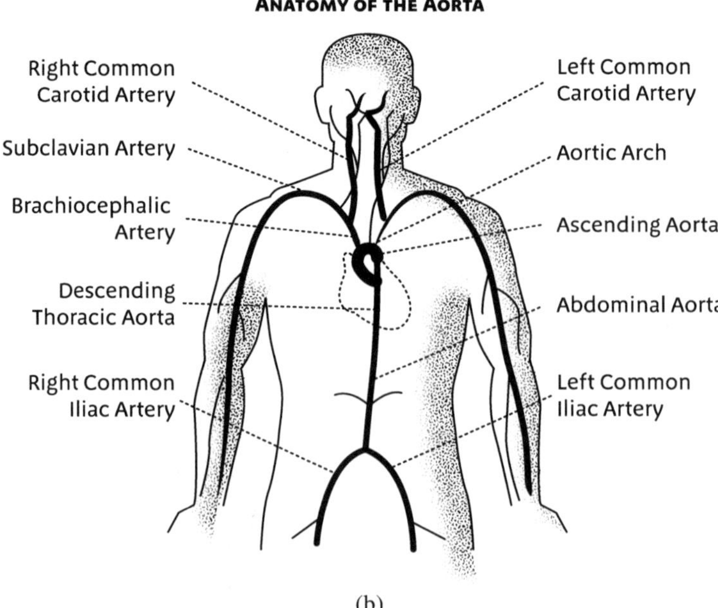


**Figure 19.1.**   (a) Schematic of the heart; (b) Schematic of the major arteries.

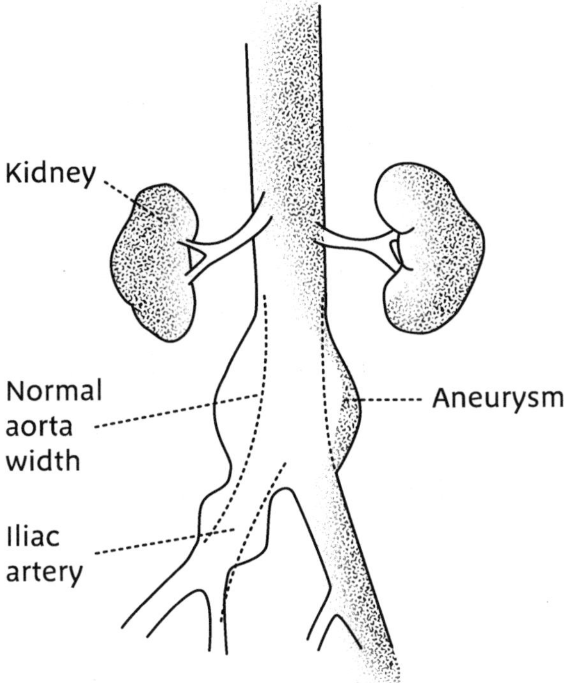


**Figure 19.2.**  Dilatation of an arterial wall.

genetic disorder Marfan syndrome lack the genetic coding that produces fibrillin, which is necessary to make the connective tissue sufficiently elastic to function effectively. They are especially affected by diseases attacking the connective tissues, particularly the aorta, ligaments, and the ciliary zonules of the eye.[2]

An aneurysm is a dilation or "ballooning" of the wall of an artery (Fig. 19.2). An aneurysm develops when the aortic artery begins to loss its elasticity, the arterial walls are weakened, and blood pressure causes the walls to dilate. Three types of aneurysms can occur (Fig. 19.3):

1. symmetric (fusiform): the artery wall dilates symmetrically;
2. asymmetric (saccular): the wall of the artery bulges on one side only; and

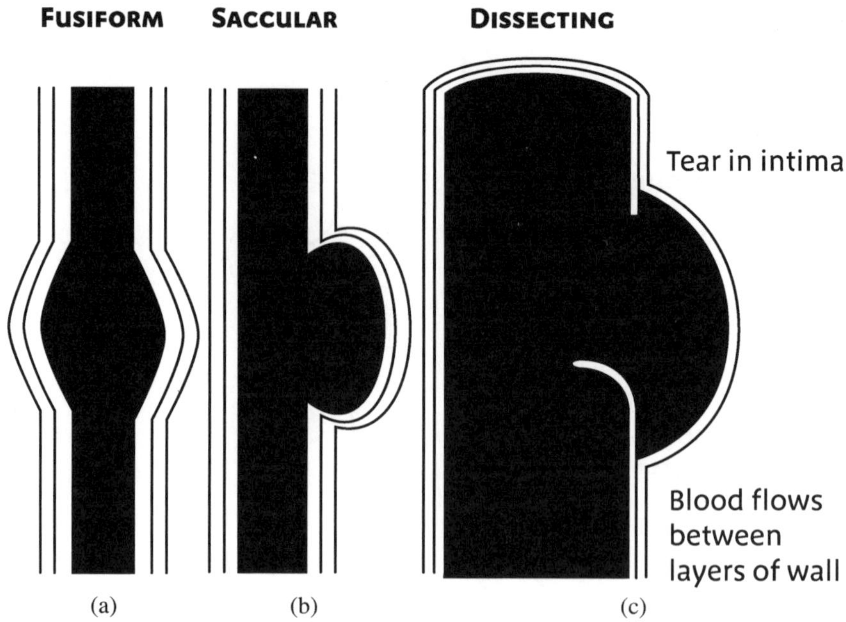


**Figure 19.3.**   Types of aneurysm.

3.  dissection (pseudoaneurysm): affects the three layers that make up the aorta, i.e. the thin inner layer (intima), a thick, elastic middle layer (media), and the thin outer layer (adventitia). The walls of the artery weaken and allow blood to penetrate the walls of the artery, causing bleeding within the walls.

Fusiform and saccular are considered true aneurysms, that is, can be further defined as a dilation that is twice the size of the normal diameter. For all types of aneurysm the individual has a low probability of survival if repair is not performed in a timely manner.

## 19.2.  Present Surgical Techniques

In the late 19th century, doctors used techniques called the Galvano-puncture and Macewan method to reduce the size of an aneurysm. These methods depended on creating blood clots within the sac. The Galvano-puncture procedure was used to treat aneurysms by inserting fine needles that caused the blood to clot by applying current to the blood near the sac. The positively charged needles were inserted into the artery while the negative pole was connected to a wet pad located on the skin near the aneurysm.[3] The current applied would start at 50 mA, increasing until the sac became firm; this would be repeated every week. The Macewan method also used needles that were made of stainless steel, but no current was applied. The needles would be used to scratch the inner surface of the sac to induce blood clot formation, and this procedure would also be repeated every seven days if necessary. Both procedures were improvements in eliminating aneurysms but the chances of rupture were still high.

The experiments continued but it was not until the 1950s that the first successful open surgery was performed by Dubost and colleagues, who excised the diseased portion of the artery and reconnected the section with homografts.[2] But the real breakthrough came with Gross's method of using a new segment of material to reconnect the sections when a large portion of the artery was diseased. Although the success rate was high, finding donors limited the number of procedures possible. It was not until 1957 that the Society of Vascular Surgery concluded its survey of the available synthetic material and recommended Dacron® and Teflon® as reliable grafting materials to use in replacing the diseased portion of the artery.[2] These new materials and surgical techniques made it possible to treat many more patients.

Improved imaging techniques have also made it possible for surgeons to view the damaged artery to aid the decision for surgery. Five imaging techniques currently used to detect and determine the size of the aortic defect are computed topography (CT), magnetic resonance imaging (MRI), echocardiography, angiography, and chest X-ray.

Surgical repair of the aneurysms is an invasive procedure. In general, three alternative procedures are utilized when accessing the diseased sections of the artery, depending upon the type of repair needed. In the case of the ascending artery, the surgeon accesses the diseased portion by making an incision through the breastbone (median sternotomy), exposing the upper portion of the heart. For the descending aorta, a larger incision is made from the back of the shoulder blade around the patient's side to the rib cage under the breastbone (thoracotomy).[3] For abdominal aneurysms an incision is made in the abdominal muscle for the surgeon to pass the graft through and then to sew the graft in place. As the surgeon is excising the diseased portion of the artery, the patient is put in a hypothermic state and placed on a cardiopulmonary bypass machine. Repair of a descending thoracic aortic aneurysm consists of the following: the diseased section of the artery is identified for removal, then the aorta is clamped, the aneurysm section is removed, and finally the graft is sewn in place.[3]

These are very invasive techniques that typically require the patient to stay in hospital for at least ten days with full recovery expected after 2–3 months. Complications can arise from this type of invasive surgery, such as myocardial infarction (MI), stroke, bowel necrosis, and limb ischemia. However, without these procedures there would be thousands of patients with nothing to prevent a rupture from occurring.

## 19.3. New Surgical Methods: Endovascular Surgery

As medical techniques and technology have continued to progress, so have the procedures for repairing aortic aneurysms. New techniques, now termed endovascular surgery, have been developed which still use the same synthetic polymeric graft materials (Dacron® and Teflon®), but the means of attaching the graft has changed dramatically. The synthetic material now encases a wire cage made of stainless steel or Nitinol shape-memory alloy wire. There are currently three companies in the United States that are clinically testing the new devices under U.S. Food and Drink Administration (FDA) Review (Table 19.1). The technique for repairing the diseased artery has changed from

**Table 19.1.**  Devices undergoing FDA review for TAA repair.

| Device (company) | Stent/graft material | Unique properties |
| --- | --- | --- |
| TAG® (Gore) | Nitinol/PTFE | Flexible |
| Talent® (Medtronic) | Nitinol/polyester | Uncovered proximally |
| TX2® (Cook) | Stainless steel/polyester | Fixation barbs |

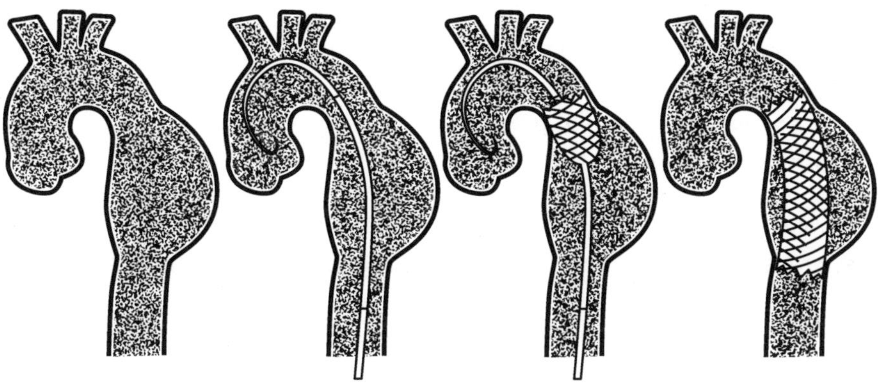

**Figure 19.4.**  Arterial repair. Endovascular repair begins with advancing the guide wire. The stent-graft is then inserted and deployment begins at the proximal portion of the aorta until there is full expansion down to the distal portion of artery. During deployment of the graft, blood flow is not interrupted.

the traditional, more invasive open surgery, as discussed above, to a less invasive procedure (Fig. 19.4), with the following steps:

1. Patient is prepared for surgery.
2. Incision to the iliac or femoral artery (located in the groin area).
3. A guide wire, usually made of stainless steel, is routed to the aorta.
4. A catheter, with the compressed graft, follows the guide wire to the affected area.
5. The graft is deployed by pulling the rip cord of the graft.
6. The catheter is removed.
7. A balloon is inserted and expanded; to ensure good compression between the graft and the artery wall, the balloon is rotated 60° at least twice, expanding the balloon at each rotation.

During the procedure, contrast material and X-rays are used to help monitor and ensure the location of the device and graft. The required landing zone for the stent graft to be successfully deployed is 15–20 mm of healthy/normal artery wall. In the case of a severe aneurysm, a second graft may be inserted and deployed. The total time for the procedure is typically 1–3 hours.

Important advantages of this new technique are:

1.  patients are able to return home within an average of three days;
2.  there is less blood loss; and
3.  faster recovery.

One very important advantage is that the patient's blood supply is never interrupted as it is with open surgery, which requires placing the patient on a cardiopulmonary bypass machine. The new stent grafts are more flexible, allowing surgeons to maneuver and access curved arches with greater success.[4]

At present, the concerns are:

1.  long-term effects are unknown, since this is a new technique in the early clinical trial stages;
2.  re-grafting of the artery due to migration of the graft or development of endoleaks;
3.  potential mechanical failures of the grafts; and
4.  not all patients are candidates for this procedure.

Patients are required to have healthy femoral or iliac arteries to allow for a 26–40 mm diameter device to be inserted and guided to the aorta.[5] The patient's aorta must also have a sufficient landing zone for deployment to securely anchor the stent graft. Endoleaks occur when blood begins to penetrate between the stent graft and the artery wall. In these cases re-grafting is necessary, which can damage the artery if repetitive access is made. There have also been mechanical failures of the devices such as collapsing of the stent graft, suture disruption, metal fracture, or fabric erosion.[6,7] With these types of failure re-grafting is necessary.

## 19.4. Ethics

As with any type of surgery, the patient, patient's family, and surgeon must weigh the risks of the operation *versus* potential rupture of the aorta with time (Fig. 19.5).[8] In treating aortic aneurysms, there are many factors which help to determine when surgery is necessary. Some of the more important of these include:

- Age of the patient.
- Genetic disorders.
- Size and rate of growth of the aneurysm.
- Pain experienced by the patient.
- Physiology of the arteries.

Age is important, as older patients recover more slowly and are more susceptible to complications (MI, renal failure, and ischemia). Their arteries are also weaker, making them less suitable for endovascular surgery. In younger patients the unknown long-term effects of endovascular surgery may limit use at present. Younger patients have the option of regular monitoring of the aneurysm and its growth rate. When the aneurysm exceeds 50% of the normal diameter or experiences a growth rate faster than 1 cm per year, they can than elect to have surgery. Older patients can do the same, but deterioration of their arteries may make it impossible to elect surgery. This is a difficult decision to face because of the

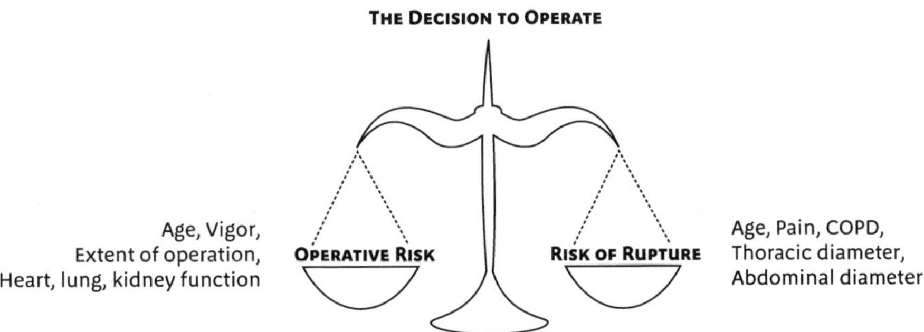


**Figure 19.5.**  Weighing the risks of surgical intervention *versus* rupture of aorta.

possible outcome: if surgery is not performed, rupture is most likely to occur and only 6%[9] of patients survive, if they make it to the hospital. If a patient suffering from Marfan syndrome experiences an aneurysm, the surgeon is more likely to perform surgery when the aneurysm is detected, because of a higher probability of rupture.

## 19.5.  Cost

Cost comparison has been estimated between traditional open-heart surgical repairs of aneurysms *versus* endovascular surgery.[10] It is estimated that in the United States, the cost to Medicare would be higher by approximately 20% (US$131,000 *versus* US$106,000) for an open-heart surgical procedure. Longer stays in the hospital and use of the cardiopulmonary bypass machine for open-heart procedures account for most of the cost differences. Although endovascular surgery is a fairly new procedure, the additional staff required and limited number of surgeons able to perform the surgery, at present, could increase the cost for the first group of patients. Until more surgeons are trained in performing endovascular surgery and the staff can support the surgeon at an equivalent scale to open-heart surgery, valid economic comparisons are difficult.

## 19.6.  Recent Developments

Surgical techniques for aortic aneurysm repair, especially the technology and devices for endovascular aortic repair (EVAR), have advanced significantly. In large part, much of the progress has been observed in stent-graft development, particularly with the introduction of a new generation of fenestrated and branched stent grafts for management of difficult aortic neck and branched vasculature.[16] EVAR has become an established treatment for repair of abdominal and aortic aneurysms due to many advantages over open repair including decreased hospitalization time, blood loss, morbidity, and mortality.[11] Although short-term results for EVAR have improved patient outcomes, the long-term results have shown similar survival rates to open repair, and significantly higher rates of secondary intervention for EVAR patients.[12,13] Despite advances in current surgical techniques, a great need still exists for developing a reliable approach which can provide patient-specific customization and consistent long-term results.

Stem cells are capable of producing angiogenic structures, most importantly smooth muscle vascular tissue, which could potentially be used for vascular repair of such defects as aortic aneurysms.[14,15] The best source for obtaining stem cells for vascular tissue regeneration has yet to be agreed upon. Many studies have derived cells from either bone marrow (BMCs) or adipose tissue (ASCs) due their high rates of proliferation and availability.[14] Both adipose and bone marrow-derived stem cells have demonstrated the ability to differentiate into smooth muscle cells (SMCs) and endothelial cells.[15] It has been argued that adipose-derived stem cell harvest is the better choice due to lower co-morbidities and the decline in bone marrow-derived stem cells with age; however, the optimal cell source is still a point of contention.[15]

Recently, Harris *et al.* demonstrated that ASCs seeded onto a vascular graft consisting of decellularized saphenous vein cultured in a bioreactor allowed for cell attachment and differentiation and proliferation of SMCs.[15] In another study by Cho *et al.*, BMCs were seeded onto decellularized porcine aortic explants and implanted in the bone marrow of donor pigs.[16] This study showed that autologous BMCs seeded onto vascular tissue scaffolds expressed extra-cellular matrix precursors, cell proliferation, and *in vivo* vascular tissue remodelling. These studies demonstrate the potential for developing tissue-engineered autologous arteries for vascular repair such as aortic aneurysm-induced deficits.

Prosthetic vascular grafts constructed of polymeric scaffolds seeded with cells have been demonstrated to allow for vascular cell proliferation and mitigate inflammatory response associated with tissue-derived graft scaffolds. Mirensky *et al.* assembled a tubular vascular graft scaffold from poly-L-lactic acid mesh coated with ε-caprolactone and L-lactide copolymer.[17] The polymer-mesh scaffolds were then seeded with human aortic endothelial and smooth muscle cells (SMCs) and implanted as abdominal aortic grafts in mice. The implanted grafts showed extensive *in vivo* remodeling, and the formation of neo-vessels with an endothelial lining and extracellular matrix similar to native tissue.

In another study by Tillman *et al.* an electrospun polycaprolactone-collagen scaffold was shown to facilitate vascular cell growth and proliferation *in vivo*.[18] The electrospun scaffold was demonstrated to retain good biomechanical properties after implantation and the electro-spinning method also allowed for consistent composition and controlled degradation of the graft. The development of polymer-based vascular scaffolds has the potential to allow improvement to previous vascular grafts, as there is prospective reduction of inflammatory and immune responses, and reproducibility of the graft scaffolds can be controlled during construction. With the broad range of possibilities for vascular tissue repair and regeneration, a considerable amount of research is still needed to establish an ideal methodology for repairing aortic aneurysms using tissue-engineering technology.

## 19.7. Conclusion

The repair of the aortic artery using stem cells to regenerate the diseased section of the artery would be an important step forward but is still far from being a clinical procedure. The scientific and technical issues of stem cell research discussed in Chapter 2 need to be addressed before this advancement becomes routine. Another approach is to repair the diseased artery using tissue-engineering techniques for developing new grafts, as is being achieved for bone grafts (Chapter 9). Other examples include impregnating the graft material with controlled-release pharmaceuticals to reduce the chances of blood clot formation, or the use of bioactive materials to regenerate and reinforce the anchoring area of the graft.

Aortic aneurysms have at times have been referred to in the literature as an "Achilles heel" for surgeons. Determining who is a candidate for surgery, knowing death is possible if no surgical intervention occurs, is a routine dilemma for vascular surgeons. With 15,000 fatalities each year due to aortic rupturing, improved diagnosis and repair of the aneurysm is necessary to reduce the mortality rate. Although the present open surgery technique is very aggressive and invasive, it has prevented many deaths. With the advances of endovascular surgery, patients now have a new alternative for reducing the chance of aortic rupture and death.

## References


1. Aortic Aneurysm Facts Sheet (2010). *Centers for disease control and prevention, national center for health statistics.* [Online]. http://www.cdc.gov/dhdsp/data_statistics/fact_sheets/docs/fs_aortic_aneurysm.pdf. [Accessed 19 February 2011].
2. Shumacker, H.B. (1980). A history of modern treatment of aortic aneurysms, *World J. Surg.*, **4**, 503–509.
3. Isselbacher, E.M. (2005). Thoracic and abdominal aortic aneurysms, *Circulation*, **111**, 816–828.
4. Appoo, J.J., Moser, W.G., Fairman, R.M., *et al.* (2006). Thoracic aortic stent grafting: improving results with newer generation investigational devices, *J. Thoracic Cardiovasc. Surg.*, **13**, 1087–1094.
5. Wheatley, G.H. III, McNutt, R. and Diethrich, E.B. (2007). Introduction to thoracic endografting: imaging, guidewires, guiding catheters, and delivery sheaths, *Ann. Thoracic Surg.*, **83**, 272–278.
6. Jacobs, T.S. and Won, J. (2003). Mechanical failure of prosthetic human implants: a 10-year experience with aortic stent graft devices, *J. Vascular Surg.*, **37**, 16–26.
7. Steinbauer, G.M., Stehr, A.S., Pfister, K., *et al.* (2006). Endovascular repair of proximal endograft collapse after treatment for thoracic aortic disease, *J. Vasc. Surg.*, **43**, 609–612.
8. Juvonen, T., Ergin, M.A., Galla, J.D., *et al.* (1994). Prospective study of the natural history of thoracic aortic aneurysms, *Ann. Thoracic Surg.*, **63**, 1533–1544.
9. Katzen, B.T., Dake, M.D., MacLean, A.A., *et al.* (2005). Endovascular repair of abdominal and thoracic aortic aneurysms, *Circulation*, **112**, 1663–1675.

10. Wilt T.J., Lederle F.A., MacDonald R., *et al.* (2006). Comparison of endovascular and open surgical repairs for abdominal aortic aneurysm, *Agency for Healthcare Research and Quality Evidence Report/Technology Assessment*, **144**, 1–210.

11. Ricotta, J.J. and Oderich, G.S. (2008). Fensestrated and branched stent grafts, *perspect. Vasc. Surg. Endovasc. Ther.*, **20**,174–187.

12. de Bruin, J.L., Baas, A.F., Buth, J., *et al.* (2010). Long term outcome of open or endovascular repair of abdominal aortic aneurysm, *New Engl. J. Med.*, **362**, 1881–1889.

13. Moise, M.A., Woo, E.Y., Velazquez, O.C., *et al.* (2006). Barriers to endovascular aortic aneurysm repair: past experience and implications for future device development. *Vasc. Endovascular. Surg.*, **40**, 197–203.

14. Ugurlucan, M., Yerebakan, C., Furlani, D., *et al.* (2008). Cell sources for cardiovascular tissue regeneration and engineering, *Thorac. Cardiovasc. Surg.*, **57**, 63–73.

15. Harris, L.J., Abdollahi, H., Zhang, P., *et al.* (2009). Differentiation of adult stem cells into smooth muscle for vascular tissue engineering. *J. Surg. Res.*, [Online]. Available at: http://www.ncbi.nlm.nih.gov/pubmed/19959190. [Accessed 18 July 2010].

16. Seung-Woo, C., Il-Kwon, K, Kang, J.M., *et al.* (2009). Evidence for *in-vivo* growth potential and vascular remodeling of tissue engineered artery, *Tissue Eng. A.*, **15**, 901–912.

17. Mirensky, T.L., Nelson, G.N., Brennan, M.P., *et al.* (2009). Tissue engineered arterial grafts: long-term results after implantation in a small animal model, *J. Pediatr. Surg.*, **44**, 1127–1133.

18. Tillman, B.W., Yazdani, S.K., Lee, S.J., *et al.* (2009). The *in vivo* stability of electrospun polycaprolactone-collagen scaffolds in vascular reconstruction, *Biomaterials*, **30**, 583–588.

# Abdominal Aortic Aneurysms: Treatment and Repair

CHAPTER **20**

John Ashton

## 20.1. Introduction

An abdominal aortic aneurysm (AAA) is a permanent local dilation of the aorta in the abdominal region (Fig. 20.1). The aorta is usually considered aneurysmal if the diameter of the permanent dilation is 50% greater than the expected diameter.[1,2] This dilation occurs over a period of years and is characterized by chronic inflammation, destructive outward remodelling of the vascular wall, and decrease in smooth muscle cells. Rupture occurs when blood pressure in the aneurysmal sac exceeds the wall strength, and takes place in one-third of all AAAs,[3] having a 67–89% mortality rate.[1] AAAs are usually asymptomatic until they rupture. Symptoms of a rupture include intense abdominal and back pain, shock, and a pulsatile abdominal mass.

The success of elective open surgical repair (OSR) has not changed much over the past 20 years. Three types of treatment are currently available: non-invasive monitoring of progression, OSR, and endovascular aneurysm repair (EVAR).

The exact causes of the vascular wall degeneration in AAA are not known. However, there are several associated risk factors, which include older age, cigarette smoking, male gender, Caucasian race, hypertension, coronary artery disease, atherosclerosis, high cholesterol, chronic obstructive pulmonary disease, and a family history of AAA.[4–6]

AAA is a significant disease in developed countries, and its incidence is believed to be growing as life expectancy increases.

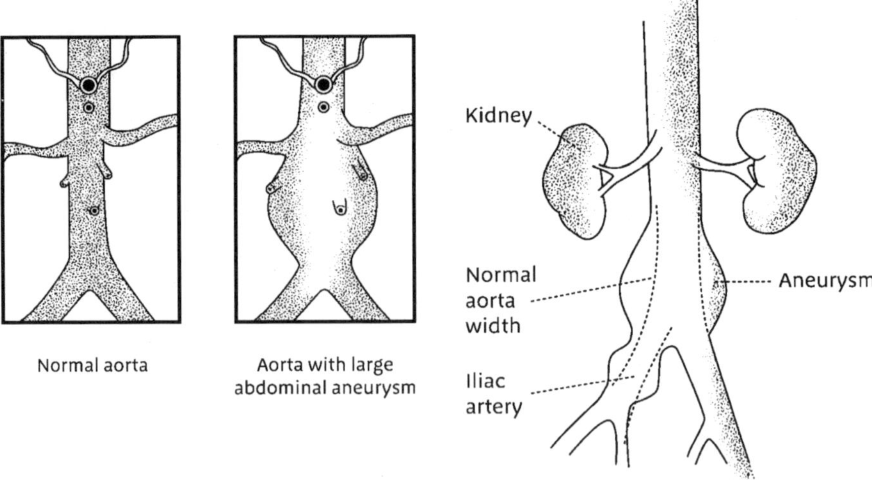


**Figure 20.1.**   Abdominal aortic aneurysms.

In people 60 years and older, AAAs are present in 4–8% of men and 1–3% of women.[1] Aortic aneurysm and dissection is the 15th leading cause of death in the United States, and about half of these aneurysms are AAAs.[8]

## 20.2. Treatment

Currently, three types of AAA treatment are clinically available in the United States. One is for the physician to non-invasively monitor the growth of the AAA over time. This can be accomplished by the physician palpating the abdominal region or through imaging the area by ultrasound, CT scan, or MRI. The other two treatments are open surgical repair (OSR) and endovascular aneurysm repair (EVAR). Elective OSR and EVAR refers to AAA patients undergoing treatment before rupture in order to reduce the risk of rupture, whereas rupture OSR and EVAR refers to patients undergoing treatment after rupture has occurred. When deciding upon elective repair, the one of the key question that must be asked is: when does the risk of rupture outweigh the risk of the procedure?

Generally, physicians will encourage elective treatment when AAA diameter exceeds 5–5.5 cm.

### 20.2.1. *Open surgical repair*

Dubost and colleagues[9] first reported treating AAA with OSR in 1951. In OSR, a large anterior incision is first made in the abdominal region. Next, the aorta is clamped above the aneurysm to stop blood flow; the aneurysm sac is cut open and any present thrombus is removed; and the aneurysm is butterfly-opened (vessel wall is not completely removed). Then a graft is sewn in place; blood flow is restored; and the AAA wall is sewn back around the wall (which provides reinforcement). Finally, the patient's abdomen is closed. OSR grafts are usually made of polytetrafluoroethylene (PTFE), Dacron®, or polyester.

The success of elective OSR has not changed much over the past 20 years. Thirty days after elective OSR there is a 5% mortality rate and a 15–30% major complication rate,[10] and ruptured OSR has a 45.7% mortality rate.[3] The long-term outcome of OSR has proven to be quite successful. In one study, late graft-related complications occurred in 3.2% of OSRs and in another study graft-related complications occurred in 2.5% of patients 12 years after surgery.[9] Long-term graft-related complications and mortality of OSR include infection, psuedoaneurysm development, graft thrombosis, and graft erosion and breakage. Since OSR has a proven record of survivability, it is considered the gold standard for AAA treatment.

### 20.2.2. *Endovascular aneurysm repair*

EVAR was introduced as a less invasive treatment for AAA in 1991.[9] In this procedure, a delivery catheter containing a stent graft is inserted into a small incision in the femoral artery and guided up to the AAA using X-ray imaging. The stent graft is deployed in position and a balloon may also be used to open and fix the ends into place. Friction with the vessel wall is the main mechanism of attachment, but metal hooks or anchors may be present on each end of

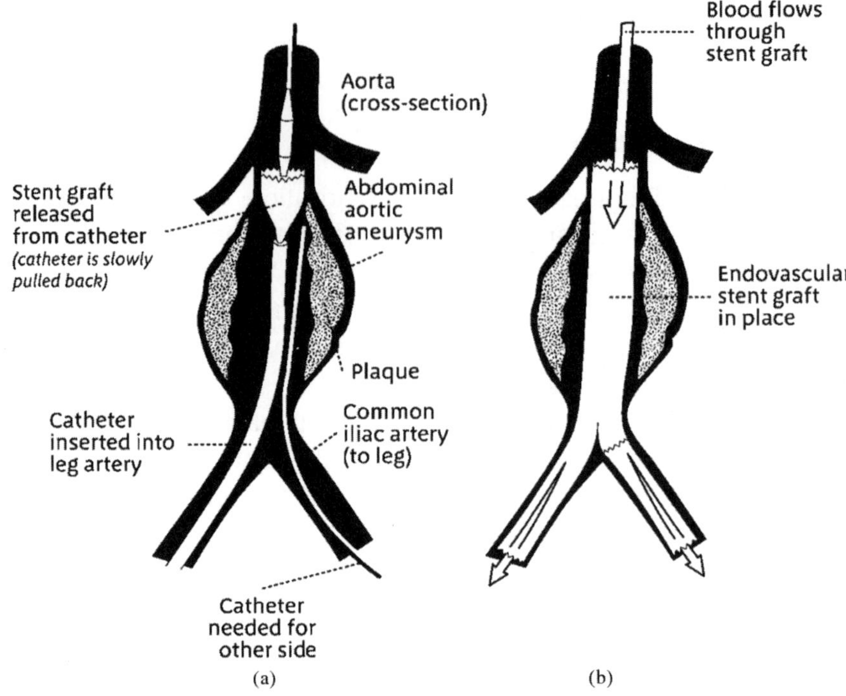



**Figure 20.2.**   Schematic of EVAR procedure.

the stent graft to help secure it in place. Figure 20.2 is a schematic of the EVAR procedure.

Stent grafts consist of a polymer tube surrounded by metal struts to reinforce it. The stent graft channels blood flow through the aneurysm with the purpose of eliminating expansion of the aneurysmal sac. Table 20.1 is a list of available stent-grafts and the biomaterial in the device. Due to the nature of stent-graft deployment and fixation, only about 50% of AAA patients are suitable for EVAR.[10]

Most physicians would consider the following AAA geometric criteria unsuitable for EVAR:[12]

- Proximal neck <15mm
- Infrarenal aortic diameter >26 mm
- External iliac diameter <7mm or >16 mm
- Bilateral internal and external iliac aneurysms

**Table 20.1.**  Available stent grafts.[11]

| Name | Company | Graft material | Stent material |
|---|---|---|---|
| Ancure® | Guidant | Polyester | Elgiloy |
| AneuRx® | Medtronic AVE | Polyester | Nitinol |
| Talent® | Medtronic AVE | Polyester | Nitinol |
| Excluder® | WL Gore | PTFE | Nitinol |
| Zenith® | Cook | Polyester | Stainless Steel |
| LifePath® | Edwards Lifesc. | Polyester | Elgiloy |
| PowerLink® | Endologix | PTFE | Stainless Steel |
| Quantum LP® | Cordis | Polyester | Nitinol |

Thirty days after EVAR, mortality rate ranges from 0.8% to 5.6%[1] with major complications significantly lower than that of OSR due to the less invasive procedure.[9] However, long-term outcome of EVAR has not been shown to be quite as successful. There is a 14% secondary intervention rate four years after the operation.[10]

The most common cause of long-term failure and need for secondary intervention after EVAR is endoleak, which is the persistence of blood flow into the aneurismal sac. When endoleak occurs, risk of rupture increases over time as the sac expands. Endoleak is divided into four types:

1. Type I: inadequate seal at either the proximal or distal end of the stent graft.
2. Type II: retrograde flow from collateral arteries around the sac.
3. Type III: fabric tears or component disconnection.
4. Type IV: blood flow through an intact stent graft (for example, due to porosity).

Another common cause of failure is stent-graft migration down the aorta.

### 20.2.3. *OSR versus EVAR*

Whether an AAA patient should undergo OSR or EVAR is based on several factors; some patients may be more suitable for the former.

For example, it may be too difficult to effectively deploy a stent graft into the lumen of an AAA if the patient's AAA geometry is highly tortuous. For other patients, EVAR may be more appropriate than OSR. For example, if the patient has several other health complications, the risk of mortality due to open heart surgery may be very high.

For patients eligible for either procedure, EVAR has advantages over OSR in that the procedure is less invasive, there is quicker patient recovery, and potential cost could be lower due to shorter hospitalization. However, since EVAR is a newer treatment, it has higher uncertainty of success and the lower cost at the time of procedure is offset by the need for more physician follow-up visits to monitor progress. In addition, there is a higher occurrence of secondary interventions in EVAR due to endoleak and stent-graft migration.

In the EVAR 1 study, an ongoing study that started in 1999, the success rate of elective EVAR *versus* OSR is compared for patients who were medically suitable for both procedures. In 2004, the 30 day post-operation mortality rate was reported to be 1.7% in EVAR compared to 4.6% in OSR,[13] and in 2005, the three-year aneurysm-related mortality rate was reported to be 4% in EVAR compared to 7% in OSR.[14] However, other studies have found the difference in success rates between the two procedures to be insignificant.[10]

## 20.3. Economic Issues

The costs associated with some of the treatments have been briefly addressed. Non-invasive monitoring of the AAA is the cheapest if it only involves the physician palpating the abdominal area. However, monitoring is more expensive if it includes ultrasound, CT scan, or MRI.

If it is determined that the risk of rupture outweighs the risk of open repair, OSR and EVAR are considered. Table 20.2 summarizes several studies that have evaluated the cost of EVAR compared to OSR. EVAR has the potential of being cheaper due to

**Table 20.2.** Is EVAR cost-effective?.[11]

| Author | Journal (year) | |
| --- | --- | --- |
| Holzenbien, J. | *Eur. J. Vasc Endovasc. Surg.* (1997) | Yes |
| Ceelen, W. | *Acta Chir. Belg.* (1999) | Equal |
| Patel, S.W. | *J. Vasc. Surg.* (1999) | Yes |
| Seiwert, A.J. | *Am. J. Surg.* (1999) | Equal |
| Quinones, W.J. | *J. Vasc. Surg.* (1999) | No |
| Sternbergh, W.C. | *J. Vasc. Surg.* (2000) | No |
| Clair, D.G. | *J. Vasc. Surg.* (2000) | No |
| Birch, S.E. | *Aus. N.Z.J. Surg.* (2000) | No |
| Turnipseed, W. | *J. Vasc. Surg.* (2001) | No |
| Bosch, J.L. | *Radiology* (2001) | No |

the reduced requirements of blood transfusion, shorter intensive care unit and hospital stay, lower 30-day mortality, and lower complication rate. However, the cost of EVAR is currently similar to OSR due to the offsetting costs of more expensive devices and the need for close post-intervention surveillance and secondary interventions.[11] As improvements are made on current EVAR devices and more become available to compete in the market, the cost of EVAR should drop significantly below that of OSR.

## 20.4. Ethical Issues

There are uncertainties as to when an AAA will rupture and the outcome of OSR or EVAR if elective repair is considered. If the patient has health insurance which has agreed to pay part of the cost, the treatment decision must be made in conjunction with the provider according to their contract. In the decision process, the patient (and provider) should understand to the best of their ability the associated risks and costs of the three treatments. Ultimately, the responsibility is upon the patient to educate herself about the options. However, the patient's physician, who understands the risks and costs, needs to take the time to explain the options available.

## 20.5. Current Biomaterials Research

### 20.5.1. *OSR grafts*

The long-term outcome of OSR is quite successful. The numbers presented show that there is a higher chance of mortality due to the surgery than to device failure. However, improvements can still be made on the graft.

Research to improve synthetic grafts includes:

1.  Embedding them with antithrombogenic drugs.
2.  Seeding them with endothelial cells.
3.  Developing more biocompatible materials.[15]

For example, in one study of small-diameter vascular grafts, those coated with heparin have led to better results than standard prostheses.[16]

Another potential improvement in biocompatibility and success of OSR grafts would be to develop autologous tissue-engineered grafts. Tissue-engineered arterial autografts have been successfully grown and implanted in swine.[17] In this procedure, autologous arteries were grown *in vitro* from vascular smooth muscle cells and endothelial cells from the aorta on a biodegradable polymer matrix (PGA) by means of a pulsatile perfusion system. The grafts were implanted in the saphenous artery and remained open for three weeks before developing thrombosis. The study also showed that similarly tissue-engineered bovine grafts ruptured at a pressure of 2150+/−709 mm Hg after eight weeks, which is much higher than the strength needed in the aorta (120 mm Hg).

Another group has been able to successfully culture a tissue-engineered autologous blood vessel from human smooth muscle cells and endothelial cells taken from the saphenous vein.[18] Mechanical testing and implantation viability were not reported in the study.

These tissue-engineered vessels are under investigation for small-diameter grafts (such as coronary artery bypass grafts), but also show potential for large-diameter grafts. These engineered grafts were grown over several weeks, and thus have potential for use in elective, but not ruptured, OSR.

## 20.5.2. *EVAR stent grafts*

As discussed above, failure of stent grafts is usually due to an endoleak or migration. Many factors may cause these, including:

1. Improper positioning at deployment.
2. Insufficient material mechanical strength.
3. Insufficient friction to hold and seal the stent graft in place.
4. Improper size of stent graft chosen.

In addition, many AAA patients do not have the option of EVAR due to restrictions such as highly tortuous vessel lumens or critical arterial branches in the AAA area.

One method of decreasing the failure of EVAR is to customize the stent graft to the patient's anatomy. The geometry of the patient-specific AAA can be imaged by computed tomography (CT) in conjunction with software. One study reported successful implantations of customized stent grafts after the geometry of the AAA was analyzed in two patients.[19]

A second approach to improve stent grafts is to provide openings at the sites of critical branching artery areas.[20] In many cases, additional smaller stent grafts can be deployed through those fenestrations for additional support in holding the stent graft in place and sealing off the aneurysmal sac to reduce endoleak and migration.

A third approach is to develop a bioactive layer that can be coated on the outside surface of the stent grafts. The purpose of the coating is to encourage adherence and growth of smooth muscle cells, endothelial cells, and fibroblasts onto the outer surface of the stent graft. This would provide a better seal between the proximal and distal ends of the stent graft and the vascular wall, thus reducing type 1 endoleak and stent-graft migration. In addition, it could encourage wound-healing cells to migrate down the side of stent graft to heal the aneurysm. Lerouge and associates are investigating the potential of doing this with a nitrogen-rich plasma-polymerized coating that they have developed.[21] Depictions of fenestrations and addition stent grafts deployed involving the renal arteries and external and internal iliac arteries can be seen in detail in reference 12.

The Anaconda system[22] is a new stent-graft design undergoing clinical trials. The graft material is a third of the thickness of most graft materials. In addition, there is a novel proximal ring stent to provide radial force to an area above the renal arteries without the graft covering renal artery inlets, which helps reduce migration and provide a better seal to reduce endoleak.

The Enovus graft is another novel EVAR device that is currently undergoing testing.[22] This graft does not contain metallic stents along the length of the graft, so technically it is not considered a stent graft, but the graft is designed to be used in EVAR. Instead of a stent, it has channels that run along the graft which, after graft placement, can be filled with a polymer that later hardens to provide radial and longitudinal reinforcement. Since there is no stent on the graft, the graft can be deployed with a delivery catheter that has a smaller diameter.

Other areas in which stent-grafts could be improved are:

1. more flexible biomaterials, which would reduce kinks and facilitate deployment; and
2. developing a way to cost-effectively manufacture stent grafts customized to each patient to fit their anatomy. One source reported that a spokesman for one company estimated that building a patient-specific stent graft would take 20 hours of hand labor plus computer aided design (the real cost of building the device is proprietary information).[23]

With the ever-increasing use and improvement of EVAR devices and surgical procedures for AAA repair, the role of OSR may decrease, although it will remain essential and its success will depend on the surgeon's experience. Surgeons will be challenged to continue to hone their skills, and to seek new biomaterials, and better techniques for OSR — most importantly for the acute care of ruptured aortic aneurysms.[24] In the short term (30 days), it has been shown that EVAR yielded a significantly decreased mortality over OSR for ruptured AAA.[25] For either EVAR or OSR procedures, the institutional and surgical experience is critical to the outcome of patients with a ruptured AAA. One of the next important steps in aortic aneurysm repair will most likely involve modifications of current EVAR techniques for implementation in repair of thoracic aneurysms. Thoracic endovascular aneurysm repair (TEVAR) was approved by the US Food and Drug Administration (FDA) in 2005 and has since experienced a boom in use for aneurysm repair of sections of the aorta previously only treatable by highly invasive OSR methods.[26]

## 20.6. Recent Developments

As the cause of aortic aneurysm is still unknown, another approach of increasing interest is the study of the biomechanical factors which contribute to formation and rupture of AAA. Developing an engineering-based understanding of the biomechanical factors involved could potentially allow for a greatly improved clinical assessment of the risk of AAA rupture. Experimental data and numerical models have allowed development of biomechanical assessment criteria including wall stress, finite element analysis rupture index, rupture potential index, and anatomical geometric considerations.[27] McGloughlin and Doyle reviewed these and other possible biomechanical indicators and concluded that some of the previously developed parameters could be close to implementation in the clinical setting, although all of these factors will require additional research, testing, and validation prior to use as clinical tools.[27]

Chapter 19 covers many of the potential advances in aneurysm repair to come in the future including the use of stem cells, tissue-engineered graft stents and a wide spectrum of new or improved materials for stent construction. A foreseeable development in aneurysm repair are improvements in current stent technology using bioactive material coatings. The next generation of stent technology may include bioactive coatings consisting of biological molecules and advanced polymeric formulations, although some of the new bioactive coatings may be inorganic materials. For example, Chai *et al.* demonstrated that sol-gel derived oxides ($TiO_2$, $Nb_2O_5$, and $TiO_2/SiO_2$) with nanoporous structures coated on titanium led to improved human vascular endothelial cell growth and proliferation over un-coated titanium.[24] This study is one of many which seek to improve current stent technology with a goal to replace artificial stents with biologically derived tissue-engineering constructs and cell-seeding techniques.

## 20.7. Summary

Non-invasive monitoring, OSR, and EVAR are the current options for treatment of AAAs. The patient, the physician, and the insurance provider should all understand the different risks and costs associated with each before making a treatment decision. Current research is aimed at increasing the success rates of devices, which will also potentially reduce treatment cost without a sacrifice of patient survival.

## References

1. Al-Omran, M., Verma, S., Lindsay, T.F., *et al.* (2004). Clinical decision making for endovascular repair of abdominal aortic aneurysm, *Circulation*, **110**, e517–e523.

2. Vande Geest, J.P. (2005). "Towards an improved rupture potential index for abdominal aortic aneurysms: anisotropic constitutive modeling and non-invasive wall strength estimation", PhD thesis, University of Pittsburgh, pp. 317.

3. Dalman, R.L., Tedesco, M.M., Myers, J., *et al.* (2006). AAA disease: mechanism, stratification, and treatment, *Ann. NY. Acad. Sci.*, **1085**, 92–109.

4. Alcorn, H.G., Wolfson S.K. Jr, Sutton-Tyrrell, K., *et al.* (1996). Risk factors for abdominal aortic aneurysms in older adults enrolled in the

cardiovascular health study, *Arterioscler. Thromb. Vasc. Biol.*, **16**, 963–970.

5. Lederle, F.A., Johnson, G.R., Wilson, S.E., *et al.* (1997). Prevalence and associations of abdominal aortic aneurysm detected through screening, *Ann. Intern. Med.*, **126**, 441–449.

6. Cole, C.W., Barber, G.G., Bouchard, A.G., *et al.* (1989). Abdominal aortic aneurysm: consequences of a positive family history, *Canadian J. Surg.*, **32**, 117–120.

7. Hoyert, D.L., Arias, E., Smith, B.L., *et al.* (2001). Deaths: final data for 1999, *Nat. Vital. Stat. Rep.*, **49**, 1–113.

8. Gillum, R.F. (1995). Epidemiology of aortic aneurysm in the United States, *J. Clin. Epidemiol.*, **48**, 1289–1298.

9. Krupski, W.C. and Rutherford, R.B. (2004). Update on open repair of abdominal aortic aneurysms: the challenges for endovascular Repair, *J. Am. Coll. Surg.*, **199**, 946–960.

10. Norwood, M.G., Lloyd, G.M., Bown, M.J., *et al.* (2007). Endovascular abdominal aortic aneurysm repair, *Postgrad. Med. J.*, **83**, 21–27.

11. Hallett, J., Mills, J., Earnshaw, J., *et al.* (2004). *Comprehensive Vascular and Endovascular Surgery*, Mosby, London, p. 728.

12. Chane, M. and Heuser, R.R. (2006). Review of interventional repair for abdominal aortic aneurysm, *J. Interv. Cardiol.*, **19**, 530–538.

13. Greenhalgh, R.M., Brown, L.C., Kwong, G.P., *et al.* (2004). Comparison of endovascular aneurysm repair with open repair in patients with abdominal aortic aneurysm (EVAR Trial 1), 30-day operative mortality results: randomised controlled trial, *Lancet*, **364**, 843–848.

14. EVAR trial participants (2005). Endovascular aneurysm repair versus open repair in patients with abdominal aortic aneurysm (EVAR Trial 1): randomised controlled trial, *Lancet*, **365**, 2179–2186.

15. Hoenig, M.R., Campbell, G.R., Rolfe, B.E., *et al.* (2005). Tissue-engineered blood vessels: alternative to autologous grafts?, *Arterioscler Thromb. Vasc.*, **25**, 1128–1134.

16. Begovac, P.C., Thomson, R.C., Fisher, J.L., *et al.* (2003). Improvements in GORE-TEX® vascular graft performance by carmeda bioActive surface heparin immobilization, *Eur. J. Vasc. Endovasc. Surg.*, **25**, 432–437.

17. Niklason, L.E., Gao, J., Abbott, W.M., *et al.* (1999). Functional arteries grown *in vitro*, *Science*, **284**, 489–493.

18. Poh, M., Boyer, M., Solan, A., *et al.* (2005). Blood vessels engineered from human cells, *Lancet*, **365**, 2122–2124.

19. Uflacker, R., Robison, J.D., Schonholz, C., *et al.* (2006). Clinical experience with a customized fenestrated endograft for juxtarenal abdominal aortic aneurysm repair, *J. Vasc. Interv. Radiol.*, **17**, 1935–1942.

20. Kaviani, A. and Greenberg, R. (2006). Current status of branched stent-graft technology in treatment of thoracoabdominal aneurysms, *Semin. Vasc. Surg.*, **19**, 60–65.

21. Lerouge, S., Major, A., Girault-Lauriault, P.L., *et al.* (2007). Nitrogen-rich coatings for promoting healing around stent-grafts after endovascular aneurysm repair, *Biomaterials*, **28**, 1209–1217.

22. Faries, P.L., Dayal, R., Rhee, J., *et al.* (2004) Stent graft treatment for abdominal aortic aneurysm repair: recent developments in therapy, *Curr. Opin. Cardiol.*, **19**, 551–557.

23. Williams, D.M. (2006). Engineering improvements in endovascular devices: design and validation, *Ann. NY Acad. Sci.*, **1085**, 213–223.

24. Chai, F., Oshsenbein, A., Traisnel, M., *et al.* (2010). Improving endothelial cell adhesion and proliferation on titanium by sol-gel derived oxide coating, *J. Biomed. Mater. Res.*, **92A**, 754–765.

25. Desal, M., Eaton-Evans, J., Hillery, C., *et al.* (2010). AAA stent-grafts: past problems and future prospects, *Ann. Biomed. Eng.*, **38**, 1259–1275.

26. Coselli, J.S. and Gopaldas, R.R. (2010). Ruptured thoracic aneurysms: to stent or not to stent?, *Circulation*, **121**, 2705–2707.

27. McGloughlin, T.M. and Doyle, B.J. (2010). New approaches to abdominal aortic aneurysm rupture risk assessment: engineering insights with clinical gain, *Arterioscl. Throm. Vas.*, **30**, 1687–1694.

# Heart Valve Stenosis and Surgical Repair: Present and Future

CHAPTER **21**

Scott Cooper

## 21.1. Introduction and Statement of Clinical Need

Heart failure is the number one cause of death in most developed countries. Both men and women face a 1 in 5 lifetime risk of heart failure.[1] Heart valve stenosis is a narrowing of the valve which can ultimately lead to heart failure. The primary method of treating stenosis is to replace the problematic valve with a prosthetic, and approximately 250,000 such procedures are performed every year in the US.[2]

There are currently two main types of replacement valves. The first is a completely man-made, mechanical valve. These are durable; however anticoagulants must be taken for the entire time that the mechanical valve is in place. The other option is a tissue valve. These valves usually come from porcine or bovine sources. Tissue valves are fixed in glutaraldehyde to kill the native cells and crosslink the extracellular matrix (ECM). Tissue valves have a low risk of rejection and anti-coagulants do not need to be taken long-term. However, they also have a limited lifetime and wear out within 10–15 years.

Average life expectancy is increasing throughout the world. As the world population ages, more body parts will become worn out and need to be replaced. This is especially critical for organs such as the heart, whose function is essential to maintain life. Therefore, in the coming decades, the demand for replacement heart valves will increase. Currently, patients only have two choices for a replacement valve: mechanical or tissue — neither of which is ideal. There is a need for valves that have a long lifetime, have a minimal risk of rejection, and do not require anti-coagulant drugs.

## 21.2. Keywords and Definitions

- *Mitral valve*: the heart valve that regulates the flow of blood between the left ventricle and left atrium. Also known as the bicuspid or left atrioventricular valve. It commonly requires replacement.
- *Aortic valve*: the heart valve that regulates blood flow between the left ventricle and the aorta. It is also common for this valve to need replacement.
- *Stenosis*: a condition in which the valve does not open completely, causing the heart to work harder. This can ultimately lead to heart failure.
- *Calcification*: hardening of the heart tissue, in this case of the valve leaflets. This reduces the flexibility of the valve, causing stenosis.
- *Thrombus*: a blood clot.
- *Thrombosis*: the coagulation of the blood in a vessel, resulting in a blood clot.
- *Thromboembolism*: a blood clot which has broken away from its initial site of formation. Thromboemboli are dangerous because they may create a blockage in another vessel, causing a heart attack or stroke.
- *Pericardium*: the membrane that surrounds the heart.
- *Congenital heart disease*: a malformation of the heart that occurred during prenatal development, often caused by genetic defects.
- *Cell differentiation*: the process by which cells change from non-specific to a particular type of cell such as muscle, blood, or skin.

## 21.3. Physiology of the Heart

The mitral valve regulates the flow of blood between the left atrium and the left ventricle. The left atrium is the chamber that

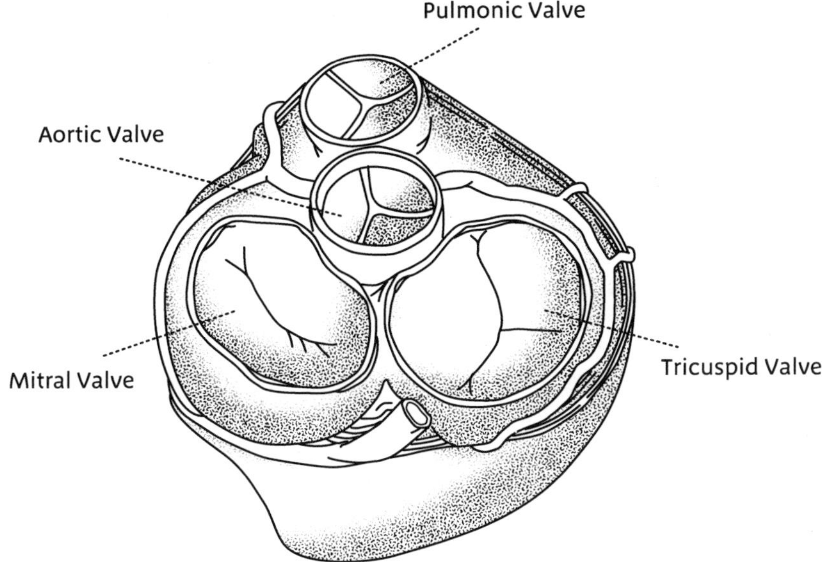



**Figure 21.1.**   Superior view of heart showing the mitral valve location.

holds the blood which returns from the lungs. The left atrium pumps blood through the mitral valve into the left ventricle. The left ventricle then pumps the blood through the aortic valve into the aorta, which delivers blood throughout the body (see Fig. 21.1 to observe position of the mitral and other valves). As stated above, stenosis is a condition in which the valve narrows and does not let blood adequately flow into the left ventricle. See references 3, 4, and 5 for high-quality images and detailed explanations of stenosis.

Mitral stenosis causes blood to build up in the left atrium due to inadequate opening of the mitral valve. Blood may also regurgitate from the left ventricle back into the left atrium. This build up of blood requires the heart to work harder to pump blood. Some symptoms include shortness of breath, fatigue, and heart palpitations. If not treated, a stenotic valve can cause a heart attack. Mitral stenosis is most frequently caused by rheumatic fever but can also be caused by congenital heart disease.[4]

## 21.4.  Current Methods of Treatment for Valve Stenosis

There are currently two types of replacement valves that are used for treating stenosis: mechanical and tissue valves. Each has its own advantages and disadvantages, and the patient must choose what best suits his or her needs.

### 21.4.1. *Mechanical valves*

Mechanical valves have been in use for over 50 years. In that time, various designs have been used clinically. The first mechanical valves were the caged-ball design; of these, only the Starr–Edwards valve is still in production today. Other designs used include the non-tilting disc and the tilting disc. Today, the tilting-disc valves are the most common. These valves are typically made of pyrolytic carbon-coated metal parts embedded in a plastic ring. Table 21.1 lists approximate usage of mechanical valves through 2003. Images of the caged ball mechanical valve design, the non-tilting disc mechanical valve design, and the tilting disc mechanical valve design can be seen in references 5 and 6.

In the early days of mechanical valve replacement, the operative mortality rate was 15–20%. Today, that number is below 2%.[6] Complications arising from catastrophic failure of mechanical valves are extremely rare. Rather, most complications of mechanical valves are due to interactions between the patient's living tissue and the foreign material. Thrombosis is a common tissue–material interaction that can occur at the surface of the valve. Many valves

**Table 21.1.**   Approximate number of mechanical valves implanted in mitral and aortic positions valves in the United States through 2003.[6]

| Mechanical valve design | Mitral valve replacements | Aortic valve replacements |
|---|---|---|
| Caged-ball | ≈ 143,000 | ≈ 128,000 |
| Non-tilting disc | ≈ 18,300 | ≈ 1,500 |
| Tilting disc | ≈ 367,000 | ≈ 504,000 |

today are coated in pyrolytic carbon, which prevents thrombus formation. Yet even with advanced surface coatings, blood will still form clots on man-made devices. For this reason, patients with mechanical valves must take anti-coagulant (blood-thinning) drugs to reduce the risk of thrombosis. With anti-coagulant treatment the chance of thrombosis on a mechanical valve is less than 1% per year.[7]

While anti-coagulant therapy does reduce the many of the risks associated with mechanical valves, patients face other complications. Anti-coagulants reduce the body's ability to stop bleeding, so patients who suffer traumatic injuries have a risk of bleeding to death. Living with such a serious risk may force patients to change their lifestyle. For example, those taking anti-coagulants may choose not to participate in sports because of the risk for traumatic injury. This lowers the quality of life for that person and may make a mechanical valve an undesirable choice.

However, for some patients mechanical valves are the best choice due to their durability. Modern mechanical valves are designed to withstand wear from decades of use and rarely undergo catastrophic failure. These valves are expected to last longer than the life of the patient. For patients in their 30s and 40s, the durability of mechanical valves may make them the best option.

### 21.4.2. *Tissue valves*

After the development of mechanical valves, biological sources began to be used for replacement valves. The first biological valves were transplanted from human cadavers, and these are still the preferred source of valves for infants and young children. However, adult heart valve prostheses typically come from animal tissue. Processed porcine heart valves became available with the advancement of tissue preservation. These tissue-derived devices are called *bioprosthetics*.[8] Later, heart valves made from bovine pericardium were developed. In this case, the bovine pericardial tissue is formed by hand into a valve shape — the bovine valve itself is not used. Both porcine and bovine sources are currently used as bioprostheses. The Carpentier–Edwards supra-annular valve can be used in either mitral or aortic positions,[8] while the Carpentier–Edwards perimount pericardial valve is typically used in the aortic position.[8]

Tissue valves have an advantage over mechanical valves because most patients do not require long-term anti-coagulant therapy. This allows them to maintain an active lifestyle without risk of bleeding. However, tissue valves can fail structurally over time. The average time for reoperation or structural failure of mitral valve is approximately 14 years.[9] However, patient age is an important factor as valves wear out more rapidly in younger patients than in older patients.[7] A 70-year-old patient can expect a low chance of having the valve fail during his remaining life. However, a patient in her 30s, 40s, or 50s should reasonably expect the tissue valve prosthesis to require replacement. For this reason, tissue valves tend to be the implant of choice for older patients. This creates a dilemma for young patients who do not want to take anti-coagulants regularly, but face a significant risk of a tissue valve failing in their lifetime.

The major drawback with tissue valves is their limited lifetime. Replacing a heart valve requires open-heart surgery and comes with significant risks. As a patient gets older, the revision surgeries become more risky. The heart valves of the future need to last longer than current tissue valves so that revision surgeries are not required.

For all heart valves — mechanical or tissue — thromboembolism remains a problem. In this case, a blood clot comes loose and may block a critical vessel in the heart or brain, resulting in a heart attack or stroke. Thromboembolism rates currently average approximately 2–3% per year for mitral valves and 1.5–2% per year for aortic valves.[7]

## 21.5. A Third-Generation Biomaterials Solution

Advances in genomics and stem cell research are paving the new way to solutions of stenosis. Two next generation solutions are presented below: early detection and tissue engineering.

### 21.5.1. *Early detection of risk for valve dysfunction*

Early detection offers the best chance of survival for any patient. As stated before, the main causes of stenosis are rheumatic fever and congenital heart defects. Because congenital heart defects are often caused by genetic insufficiencies, those who have congenital heart disease may not know that they are at risk for valve failure. By understanding the genes that regulate healthy valve formation, tests can be created to see if patients are at risk. Insight into the stages of heart formation have shed light on what can go wrong if certain stages of development are interrupted. For example, it is known that disruption during development of the heart can cause stenosis. However, more work needs to be done to truly understand the genetic and molecular mechanisms for the formation of these body parts. Studies show that absence of the gene *Ptpn11* can cause stenosis of heart valves. Similarly, if the *Smad6* protein is not present in mice, mitral valves become thick and gelatinous.[10,11]

These are just two examples of biological factors that can lead to stenosis. In reality, there are many genes, proteins, and pathways of intracellular communication that dictate healthy development of heart valves in the fetus. When these are identified, patients can be tested for deficiencies of these indicators. If key genes or proteins are lacking, both the patient and doctor will be better informed of an elevated risk for valve failure, and preventative measures can be taken.

### 21.5.2. *Tissue engineering a replacement valve*

Another way to replace the stenotic valve is to grow a replacement valve from human cells which have been stimulated to differentiate into the proper phenotypes and 3D architecture. This is the goal of tissue engineering.

The conventional tissue engineering paradigm is to grow cells *in vitro* on a substrate called a scaffold. (See Chapters 2 and 9 for additional discussion of tissue-engineering technologies.) The scaffold is seeded with cells and placed in a bioreactor[12] which provides the right chemical and mechanical environment for the cells to grow. Proteins and other biological signals are fed into the bioreactor, causing differentiation into cells which will form a heart valve. As the cells grow, the scaffold dissolves, requiring the extra-cellular matrix of the proliferating cells to provide strength to the construct. The scaffold is then implanted into the patient and growth continues *in vivo*.

Complete regeneration of cardiac tissue such as heart valves is one of the loftiest goals in tissue engineering because it has the possibility to affect so many people. However, heart valves present their own unique challenges.[13] They are difficult tissues to regenerate because the valve must withstand cyclic biomechanical forces. Design of scaffolds for engineering of valves is an intensely studied area; the scaffold must provide a structure which allows for cell proliferation while withstanding the forces due to vascular pressure and pulsatile flow.

Materials used for creating heart value scaffolds include: synthetic polymers, natural polymers, and decellularized tissue.[14] The benefit of synthetic scaffolds is that they are reproducible and allow for structure and properties to be tuned. However, the challenges facing them include control over cell adhesion, resorption rate, and diffusion of cells into the scaffold.[15–17]

An example of a synthetic polymer scaffold for heart valve regeneration is the electrospun polydioxanone (PDO), as shown by Kalfa *et al.*[18] In this study PDO scaffolds were seeded with mesenchymal stem cells, labeled with quantum dots, and implanted into the right ventricular outflow tract of sheep. It was observed that the scaffold completely resorbed and was replaced by viable multi-layered endothelial tissue with ECM similar to that of native tissue. The quantum dots were used to verify that some of the original stem cells remained in the grafted transannular valve region. Diagnostic imaging, histology, and biochemical analysis all displayed encouraging evidence that the scaffold provided a viable tissue engineered heart valve. This study by Kalfa *et al.* demonstrates the multidisciplinary approach that must be taken to achieve a device that performs well *in vivo*.

Recently, synthetic polymers have been combined with ligands from the ECM to provide better cell attachment, differentiation, and scaffold remodeling.[19] This combines the benefits of synthetic and naturally derived polymers. Naturally derived polymer scaffolds come from a number of human, animal, plant, or bacterial sources[20] and are composed of purified polymers from the ECM (for instance collagen or fibrin) which is processed into a desired shape. The advantage is that extracellular ligands can provide cell-binding sites, which are critical for proliferation and differentiation. While the chemistry of natural polymers is not always well defined, their properties can be adjusted to give the desired response. An example is a study by Masters *et al.*, in which hyaluronic acid was used to encapsulate valvular interstitial cells.[21] Hyaluronic acid is a critical polysaccharide in the ECM of human tissue, including heart valves.[22] In this case, the hyaluronic acid was modified with methacrylate groups and crosslinked using UV polymerization. Cells remained viable after being encapsulated for six weeks in these gels and produced a significant amount of the ECM protein elastin.

Scaffolds can also be created by decellularizing tissue, which preserves the 3D architecture of the tissue. One advantage of these materials is that it maintains many of the ECM molecules, although some may be rendered inactive in the decellularization process. The disadvantages include the possibility of an immune response, potential change in properties upon decellularization, and risk of calcification. One example of this type of material is subintestinal submucosa (SIS), which has shown promise as a valve in animal models.[23] Decellularized porcine pericardium is another example. In a study by Tedder *et al.*,[24] the decellularized pericardium was coated with penta-galloyl glucose which allowed for collagen binding and crosslinking. The scaffolds from this study promoted fibroblast growth and scaffold remodeling. This gave the tissue construct correct anatomical structure, excellent biaxial mechanical properties, and support of *in vivo* cell proliferation without calcification.

The cells grown on scaffolds can come from either differentiated heart cells or stem cells. Differentiated heart cells can be extracted by biopsy of heart tissue; however they do not reproduce as easily as stem cells. While differentiated cells are "committed" to a specific lineage, stem cells are pluripotent — that is, they can become a number of different types of cells.[25] Stem cells can be found in bone marrow of adults. These cells are often used, because they can be easily accessed and cultured.

Another paradigm of tissue engineering is to bypass the *in vitro* growth step altogether. In this case, a scaffold is placed directly into the patient and "captures" the patient's own circulating cells. This is typically achieved by ligands attached to the scaffold which capture cells and induce differentiation. These cells are called circulating progenitor cells and share many traits with mesenchymal stem cells found in bone marrow. Benefits of using the patient's circulating cells is that the therapy can be applied quickly with little chance of tissue rejection. This method also has the possibility of treating the patient in a non-invasive procedure. This is a significant consideration, since current valve replacement surgery is extremely invasive and requires a long recovery time. Recent studies have shown that injection of a patient's own stem cells shortly after a heart attack can help to repair the wall of the heart.[26–28]

## 21.6. Ethical and Economic Considerations of Tissue Engineering of Mitral Valves

There are always ethical considerations regarding the manipulation of life — even on the cellular level — and tissue engineering is a field which often raises these issues.[29] Further ethical and economic implications are posed when applying tissue engineering in a clinical setting. The first consideration is whether or not tissue-engineered heart valves are safe to test in humans. As with most medical devices, animal testing must first show that results justify human testing. Liability for patient safety during these early stages of human testing must be established between doctors, biomedical engineers, and companies sponsoring the studies. Patient selection must be done to minimize risk, yet identify those patients who would benefit from the potentially risky early stages of testing.

The second ethical question, which also becomes an economic issue, is who shall receive a tissue-engineered heart valve? In developed countries with publicly funded health-care, treatments are allocated to patients on the basis of need. However, in the United States patients pay for their health-care and devices are allocated based on what the patient can pay. Should allocation of tissue-engineered heart valves be based on need or socio-economic status? These are long-standing medical ethical issues that have yet to be resolved.[30]

A lack of supply for tissue-engineered heart valves could create a "tragedy of the commons," causing the price of such heart valves to be extremely high. This would allow access to these replacement tissues only to the wealthy. Patients in developing countries would not have access to such treatment, despite the fact that early-stage testing of new medical devices are often done in these countries. Businesses producing engineered heart valves should be aware of these issues and seek to enter the market-place only if sufficient supply can be made available to a wide range of patients.


## 21.7. Recent Developments

Over the past ten years, surgical procedures for treating stenosis have progressed from valve *repair* to valve *replacement*. Valve repair offers a number of advantages over valve replacement including a reduction in operative mortality, decreased risk of infection and stroke, and greatly improved long-term survival. In the same period there has also been a progression towards the use of tissue valves over mechanical valves.[31] This increased popularity of tissue is because of improved hemodynamics and hemocompatibility. However, tissue valves still have serious shortcomings including calcification of valve leaflets, mechanical fatigue, and tissue failure.[32] Both mechanical and tissue valve replacements continue to suffer from numerous limitations. The most pressing issue is that both replacement and repair procedures require open-heart surgery.

Recent advances in endovascular stent technology have led to the development and successful implementation of a significantly less invasive procedure known as percutaneous valve replacement surgery (PVR). PVR is an endovascular technique which delivers a stent-graft construct, which contains an expandable heart valve prosthesis, to the defective valve. A wide variety of materials for PVR have been studied including synthetic and tissue-derived valve prosthetics or valved stents.[32] In a recent study, porcine heart valves and SIS were used to assemble a valved stent, followed by culture in a bioreactor with ovine endothelial cells and myofibroblasts.[33] After culturing the cells *in vitro*, the valved stent was percutaneously deployed in the pulmonary valve of sheep. Angiography of the valve after four weeks revealed successful implantation, excellent valvular mechanics, and smooth muscle growth around the stent. While PVR is still in the early stages of development it has been mainly restricted to replacement of the aortic valve. In the future this method has the potential to become the technique of choice for mitral or aortic valve replacement in the future.

## 21.8. Summary

There are approximately 250,000 heart valves replaced in the US every year.[2] Two options are currently available to patients: mechanical or tissue valves. A mechanical valve will rarely fail, but requires continual anti-coagulant medication. The second option is a tissue valve from an animal source. The tissue valves require no blood thinners, but have a projected survivability of approximately 14 years, depending on the age of the patient. This is especially problematic for young patients who do not wish to have multiple revision surgeries in their lifetime.

Tissue engineering offers a new method to solving the valve replacement problem. If a new valve can be regenerated from the patient's own cells, the patient could have a long-lasting valve with no risk of rejection that does not require immunosuppressant drugs. A better understanding of how cells develop into tissues, how the body heals itself, and how materials influence cell behavior will be crucial to this effort. Ultimately, regenerative medicine holds the promise for better health treatment for patients suffering from heart valve stenosis.

## References


1. Ahmad, F., Seidman, J.G. and Seidman, C.E. (2005). The genetic basis for cardiac remodeling, *Ann. Rev. Genomics Hum. Gene*, **6**, 185–216.

2. Hench, L.L. and Jones, J.R. (2005). *Biomaterials, Artificial Organs, and Tissue Engineering*, Woodhead Publishing Ltd, Cambridge, UK.

3. National Institute of Health, Medline Plus. (2010). *Heart valves — anterior view*. [Online]. Available at: http://www.nlm.nih.gov/medlineplus/ency/imagepages/18092.htm. [Accessed 8 November 2010].

4. Mayo Clinic (2006). *Mitral stenosis — diagnosis and treatment options at Mayo Clinic*. [Online]. Available at: http://www.mayoclinic.org/mitral-valve-disease/mitral-stenosis.html. [Accessed 8 November 2010].

5. National Institutes of Health (2006). *MedlinePlus medical encyclopedia: mitral stenosis*. [Online]. Available at: http://www.nlm.nih.gov/medlineplus/ency/imagepages/18147.htm. [Accessed 8 November 2010].

6. Gott, V.L., Alejo, D.E. and Cameron, D.E. (2003). Mechanical heart valves: 50 years of evolution, *Ann. Thorac. Surg.*, **76**, S2230–S2239.

7. Starr, A., Fessler, C.L., Grunkemeier, G., *et al.* (2002). Heart valve replacement surgery: past, present, future, *Clin. Exp. Pharm. Phys.*, **29**, 735–738.

8. Grunkemeier, G., Li, H., Naftel, D.C., *et al.* (2000). Long-term performance of heart valve prostheses, *Curr. Probl. Cardiol.*, **25**, 76–154.

9. Khan, S.S., Chaux, A., Blanche, C., *et al.* (1998). A 20-year experience with the hancock porcine xenograft in the elderly, *Ann. Thorac. Surg.*, **66**, S35–S39.

10. Srivastava, D. (2006). Genetic regulation of cardiogenesis and heart disease, *Ann. Rev. Pathol. Mech. Dis.*, **1**, 199–213.

11. Srivastava, D. and Olson, E.N. (2000). A genetic blueprint for cardiac development, *Nature*, **407**, 221–226.

12. Karim, N., Golz, K. and Bader A. (2006). The cardiovascular tissue-reactor: a novel device for the engineering of heart valves, *Artif. Organs,* **30**, 809–814.

13. Sack, M.S. Schoen F.J. and Mayer J.E. (2009). Bioengineering challenges for heart valve tissue engineering, *Annu. Rev. Biomed. Eng.*, **11**, 289–313.

14. Dohmen, P.M. and Konertz, W. (2009). Tissue engineered heart valve scaffolds, *Ann. Thoracic Cardiovasc. Surg.*, **15**, 362–367.

15. Mendelson, K. and Schoen, F.J. (2006). Heart valve tissue engineering: concepts, approaches, progress and challenges, *Ann. Biomed. Eng.*, **34**, 1799–1819.

16. Polak, J. and Hench L.L. (2005). Stem cells and tissue engineering: past, present, and future, *Gene Ther.*, **12**, 1725–1733.

17. Neuenschwander, S. and Hoerstrup, S.P. (2004). Heart valve tissue engineering, *Transplant Immunol.*, **12**, 359–365.

18. Kalfa, D., Bel A., Chen-Tournoux, A., *et al.* (2005). A polydioxanone electrospun valved patch to replace the right ventricular outflow tract in a growing lamb model, *Biomaterials*, **31**, 4056–4063.

19. Lutolf, M.P. and Hubbell, J.A. (2005). Synthetic biomaterials as instructive extracellular microenvironments for morphogenesis and tissue engineering, *Nat. Biotechnol.*, **23**, 47–55.

20. Cebotari S., Lichtenberg A., Tudorache I., *et al.* (2006). Clinical application of tissue engineered human heart valves using autologous progenitor cells, *Circulation*, **114**, I132–I137.

21. Masters, K.S., Shah, D.N., Leinwand, L.A., *et al.* (2005). Crosslinked hyaluronan scaffolds as a biologically active carrier for valvular interstitial cells, *Biomaterials*, **26**, 2517–2525.

22. Grande-Allen, K.J. Calabro, A., Gupta, V. *et al.* (2004). Glycosaminoglycans and proteoglycans in normal mitral valve leaflets and chordae: Association with Regions of tensile and compressive loading, *Glycobiology*, **14**, 621–633.

23. Pavcnik, D., Uchida, B.T., Timmermans, H.A., *et al.* (2002). Percutaneous bioprosthetic venous valve: a long-term study in sheep, *J. Vasc. Surg.*, **35**, 598–602.

24. Tedder, M.E., Liao. J., Weed, B., *et al.* (2009). Stabilized collagen scaffolds for heart valve tissue engineering, *Tissue Eng. A*, **15**, 1257–1268.

25. Engler A.J., Sen, S., Sweeney, H.L., *et al.* (2006). Matrix elasticity directs stem cell lineage specification, *Cell*, **126**, 677–689.

26. Schachinger, V., Assmus, B., Britten, M.B., *et al.* (2004). Transplantation of progenitor cells and regeneration enhancement in acute myocardial infarction: final one-year results of the TOPCARE-AMI Trial, *J. Am. Coll. Cardio.*, **44**, 1690–1699.

27. Wollert, K.C., Meyer, G.P., Lotz, J., *et al.* (2004). Intracoronary autologous bone-marrow cell transfer after myocardial infarction: the BOOST randomised controlled clinical trial, *Lancet*, **364**, 141–148.

28. Fernandez-Aviles F., San Román J.A., García-Frade J., *et al.* (2004). Experimental and clinical regenerative capability of human bone marrow cells after myocardial infarction, *Circ. Res.*, **95**, 742–748.

29. Hench, L.L. (2001). *Science, Faith, and Ethics*, Imperial College Press, London, pp. 132–151.

30. McConnel, C. and Turner, L. (2005). Medicine, ageing, and human longevity, *EMBO Reports*, **6**, S59–S62.

31. Gammie, J.S., Sheng, S., Griffith, B.P., *et al.* (2009). Trends in mitral valve surgery in the united states: results from the society of thoracic surgeons adult cardiac database, *Ann. Thorac. Surg.*, **87**, 1431–1439.

32. Kidane, A.G., Burriesci, G., Cornejo, P., *et al.* (2008). Current developments and future prospects for heart valve replacement therapy, *J. Biomed. Mater. Res.*, **88B**, 290–303.

33. Metzner, A., Stock, U.A., Iino, K., *et al.* (2010). Percutaneous pulmonary valve replacement: autologous tissue engineered valved stents, *Cardiovasc. Res.*, **88**, 453–461.

# Prosthetic Vascular Grafts for Treatment of Peripheral Arterial Disease

CHAPTER 22

Margo Ellis

## 22.1. Introduction

Peripheral arterial disease (PAD) is a progressive and destructive condition that affects the arterial wall of the aorta and its branches.[1–4] The aorta is the main artery that spans the trunk of the body and branches into the common iliac arteries near the umbilicus (Fig. 22.1). The common iliac arteries then branch into the external and internal iliac arteries. At the groin, the external iliac arteries become the femoral arteries that feed oxygenated blood to the lower extremities.[5] PAD occurs when a portion of the artery becomes blocked or narrowed by the build-up of fatty plaque deposits (atherosclerosis). Common symptoms of blockage include hip, buttock or thigh pain that worsens during prolonged use of limbs (intermittent claudication), cold feet and legs, and nerve damage. Left untreated, this condition can result in gangrene and possible amputation of the affected limbs.[1–4]

PAD affects 4.5–7.6 million Americans and by the year 2050 as many as 16 million adults in the United States aged 65 and over are expected to be diagnosed with PAD.[2] The risk factors for PAD increase with age and include type II diabetes, smoking, hypertension, and high cholesterol.[1–4]

There are several noninvasive techniques for the diagnosis of PAD including the Ankle-Brachial Index (ABI), the World Health Organization/Rose questionnaire on intermittent claudication, the Edinburgh Claudication Questionnaire, Doppler ultrasound examination and the Visual Flush Test.[1–4]

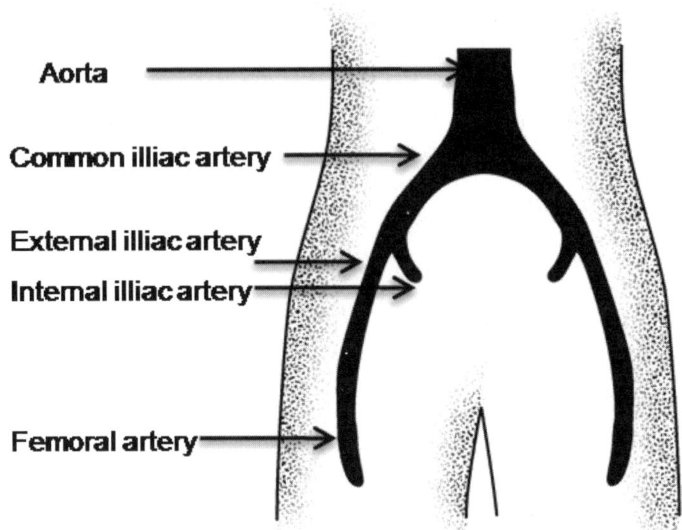



**Figure 22.1.**   Schematic of the aorta and the iliac arteries.

## 22.2. Present Medical and Biomaterials Technology and Limitations

Methods of treatment for PAD depend on the severity of the condition and range from simple lifestyle changes to surgery. Early diagnosis and risk factor management is crucial to reducing the number of cases requiring surgery. According to the American Heart Association, the American College of Cardiology, and the Society for Vascular Surgery, life-style change is the first line of treatment to halt the progression of PAD.[4] For smokers, the cessation of smoking is the cornerstone of this treatment. Exercise has been shown to reduce leg pain and increase endurance, while diet modifications may help to lower blood pressure and cholesterol.[2,4] Drug therapy may include statin drugs to lower cholesterol, antihypertensive medication to lower blood pressure, antiplatelet agents to prevent blood clots, and medications such as Cilostazol to improve exercise tolerance.[2,4] The effects on PAD of exercise; smoking cessation; control of high blood pressure; and high cholesterol and drug therapy are summarized in Table 22.1.

**Table 22.1.**  Summary of PAD treatments and their effects (adapted from[3]).

| Component | Effect on mortality or vascular events | Effect on PAD |
| --- | --- | --- |
| Exercise | Reduction in risk of cardiovascular mortality | Significant improvement in walking distance |
| Smoking cessation | Significant reduction in mortality | Significant reduction in pain at rest |
| Cholesterol control | Significant reduction in major vascular events | Slight increase in walking distance |
| Blood pressure control | Significant reduction in mortality | Unknown |
| Diabetes control | Significant reduction in mortality | Significant reduction in lower limb amputation |
| Antiplatelet agent | Significant reduction in vascular events | Possible increase in walking distance |
| Cilostazol | Unknown | Small increase in walking distance |

Percutaneous intervention (minimally invasive) or invasive surgery is reserved for patients whose symptoms are not resolved with life-style changes, who have significant arterial blockages, or who risk losing a limb due to tissue loss or gangrene. Surgical revascularization options may include surgical removal of plaque (endarterectomy), widening of the artery with or without stenting (angioplasty), or bypassing the affected blockage (invasive bypass surgery).[4]

Bypass grafting has higher patency rates than do percutaneous techniques but the severity of complications is higher and is therefore contraindicated in high-risk patients.[6,7] Table 22.2 summarizes the patency and mortality rates of revascularization options.[6] Bypass options include axillofemoral, iliofemoral, femorofemoral, obturator foramen, thoracic aortobifemoral, and aortobifemoral bypass grafting.[6,7]

The aortobifemoral bypass graft is usually considered to be the gold-standard procedure for revascularization.[7] Autologous arterial or vein grafts are the preferred choice for bypass graft procedures.

**Table 22.2.** PAD Surgical options and related mortality and patency rates (adapted from[6]).

| | Surgical options for the management of PAD | | |
|---|---|---|---|
| Procedure | Mortality (%) | 5-year patency (%) | Comments |
|---|---|---|---|
| Aortoiliac endarterectomy | 1–7 | 72–95 | Usually involves stenting |
| Aortobifemoral bypass graft | 1 | 85 | The gold standard of bypass techniques; potential sexual dysfunction in males |
| Iliofemoral bypass graft | 5 | 83 | Suitable for long external iliac artery occlusions |
| Axillofemoral bypass graft | 3–13 | 79 | Long bypass graft |
| Femorofemoral bypass graft | 1.3–1.5 | 60–92 | Groin incision only |
| Obturator foramen bypass graft | 14 | 60 | Rare application for routine procedure, used for infection, crush injuries, and malignancies |
| Thoracic aortobifemoral bypass graft | 4 | 79 | Typically used for abdominal redo or in sepsis cases |
| Hybrid open groin approach | <1 | 80 | Minimally invasive and allows for stent grafting |

However, a percentage of patients with PAD do not have veins or arteries that are suitable for grafting. Therefore, the bypass procedure is often performed using an engineered graft composed of natural or synthetic materials including: collagen; polyester; Dacron®; polytetrafluoroethylene (PTFE) or expanded PTFE ($\varepsilon$-PTFE) materials such as GoreTex®.[8–14] Dacron® emerged on the market in 1957 and was used to fabricate arterial prostheses. Complications and limitations of vascular graft bypass surgery and prosthetic vascular implants exist and are discussed below.

(See reference 8 for images and detailed explanations of the various arterial grafts.)

## 22.3. Complications and Limitations

### 22.3.1. *Graft infection*

The rate of infection for vascular prosthetic grafts has been reported to be as low as 1–3%[15] and as high as 6%.[16] Although prosthetic graft infection is uncommon, the associated mortality rates range from 25–88% of cases and amputation rates from 5–25%.[15,16] Contamination during the surgical procedure is the leading cause of graft infection. Infection may also be related to the type of prosthetic graft material used. Studies show that Dacron® grafts may show more resistance to treatment when compared to PTFE grafts.[15]

### 22.3.2. *Anastomotic intimal hyperplasia*

A thickening of the luminal wall due to cellular proliferation and extracellular matrix deposition at the site of the graft–artery interface (anastomosis) causes anastomotic stenosis and contributes to prosthetic arterial graft failure.[13,14] Anastomotic stenosis interferes with flow dynamics through the prosthetic graft leading to thrombosis.[14] Although autologous grafts are typically preferred to prosthetic grafts, autologous vein grafts can also develop intimal thickening.[17]

### 22.3.3. *Other complications*

Graft dilation has been found to occur for both Dacron® and PTFE type grafts. However, according to the study of explanted graft material, graft dilation did not appear to be directly linked to graft structural failure.[12] In addition, false aneurysms were not found to be due to graft failure but to the suture lines pulling through the natural arteries at the anastomosis site and from compliance mismatch between the graft and the natural vessel.[9,12]

### 22.3.4. *Limits to prosthetic vascular grafts*

Although vascular grafts are successfully used to bypass blockages in large-diameter blood vessels (>5–6 mm), thrombus formation in small-diameter grafts (4–5 mm) remains a problem.[10,11]


## 22.4. Summary of New Technology

Recent advances in the design of commercially available prosthetic vascular grafts have addressed some of the concerns associated with prosthetic grafts. Advances include long-term resistance to dilation, resistance to spread of infection, kink resistance, heparin bonding, type I collagen coating, external surface promotion of tissue in-growth, smooth internal surface to promote blood flow and minimize thrombus accumulation, increased burst strength, radial tensile strength, and suture pull-out strength.


### 22.4.1. *Luminal surface modification*

Current research approaches for improving the patency of prosthetic vascular grafts are varied. They include various modifications of the prosthetic graft's luminal surface to allow complete endothelial cell proliferation thus promoting a non-thrombogenic surface character.[11,14] Other approaches focus on bonding heparin to the luminal surface of εPTFE grafts, creating a bioactive surface that inhibits thrombosis formation.[8,18] One of the most recent advances for small-diameter prosthetic grafts used for treatment below the knee is the commercially available Gore® Propaten® Vascular Graft. In a proprietary design, one end of a heparin (anticoagulant) molecule is bonded to the internal luminal surface of a PTFE vascular graft. The other end of the molecule, containing the active site, is free to interact with the blood where it binds to antithrombin (AT). AT can then bind to thrombin (T) forming a neutral AT-T complex which prevents the conversion of fibrinogen to fibrin. Fibrin is the substance that contributes to the formation of thrombus. The neutral AT-T complex is released and the heparin-active site is free to react with another AT molecule. The primary patency rates for the heparin modified Gore® Propaten® vascular graft used in bypass below the knee was 72% after two years, which indicates a major improvement over conventional prosthetic grafts.[8]

## 22.4.2. *Endothelial cell seeding on prosthetic vascular grafts*

A monolayer of endothelial cells (ECs) lines the luminal wall of a normal blood vessel. This endothelial layer acts as a bio-regulator and plays a role in structural integrity and blood flow control, acts as a permeable barrier and inhibits thrombogenesis.[19] Ongoing research continues to search for solutions to prevent thrombus formation in small-diameter prosthetic vascular grafts. The importance of this research is highlighted by the 1995 comparison of three-year primary patency rates for autologous vein of 84%, *versus* a 35% patency rate for prosthetic grafts used for below-the-knee bypass.[20]

Different seeding methods have been attempted by various researchers.[19,20,21] The main source for ECs is from the donor's vein, artery, omentum, and subcutaneous fat.[19] There are two basic types of seeding techniques: two-stage and single-stage. Two-stage seeding involves the extraction of ECs from a vein or an artery. These cells then undergo an extended cell culture period (2–4 weeks) in order to increase the number of seeding cells.[19] Although some cells are lost on exposure to pulsatile blood flow, this technique results in a high density of seeded ECs remaining on the luminal surface and has led to excellent patency rates in animals. However, animals have an inherent ability to self-endothelialize, whereas human lack this ability. Therefore, clinical trials in humans have been less successful. The major drawback for clinical application of this method is the prolonged cell-culture time and increased cost related to the maintenance of a cell-culture laboratory.[19]

In comparison, single-stage seeding uses ECs harvested from microvascular sources such as the omentum and subcutaneous fat. These cells are used to line the graft at the time of the surgery.

Animal trials for single-stage seeding showed excellent results and improved patency rates.[21] Initial attempts of single-stage seeding in human clinical trials resulted in low EC density. This was determined to be due to the low number of ECs that could be collected from veins. A larger number of ECs are needed for single-stage seeding. Higher densities of ECs can be found in subcutaneous fat; however, use of these cells in the single-stage seeding technique showed inferior results when compared to two-stage seeding.[19,21] Harvesting cells from omental fat requires a laparotomy which increases risk to the patient. However, microvascular endothelial cells (MVECs) obtained from omental fat are thought to be a better model for single-stage seeding than animal cells, or ECs derived from the human umbilical vein.[21] Omental fat MVEC cells have been found to have a significantly better surface retention when seeded onto εPTFE grafts. However, clinical trials for the single-stage seeding technique have not shown high patency rates when compared to trials for two-stage seeding.[19]

Another approach for improving seeding efficiency is the use of genetically modified endothelial cells. The advantage to this approach is that the modified cells readily repopulate and secrete anticoagulant peptides, thus reducing thrombosis. Researchers have been successful with *in vitro* and *in vivo* animal studies using these cells. However, unmodified cells have performed significantly better than genetically modified cells upon exposure to flow when placed *in vivo*.[19]

### 22.4.3. *Tissue engineering*

Tissue engineering of small-diameter blood vessels presents many challenges (see Chapter 23 for additional discussion of this topic). The mechanical properties of the engineered vessel must be comparable to the natural vessel.[22] In addition, the luminal surface of the graft must have the antithrombogenic properties of ECs.[19] The ideal construct would be composed of a biodegradable scaffold with biologically active ECs lining the lumen, a vascular smooth muscle cell (SMC) middle layer, and an external fibroblast layer.[23]

Ideally, cell adhesion, function, and growth would be supported by the construct during the generation of new ECs and SMCs to produce a new basal lamina. The end goal is to obtain a fully functional vessel *in vivo* with properties comparable to a natural vessel.[23]

Currently, the three types of tissue-engineered grafts that are available for vascular vessel replacement are the biodegradable, bioresistant, and decellularized tubular scaffolds. The inner surface (lumen) of these scaffolds is lined with smooth muscle cells, chondrocytes, myofibroblasts or extracellular basement membrane. The surface is then seeded with ECs and cultured in a bioreactor or perfusion culture chamber.

The first tissue-engineered vascular vessels used bovine ECs, SMCs, and fibroblasts on a permanent Dacron® scaffold. However, the graft was unable to withstand *in situ* conditions.[19] Subsequently, engineered vascular grafts have been modified to withstand the high venous and arterial pressures found *in situ*. One such device is composed of a marine source collagen scaffold with a polymer coating of poly(lactide-co-glycolide) (PLGA) fiber deposited by electrospinning.[24] *In vitro* co-culture of ECs and SMCs on the collagen/PLGA scaffold are performed in a pulsatile perfusion bioreactor designed to mimic the *in vivo* environment. Cellular alignment, enhanced development of vascular ECs, and the retention of differentiated cell phenotype were achieved using the pulsatile perfusion bioreactor.[24] Improvements have also been made in cell-seeding methods. Recently, a new seeding device and technique has been used that imparts the effects of vacuum, centrifugal force and flow on the structure, which allows for rapid seeding of both surface and bulk scaffold material.[25] These new techniques in tissue engineering represent a significant advancement in the progress towards the development of a structurally sound, thrombus resistant, biologically functional, engineered vascular replacement.


## 22.5. Recent Developments

Cell-based therapeutic techniques have recently gained rapidly growing acceptance for clinical treatment of patients suffering from

PAD. Currently, there are three major modes of cell therapy for treating PAD undergoing clinical trials: direct intramuscular injection of bone marrow derived mononuclear cells, direct intramuscular injection of cytokine-mobilized peripheral blood mononuclear cells, and cytokine-induced mobilization of progenitor cells.[26] Mheid and Quyyumi review several clinical studies in which each of these three approaches was evaluated in patients with various etiologies of PAD.[26] Cell-based therapies are especially advantageous for use in patients with contraindications for traditional revascularization techniques, particularly those at high risk of infection.

Evidence for the use of cell-based therapies for PAD has been encouraging, with the considerable therapeutic advances and recent clinical trials showing positive outcomes. In a review by Fadini *et al.*, analysis of 37 clinical trials found that intramuscular injection of autologous bone marrow cells improved limb ischemia, pain, healing, and various subjective and objective indices for evaluating arterial remodeling and repair.[27] This review also showed that intramuscular bone marrow cell therapy were well tolerated and safe, although it is mentioned that further large-scale studies must be conducted in order to verify these findings. Liu *et al.* have demonstrated a novel method which utilized intra-arterial transplantation of bone marrow cells (BMCs) in a mouse model hind-limb which allowed the fate of the injected cells to be determined.[28] The results from this study revealed that after injection, adult BMCs promoted angiogenesis by the differentiation of BMCs into neovasculature and also skeletal myofibers. The regeneration of skeletal muscle provides increased blood flow but also may provide benefits to patients who have suffered from muscular atrophy. This study is an example of the intrinsic potential benefits that cell therapy offers over traditional surgical revascularization intervention.

Diabetic patients suffer a greatly increased risk for PAD, commonly presenting symptoms of critical limb ischemia, which potentially can to lead to limb amputation. It has been shown that diabetic patients who have undergone revascularization procedures (percutaneous or open) have poorer outcomes, with increased morbidity and mortality compared to non-diabetic patients, including a higher rate of amputation.[29] Cell therapy offers a potential technique which could be developed to target revascularization of diabetic patients with PAD. Again, cell therapy also has the advantage of reduced morbidity, faster healing, and decreased risk of infection — all of which are important considerations for the PAD patient population.

The use of stem cell and progenitor cells has recently been explored as a potential new therapeutic approach for inducing angiogenesis in PAD patients. The use of autologous cell therapies has been shown to be effective, feasible, and safe in preliminary studies.[30] It has also been found that a combination therapy of both stem and progenitor cells, for example mesenchymal stem cells and endothelial progenitor cells, is effective and possibly preferred for the treatment of limb ischemia, such as that observed in patients with PAD.[31] Huang *et al.* recently examined the use of embryonic stem cells (ESCs) and embryonic stem cell-derived endothelial cells (ESC-ECs) for the localization, function, viability, and the propensity to form neovasculature in ischemic hind-limbs of mice.[32] The application of ESCs is advantageous in treating patients suffering from both PAD and cardiovascular disease, since autologous adult stem cells are fewer in number and have reduced function in such patients. These studies demonstrate the rapid advancement of cell therapy for treating PAD, although they also illustrate the wide spectrum of potential therapeutic candidates and the lack of an established cell source. Not only is there a need for determining the optimal cell source, it is also very important to develop new cell-delivery techniques and large participant multi-center clinical trials.

## 22.6. Ethical Issues Related to Treatment of Preripheral Arterial Disease

Since the degree of PAD progression is directly related to early diagnosis and treatment, the role of the health-care system and the patient must be considered. Ethically, the role of the physician is to be informed about the symptoms, diagnosis and treatment of PAD and to make this information available to the patient. Once the physician has made clear the relationship between life-style choices and PAD progression, the patient has the ethical responsibility to consider a change in life-style in order to avoid progression of the disease. Ideally, the physician will guide the patient through the decision-making process if the patient's condition has already progressed to the stage where surgical intervention becomes necessary. Some patients may struggle with the ethical implications of using a prosthetic device to extend life or to improve the quality of life. Others may struggle in making the decision to accept the risks associated with surgery and with the device in order to gain a potential improvement in their quality of life. Ethical dilemmas also arise when patients are not afforded the same options for treatment. For example, those without insurance are not likely to visit a physician until their symptoms are severe. Since early diagnosis and treatment are essential to halt the progression of PAD, these patients are usually at greater risk by the time they seek treatment.

Ethical issues of stem cell research in tissue-engineered materials present concerns. However, the harvest of stem cells from bone marrow and umbilical cord reduces the ethical implications. On the other hand, gene therapy, when used in tissue engineering, does pose a debatable issue.

## 22.7. Economic Cost

The incidence of PAD is relatively common among those aged 65 and older. However, the disease is commonly undiagnosed because most cases of PAD are asymptomatic or because leg pain symptoms are attributed to other diseases of the elderly.[33] PAD is a progressive disease that leads to disability and a decreased quality of life if not detected and treated early. The economic burden for the patient and for society rises as the disease progresses.[33–35] For example, it may cost about US$650 per year to treat a case of PAD with intermittent claudication, compared to US$9,353 per year to treat a case that has progressed to critical leg ischemia.[35]

The economic burden of PAD originates from both cost of health-care and from indirect costs related to the loss of productivity of the patient. A 2004 study reported that more than 400,000 hospitalizations occur annually in the United States for treatment of PAD.[33] This number demonstrates the magnitude of the economic burden of PAD with respect to health-care costs and lost productivity. Total hospitalization costs vary depending on the type of procedure performed and on related complications.[16]

## 22.8. Summary

PAD is a progressive condition affecting 4.5–7.6 million Americans aged 65 and over. Many cases of PAD go undetected due to lack of symptoms or to leg pain symptoms incorrectly attributed to other causes. Simple methods such as the Ankle-Brachial Index can be routinely used for the early diagnosis of PAD. Cost of treatment for PAD increases proportionally with the degree of progression. Health-care cost could be greatly reduced if PAD were diagnosed and treated early. Improvement in surgical skills and in prosthetic vascular grafts has made revascularization possible in advanced PAD cases. Ongoing research is striving to make small-diameter, tissue-engineered prosthetic vascular grafts available for below-the-knee bypass procedures.

## References


1. Lefebvre, K.M. (2006). Research corner. Outcomes measures in cardiopulmonary physical therapy: Focus on the ankle brachial index (ABI), *Cardiopulm. Phys. Ther. J.*, **17**, 134–137.
2. Oka, R.K. (2006). Peripheral arterial Disease in older adults: Management of cardiovascular disease risk factors, *J. Cardiovasc. Nurs.*, **21**, S15–S20.
3. Thom, T., Haase, N., Rosamond, W., *et al.* (2006). Heart disease and stroke statistics — 2006 update. A report from the American Heart Association statistics committee and stroke statistics subcommittee, *Circulation*, **113**, E85–E151.

4.  Kolh, P. (2010). Improving quality of life in patients with peripheral arterial disease: an important goal, *Eur. J. Vasc. Endovasc. Surg.*, **40**, 626–627.

5.  Crouch, J.E. and McClintic, J.R. (1976). *Human Anatomy and Physiology*, 2nd ed., Wiley & Sons, New York, NY.

6.  Ouriel, K. (2001). Peripheral artery disease, *Lancet*, **358**, 1257–1264.

7.  Harthun, N.L., Cheanvechai, V., Graham, L.M., *et al.* (2004). Arterial occlusive disease of the lower extremities: do women differ from men in occurrence of risk factors and response to invasive treatment?, *J. Thorac. Cardiovasc. Surg.*, **127**, 318–321.

8.  Gore Propaten Literature Summary: A Multi-Year and Multi-Center Track Record in Below-Knee Bypasses (2010). *W. L. Gore & Associates.* [Online]. Available at: http://www.goremedical.com/propaten/. [Accessed 9 November 2010].

9.  Cintora, I., Pearce, D.E. and Cannon, J.A. (1988). A clinical survey of aortobifemoral bypass using 2 inherently different graft types, *Ann. Surg.*, **208**, 625–630.

10.  Huynh, T., Abraham, G., Murray, J., *et al.* (1999). Remodeling of an acellular collagen graft into a physiologically responsive neovessel, *Nat. Biotechnol.*, **17**, 1083–1086.

11.  Wissink, M.J.B., Beernink, R., Poot, A.A., *et al.* (2000). Improved endothelialization of vascular grafts by local release of growth factor from heparinized collagen matrices, *J. Controlled Release*, **64**, 103–114.

12.  Veith, F.J., Gupta, S.K., Ascer, E., *et al.* (1986). 6-year prospective multicenter randomized comparison of autologous saphenous-vein and expanded polytetrafluroethylene grafts in infrainguinal arterial reconstructions, *J. Vasc. Surg.*, **3**, 104–111.

13.  Cabrera Fisher, E.I., Bia Santana, D., Cassanello, G.L., *et al.* (2005). Reduced elastic mismatch achieved by interposing vein cuff in expanded polytetrafluoroethylene femoral bypass decreases intimal hyperplasia, *Artif. Organs*, **29**, 122–130.

14.  Stone, D., Phaneuf, M., Sivamurthy, N., *et al.* (2002). A biologically active VEGF construct *in vitro*: implications for bioengineering-improved prosthetic vascular grafts, *J. Biomed. Mater. Res.*, **59**, 160–165.

15. Swain, T.W. III, Calligaro, K.D. and Dougherty, M.D. (2004). Management of infected aortic prosthetic grafts, *Vasc. Endovasc. Surg.*, **38**, 75–82.

16. Orton, D.F., LeVeen, R.F, Saigh, J.A., *et al.* (2000). Aortic prosthetic graft infections: radiologic manifestations and implications for management, *Radiographics*, **2**, 977–993.

17. Newby, A.C. and Zaltsman, A.B. (2000). Molecular mechanisms in intimal hyperplasia, *J. Pathol.*, **190**, 300–309.

18. Laredo, J., Xue, L., Husak, V.A., *et al.* (2004). Silyl-heparin bonding improves the patency and *in vivo* thromboresistance of carbon-coated polytetrafluoroethylene vascular grafts, *J. Vasc. Surg.*, **5**, 1059–1065.

19. Seifalian, A.M., Tiwari, A., Hamilton, G., *et al.* (2002). Improving the clinical patency of prosthetic vascular and coronary bypass grafts: the role of seeding and tissue engineering, *Artif. Organs*, **26**, 307– 320.

20. Smyth, J.V., Welch, M., Carr, H.M.H., *et al.* (1995). Fibrinolysis profiles and platelet activation after endothelial cell seeding of prosthetic vascular grafts, *Ann. Vasc. Surg.*, **9**, 542–546.

21. Salacinski, H.J., Punshon, G., Krijgsman, B., *et al.* (2001). A hybrid compliant vascular graft seeded with microvascular endothelial cells extracted from human omentum, *Artif. Organs*, **25**, 974–982.

22. Stock, U.A., Vacanti, J.P., Mayer, J.E. Jr, *et al.* (2002). Tissue engineering of heart valves — current aspects, *Thorac. Cardiov. Surg.*, **50**, 184– 193.

23. Krenning, G., Dankers, P.Y.W., Jovanovic, D., *et al.* (2007). Efficient differentiation of Cd14(+) monocytic cells into endothelial cells on degradable biomaterials, *biomaterials*, **28**, 1470–1479.

24. Jeong, S.I., Kim, S.Y., Cho, S.K., *et al.* (2007). Tissue-engineered vascular grafts composed of marine collagen and PLGA fibers using pulsatile perfusion bioreactors, *Biomaterials*, **29**, 1115–1122.

25. Soletti, L., Nieponice, A., Guan, J., *et al.* (2006). A seeding device for tissue engineered tubular structures, *Biomaterials*, **27**, 4863–4870.

26. Mheid, I.A. and Quyyumi, A.A. (2009). Cell therapy in peripheral arterial disease, *Angiology*, **59**, 705–716.

27. Fadini, G.P., Agostini, C. and Avogaro, A. (2010). Autologous stem cell therapy for peripheral arterial disease: meta-analysis and systematic review of the literature, *Artherosclerosis*, **209**, 10–17.

28. Liu, Q., Chen, Z., Terry, T., *et al.* (2009). Intra-arterial transplantation of adult bone marrow cells restore blood flow and regenerates skeletal muscle in ischemic limbs, *Vasc. Endovasc. Surg.*, **43**, 433–443.

29. Lawall, H., Bramlage, P, and Amann, B. (2010). Stem cell and progenitor cell therapy in peripheral artery disease, *Thromb. Haemostasis*, **103**, 696–708.

30. Lasala, G.P., Silva, J.A, Gardner, P.A., *et al.* (2010). Combination stem cell therapy for treatment of severe limb ischemia: safety and efficacy analysis, *Angiology*, **61**, 551–556.

31. Huang, N.F., Niiyama, H.M., Peter, C., *et al.* (2010). Embryonic stem cell-derived endothelial cells engraft into the ischemic hindlimb and restore perfusion, *Arterioscl. Thromb. Vas.*, **30**, 984–991.

32. Brevetti, G. and Chiariello, M. (2004). Peripheral arterial disease: the magnitude of the problem and its socioeconomic impact, *Current Drug Targets — Cardiovasc. Hematol. Disorders*, **4**, 199–208.

33. Jansen, R.M.G., de Vries, S.O., Cullen, K.A., *et al.* (1998). Cost-identification analysis of revascularization procedures on patients with peripheral arterial occlusive disease, *J. Vasc. Surg.*, **28**, 617–623.

34. Kugler, C.F.A. and Rudofsky, G. (2003). The challenges of treating peripheral arterial disease, *Vasc. Med.*, **8**, 109–114.

# Small-Diameter Vascular Grafts

CHAPTER **23**

Kyle Marr

## 23.1. Introduction and Statement of Clinical Need

In the human anatomy, blood vessels function to transport blood and oxygen as well as vital nutrients to organs and tissues throughout the body. Atherosclerosis is a disease that affects the blood vessels and is a leading cause of morbidity and mortality worldwide, particularly for vessels of small diameter (<6 mm).[1] Atherosclerosis is a chronic inflammatory response in the walls of the arteries and is commonly referred to as a "hardening" of the arteries. There are two main problems in atherosclerosis that account for such a large death rate. First, there is plaque build-up through deposits on the vessel wall, which can lead to a narrowing of the artery, known as stenosis. Occlusion of the artery hinders proper blood flow and can cause tissue failure and organ damage.[2,3] Sometimes the plaque will rupture, causing coronary thrombosis which can lead to a myocardial infarction, more commonly known as a heart attack. The second problem with atherosclerosis is that the vessel wall can balloon out causing an aneurysm.

Replacement of damaged arteries is a common method of treatment in cardiovascular diseases. In the United States, more than 1.4 million arterial bypass operations are performed yearly.[2] Autologous grafts, using saphenous or mammary veins continue to remain the "gold standard" in coronary bypasses. However, often a patient's own blood vessels may not be available for harvesting due to vascular disease or due to a previous vessel harvest. Additionally, vascular grafts are necessary for aneurysm repair and organ transplantation, and there is a rather high cost associated with the harvest and grafting of autologous vessels.[2] Thus, there continues to be a pressing need for the development of a functional and readily available small-diameter vascular graft. Many surgeons use polymer-based synthetic vascular grafts when the patient's own vessels are not available. Polyethylene terephthalate (Dacron®), expanded polytetrafluoroethylene (ePTFE), and other polyurethanes are several of the polymers used clinically for synthetic vascular grafts to date. While these synthetic-based grafts perform rather well for large-diameter vascular grafts, they fail for smaller diameter (<6 mm) grafts, primarily because of increased thrombogenicity and intimal hyperplasia due to the low flow conditions in these vessels.[1,4] Researchers have examined ways of improving existing synthetic grafts as well as developing new options for vascular grafts, including electrospinning and tissue engineering, which attempt to re-create a more specific structure that matches the vascular basement membrane of the arterial wall.[5,6] This chapter reviews the status and condition of existing synthetic vascular grafts and presents new methods and techniques for the improvement of small-diameter vascular grafts.

## 23.2. Present Medical and Biomaterials Technology and Limitations

As mentioned above, polyethylene terephthalate (Dacron®), expanded polytetrafluoroethylene (ePTFE), and other polyurethanes are the current choice for synthetic vascular grafts. When the graft is >6 mm, these polymers have proven to be a successful and reliable option for surgeons when an autologous vessel is not available.

Polyethylene terephthalate (Dacron®) is a type of polyester yarn consisting of multiple filaments, either woven or knit into a graft. This multiple-filament, woven pattern, makes the graft soft, elastic, and easy to handle.[2,7] Woven Dacron® grafts tend to be tightly constructed due to the pattern, and have low porosity, causing the graft to have little or no stretch. The knitted Dacron® graft is structured in a warp pattern and provides more stability, so is more commonly used commercially.[7] Despite the fact that Dacron® is used successfully for large-diameter grafts, it is not used often for bypass surgeries in the lower extremities, especially those below the knee.[2]

Polytetrafluoroethylene is an inert fluorocarbon polymer and is made more micro-porous through the processes of extrusion and sintering to form expanded polytetrafluoroethylene (ePTFE).[2] ePTFE is non-biodegradable, impervious to blood, resistant to dilatation, chemically inert, highly electronegative, and highly hydrophobic with an electronegative luminal surface that is antithrombotic. It is now widely used for lower limb bypass grafts (7–9 mm) with excellent results.[2,7] Whereas autologous vein grafts have 77% patency, only 45% of ePTFE grafts are patent in femoropopliteal bypass grafts at five years. A unique use for ePTFE grafts is as an arteriovenous graft to provide vascular hemodialysis access for patients with renal failure. These grafts are interposed between the radial artery and cephalic vein in the wrist instead of having a direct arteriovenous fistula.[2,8] ePTFE grafts are also available in different wall thicknesses depending on the specific purpose and location of the graft.

Although both Dacron® and ePTFE provide excellent results, neither is able to spontaneously develop an endothelialized layer lining the interior wall of the graft.[2] This can lead to the adhesion of platelets on the synthetic vessel wall and the development of a luminal, or interior, fibrin layer which can cause thrombosis. Thrombosis is particularly dangerous in the small-diameter (<6 mm) vascular grafts where a decreased flow is detrimental. With such a small diameter, it is vital to have a graft that is antithrombotic, to keep patency within the graft.

Considerable research has been undertaken into the use of polyurethane for vascular grafts, in addition to Dacron® and ePTFE. Theoretically, polyurethane grafts would provide a relatively smooth, non-thrombogenic inner surface, and a thin wall. Although this seems appealing, research has shown that patency rates within these grafts are rather low and the degradation of the polymer, which can lead to an aneurysm, remains a persistent problem.[7] However, some research has shown that polyurethanes function well as alternatives to hemodialysis because of their ability to self-seal at a given puncture site.[2]

## 23.3. New Technology: Advantages, Limits and Unknowns

Due to the fact that current synthetic vascular grafts fail for small diameters (<6 mm), there is a demand and a need for the development of a graft that is functional at this scale. Many researchers have looked into making improvements on existing vascular grafts, with one such technique being the endothelialization of the inner arterial wall. An endothelial cell layer on the luminal surface of an artery is vital to prevent thrombosis and maintain patency within the vessel. Herring *et al.* conceived the idea of seeding the luminal layer or grafts with endothelial cells, and numerous studies in the past have shown that the endothelialization of an ePTFE graft improves patency. However, clinically, traditional methods for endothelial cell seeding have failed primarily due to the time between cell harvest and actual implantation, as well as loss of endothelial cells due to the pulsate conditions of blood flow within the graft.[9,10,11] However, Bowlin *et al.* proposed a new method that electrostatically seeds the endothelial cells on the surface. An electrostatic device gives the luminal surface of the ePTFE graft a temporary positive or less negative charge. Once removed from this device, the graft then reverts back to its natural negative charge. The authors hoped that this positive charge would attract endothelial cells and allow for better adhesion to the luminal wall of the graft. An *in vitro* study of this technique revealed an increase in the number of endothelialized cells that adhered to the vessel wall as well as an acceleration of the cells' morphological maturity.[9]

In the study by Bowlin *et al.* an *in vivo* study was performed to evaluate the persistence and source of the electrostatically seeded endothelial cells that were lining the interior walls of the ePTFE graft.[9,12] The PKH 26 labeling kit was placed in the endothelial cell membrane prior to seeding and the evaluation took place after one week of implantation using a canine femoral artery model. The procedure took place in the clinically acceptable time of 16 minutes, and results were conclusive. They revealed endothelial cell persistence and adherence to the ePTFE graft wall as well as formation of an early neointima luminal layer.[9] This research provides an excellent foundation for endothelial cell seeding *via* electrostatic charge; however, more clinical research is necessary to account for various acute and chronic complications experienced by small-diameter vascular grafts.

Another study, performed by Pislaru *et al.*, also sought to improve on the endothelialization of current synthetic vascular grafts. The authors had previously shown that blood-derived endothelial cells are effective in preventing restenosis as well as improving the vascular function of arteries in certain animal models.[10] However, clinically, previous methods for seeding vascular grafts with endothelial cells still proved inefficient and did not provide the necessary amounts of cell adhesion. Pislaru *et al.* postulated that cell capture and retention could be achieved using magnetic forces.[10] Endothelial cells were rendered magnetically attractable by loading them with superparamagnetic microspheres and a prototype vascular graft was developed. For this experiment endothelial cells were harvested from peripheral blood as opposed to tissue harvesting to harvest the cells. The magnetic approach implemented by Pislaru *et al.* attached the magnetic material around the Dacron®, 6–8 mm vascular grafts, leaving the interior unaltered. Through the procedure described in the work by Pislaru *et al.* it was demonstrated that endothelial cells are easily loaded with superparamagnetic (SPM) particles and they are rapidly captured and retained on the luminal surface of the graft under pulsatile flow conditions. This experiment, conducted in animals, revealed significant amounts of endothelial cells on the luminal surface of the grafts after a period of 24 hours. The results displayed more endothelial cells on the interior (non-magnetic) side of the graft than on the exterior (magnetic) side, possibly pointing to the fact that when magnetically attracted and bonded to the luminal surface, endothelial cells adhere better under pulsatile conditions.

In addition to endothelial cell seeding on existing vascular grafts, new nanotechnologies have become more readily available and we are seeing an increase in biomedical applications. Nanoscale synthetic biomaterials show great promise as scaffolding for regeneration of damaged cellular membranes and tissues, as well as allowing researchers to create a structure that mimics the basement membrane in many tissues and organs.[5] This new nanotechnology concept aims to develop and combine new materials by specifically and precisely making atoms and molecules yield new molecular assemblies on the nanoscale of individual cells and tissues.[13] The creation of nanofiber meshes may allow for the development of vascular grafts with superior mechanical properties to avoid patency problems common in synthetic grafts, particularly in small-diameter vascular grafts.[5]

A technique known as electrospinning has recently been explored for the formation of nanoscale diameter fibers for applications such as vascular grafts.[14] Electrospinning is a process of making nanofibers from a polymer solution through electrostatic force. These nanofibers have a large surface area per unit mass so that fabrics created of these nanofibers on a screen can be used in many biomedical applications, such as the construction of a vascular graft.[15] This advanced electrospinning technique is a promising method for fabricating a controllable continuous nanofiber scaffold, similar to the natural extracellular matrix of tissues, or the vascular basement membrane for use in vascular grafts.[15]

The concept of electrospinning is versatile; fibers of different forms and morphology, as well as different materials, can be made.[15] Thus, different polymers or polymer blends and mixtures can be engineered into fibers for specific applications. Nanotechnology provides for the creation of 3D nanofiber scaffolds made primarily of collagen to seed smooth muscle cells. This 3D scaffold would mimic the natural extracellular matrix found within the vascular basement membrane of a small-diameter blood vessel. Recent studies have shown that when implanted onto a nanofiber scaffold, smooth muscle cells can develop into a similar, functional architecture of the natural blood vessel.[14,15] Additionally, making surface modifications of electrospun fibers to existing synthetic vascular grafts such as Dacron® and ePTFE, could improve the spreading and proliferation of endothelial cells on the luminal surface of the grafts.[15]

Electrospinning provides a means for creating a structure much closer to natural blood vessels. The method involves forming a tubular scaffold made of the elecrospun nanofibers.[15] By controlling the electric field, a rotating collector can be used to form an aligned pattern of electrospun nanofibers on the outer circumference of the small-diameter tube. This tube can then be extracted to give a tubular nanofiber scaffold. Although the concept of electrospinning is still in its early stages, this technology gives rise to the hope that it may be feasible to use endothelial cells, smooth muscle cells, and fibroblasts on the vascular construct to engineer a "real" synthetic blood vessel.

Many researchers are moving in the direction of regeneration of tissues. Many biomedical applications are attempting to create scaffolds which enable the development of functional tissues specific to a certain application in the body. As far as vascular grafts are concerned, there has been an extensive amount of research into the development of tissue-engineered small-diameter vascular grafts.[16] Currently, Sarkar *et al.* are researching the development and characterization of porous micro-patterned poly-caprolactone (PCL) scaffolds using several techniques to integrate soft lithography, melt molding, and particulate leaching of polylactic-co-glycolic acid (PGLA) nanoparticles.[17] The results of this particular study reveal that it may be feasible to fabricate patterned cell sheets to build up 3D tissues that can be layered to create vessel walls with specific cellular organizations.

## 23.4. Recent Developments

Vascular-tissue scaffolds have become one the most heavily investigated and challenging areas of tissue engineering for small-diameter vascular grafts. A great deal of focus has been placed on exploring various materials, natural and synthetic, and biological factors, both *in vitro* and *in vivo*, for creating viable vascular-tissue scaffolds. Many challenges remain in the development of a suitable vascular-graft scaffold due in part to the small diameter required (<6 mm) and the biomechanical characteristics that such a scaffold must possess. Current surgical treatment of peripheral vascular disease usually consists of harvesting autologous tissues, such as the saphenous vein, followed by re-implantation in the location of diseased tissue. Limitations of such surgical procedures include donor site morbidity, increased risk of infection, and unreliable long-term patency of the graft.[18,19] Therefore it is essential to develop an optimal tissue-engineering scaffold to alleviate the obstacles associated with current surgical techniques for restoring diseased small-diameter vascular tissue.

A wide spectrum of materials has been studied for the development of synthetic vascular scaffolds, and major advances have also been made at the biological interface, particularly with *in vitro* cell-seeding approaches. Recently, scaffolds were created from compound poly(ester urethane) urea *via* a thermally induced phase separation (TIPS) technique, followed by coating with electrospun polymer to create elastomeric biodegradable vascular-tissue scaffolds.[20] The scaffolds were then seeded with muscle-derived stem cells and implanted into a rat model. The cell-seeded, electrospin-coated scaffold exhibited higher patency rates, an increased mechanical profile, and improved induction of tissue growth when compared to uncoated or unseeded control scaffolds. In another study, Soletti *et al.* developed a tubular scaffold: a polycaprolactone bi-layered structure comprising a highly porous electrospun internal layer and a strong fibrous external layer produced by a TIPS method.[21] The scaffold was seeded with adult muscle-derived stem cells and cultured in a dynamic environment under simulated mechanical stress conditions. After seven days of dynamic culture the mechanical properties were found to be consistent with native vascular tissue and the scaffolds were observed to have a high degree of cell integration.

Electrospinning has been a preferred technique for creating vascular tissue scaffolds, and recently nano-structured scaffolds were developed by electrospinning. He *et al.* constructed a tubular poly(L-lactic acid)-co-poly(caprolactone) biodegradable nanofiber scaffold coated with a collagen layer.[23] Human endothelial cells were seeded onto the nanofiber scaffolds of the tubular structure, cultured and finally implanted in rabbits to replace the superficial epigastric vein. The scaffolds exhibited mechanical properties superior to that of commercially available Teflon® grafts, retained cellular phenotype in culture, and maintained structural integrity *in vivo* for a seven-week period. All three of the studies discussed used similar processing techniques and materials but were subject to a variety of different experimental conditions; yet each study yielded promising potential for the development of a clinically functional tissue-engineered vascular graft.

Although a majority of the research into vascular graft scaffolds has focused on the use of synthetic materials, several groups have begun to investigate decellularized tissue as scaffolds for small-diameter vascular grafts. Gui *et al.* evaluated decellularized human umbilical arteries for endothelial cell compatibility, mechanical properties and *in vitro* stability.[19] The umbilical arteries provided the necessary extracellular matrix to promote endothelial cell growth and proliferation and proved similar in mechanical strength to native vascular tissue. When the decellularized umbilical arteries were implanted into nude rats it was observed that the grafts remained functional *in vivo* for up to eight weeks. Although decellularization reduces immunogenicity, the use of human-derived tissue as allograft scaffold constructs carries with it the possibility of transmitting pathogens and a lack of reproducibility of source tissue. Other groups have circumvented the use of synthetic materials by utilizing tissue sheets for developing vascular tissue scaffolds. Recently, Guillemette *et al.* developed tissue-engineered vascular adventitia with *vasa vasorum* by co-culture of endothelial cells with a fibroblast-incorporated tissue sheet which was created completely *in vitro* without the introduction of synthetic scaffold material.[18] Implantation of the tissue sheet-based construct in a mouse model demonstrated encouraging results with excellent tissue integration, vascularization, and mechanical strength similar to native arterial tissues over a long-term follow-up period. Another group developed a self-assembled fully biological scaffold-free small-diameter vascular tissue using a unique bio-printing method.[23] The bio-printed vascular tissue was not only viable, but the fabrication method allowed for intricate control over the composition and geometry of the tissue structures, which are accurate and consistent.

Further investigation of the biomechanical influences and required mechanical properties for a viable small-diameter vascular tissue-engineering scaffold must be conducted before an appropriate graft can be introduced for clinical application. Gaining a better understanding of the mechanism of cell differentiation and tissue formation will better allow for biomimetic control to be employed in tissue engineering of vascular tissue. Understanding the biological and biomechanical factors which direct the formation of vascular tissue will greatly advance tissue-engineered small-diameter vascular grafts.[24–26] Continued research by multi-disciplinary groups will be needed to conceive a fully functional optimized small-diameter vascular-graft construct.

## 23.5.  Ethical Issues

Most of the aforementioned research has been received and accepted with few ethical concerns. However, there are specific techniques pertaining to the improvements of small-diameter vascular grafts that are ethically questionable. Several research groups have sought to endothelialize the luminal surface of vascular grafts *via* bone marrow-derived cells. Although the authors have shown that bone marrow-derived stem cells evolved into functional endothelial cells and produce increased patency in synthetic grafts as well as a capacity to possibly be used in tissue engineering, there are ethical concerns regarding morbidity.[27,6] Long-term survivability must be demonstrated before new small-diameter grafts are widely used. The issues addressed in Chapter 1 regarding rapid commercialization of medical devices where a large need exists must be considered.

## 23.6. Summary

Current synthetic polymer-based vascular grafts, namely Dacron® and ePTFE have excellent results for many large-diameter vascular-graft applications, with ePTFE working rather well for grafting in lower extremities. However, these grafts have proven to fail at small diameters (< 6 mm) which prompts a need for the development of a functional and readily available small-diameter vascular graft. Research shows promise in areas of the endothelialization of the luminal surface of current synthetic vascular grafts, fabrication of electrospun nanofiber structures and tissue-engineered constructs. However, a significant amount of research is still necessary to ensure a proven and reliable small-diameter device that can be used widely throughout the human population and with long-term survivability.

## References

1. Ross, R. (1999). Mechanisms of disease — Atherosclerosis — An inflammatory disease, *New England J. Med.*, **340**, 115–126.
2. Wang, X., Lin, P., Yao, Q.H., *et al.* (2007). Development of small-diameter vascular grafts, *World J. Surg.*, **31**, 682–689.
3. Helling, T.S., Nelson, P.W., Shelton, L., *et al.* (1992). A prospective evaluation of plasma-TFE and expanded PTFE grafts for routine and early use as vascular access during hemodialysis, *Ann. Surg.*, **216**, 596–599.
4. Patel, S.H., Tsang, J., Harbers, G., *et al.* (2004). Endothelial cell function on a poly(acrylamide-co-polyetylene glycol/acrylic acid) interpenetrating polymer network: cardiovascular applications, *IEEE*, **26**, 5040–5043.
5. Buxton, D.B., Lee, S.E., Wickline, S.A., *et al.* (2003). Recommendations of the national heart, lung, and blood institute nanotechnology working group, *Circulation*, **108**, 2737–2742.
6. Cho, S.W., Lim, S.H., Kim, I.K., *et al.*, (2005). Small-diameter blood vessels engineered with bone marrow-derived cells, *Ann. Surg.*, **241**, 506–515.

7. Xu, W.L., Zhou, F., Ouyang, C.X., *et al.* (2010). Mechanical properties of small-diameter polyurethane vascular grafts reinforced by weft-knitted tubular fabric, *J. Biomed. Mater. Res.*, **92A**, 1–8.

8. Hagen, P.O., Zhang, Z.G., Mikat, E.M., *et al.* (1982). Antiplatelet therapy reduces aortic intimal hyperplasia distal to small diameter vascular prostheses (PTFE) in nonhuman primates, *Ann. Surgery*, **195**, 328–339.

9. Bowlin, G.L., Meyer, A., Fields, C., *et al.* (2001). The persistence of electrostatically seeded endothelial cells lining a small diameter expanded polytetrafluoroethylene vascular graft, *J. Biomater. Appl.*, **16**, 157–173.

10. Pislaru, S.V., Harbuzariu, A., Agarwal, G., *et al.* (2006). Magnetic forces enable rapid endothelialization of synthetic vascular grafts, *Circulation*, **114**, 314–318.

11. Chan, B.P., Liu, W., Klitzman, B., *et al.* (2005). *In vivo* performance of dual ligand augmented endothelialized expanded polytetrafluoroethylene vascular grafts, *J. Biomed. Mater. Res.*, **72B**, 52–63.

12. Bowlin, G.L. and Rittgers, S.E. (1997). Electrostatic endothelial cell transplantation within small-diameter (<6 mm) vascular prostheses: a prototype apparatus and procedure, *Cell Transplant*, **6**, 631–637.

13. Wickline, S.A., Neubauer, A.M., Winter, P., *et al.* (2006). Applications of nanotechnology to atherosclerosis, thrombosis, and vascular biology, *Arterioscler. Thromb. Vasc. Biol.*, **26**, 435–441.

14. Jin, H.J., Fridrikh, S.V., Rutledge, G.C., *et al.* (2002). Electrospinning *Bombyx mori* silk with poly(ethylene oxide), *Biomacromolecules*, **3**,1233–1239.

15. Liao, S., Li, B., Ma, Z.W., *et al.* (2006). Biomimetic electrospun nanofibers for tissue regeneration, *Biomed. Mater.*, **1**, R45–R53.

16. Heyligers, J.M.M., Arts, C.H.P., Verhagen, H.J.M., *et al.* (2005). Improving small-diameter vascular grafts: from the application of an endothelial cell lining to the construction of a tissue engineered blood vessel, *Ann. Vasc. Surg.*,**19**, 448–456.

17. Sarkar, S., Lee, G.Y., Wong, J.Y., *et al.* (2006). Development and characterization of a porous micro-patterned scaffold for vascular tissue engineering applications, *Biomaterials*, **27**, 4775–4782.

18. Guillemette, M.D., Gauvin, R., Perron, C., *et al.* (2010). Tissue engineered vascular adventitia with *vasa vasorum* improves graft integration and vascularization through inosculation, *Tissue Eng.*, **16**, 2617–2626.

19. Gui, L., Mutu, A., Chan, S.A., *et al.* (2009). Development of Decellularized human umbilical arteries as small diameter vascular grafts, *Tissue Eng. A.*, **15**, 2665–2676.

20. Nieponice, A., Soletti, L., Guan, J.J., *et al.* (2010). *In vivo* assessment of a tissue-engineered vascular graft combining a biodegradable elastomeric scaffold and muscle derived stem cells in a rat model, *Tissue Eng. A.*, **16**, 1215–1223.

21. Soletti, L., Hong, Y., Guan, J.J., *et al.* (2010). A bi-layered elastomeric scaffold for tissue engineering of small diameter vascular grafts, *Acta Biomater.*, **6**, 110–122.

22. Norotte, C., Marga, F.S., Niklason, L.E., *et al.* (2009). Scaffold-free vascular tissue engineering using bio-printing, *Biomaterials*, **30**, 5910–5917.

23. He, W., Ma, Z.W., Teo, W.E., *et al.* (2009).Tubular nanofiber scaffolds for tissue engineered small-diameter vascular grafts, *J. Biomed. Mater. Res.*, **90A**, 205–216.

24. Roh, J.D., Sawh-Martinez, R., Brennan, M.P., *et al.* (2010).Tissue engineered vascular grafts transform into mature blood vessels via an inflammation-mediated process of vascular remodeling, *Proc. Nat. Acad. Sci. USA*, **107**, 4669–4674.

25. Chan-Park, M.B., Shen, J.Y., Cao, Y., *et al.* (2009). Biomimetic control of vascular smooth muscle cell morphology and phenotype for functional tissue engineered small diameter blood vessels, *J. Biomed. Mater. Res.*, **88A**, 1104–1121.

26. O'Cearbhail, E.D., Murphy, M., Barry, F., *et al.* (2010). Behavior of human mesenchymal stem cells in fibrin-based vascular tissue engineering constructs, *Ann. Biomed. Eng.*, **38**, 649–657.

27. Jin, H.J., Chen, J.S., Karageorgiou, V., *et al.* (2003). Human bone marrow stem cell responses on electrospun *Bombyx mori* silk fibroin, *Bioinsp. Nanosc. Hyb. Sys.*, **735**, 129–133.

Julie Lockwood

# Spinal Cord Repair: Current Methods and Limitations

CHAPTER 24

## 24.1. Introduction

Spinal cord injury (SCI) often results in severe neurological impairment and disability where the affected individual sometimes requires personal care attendants and/or supervision for daily activities. The consequences of SCI are particularly severe when the spinal cord is completely severed or compressed. The spinal cord is made up of four regions, each controlling different functions of the body. The cervical region, located at the neck, is responsible for signals to the neck, arms, and hands. The upper back or thoracic region controls the torso and parts of the arms. The middle back or lumbar region controls signals to the hips and legs. Finally, the sacral region is responsible for signals to the groin, toes, and parts of the legs.

Generally, the spinal nerves below a lesion have only partial or complete lack of communication with the brain, which results in quadriplegia or paraplegia depending on the site of injury. Messages to the brain from below the level of injury will be blocked by the damage to the spinal cord and nerves above the injury will continue to exhibit normal function. Therefore, if an individual suffers damage above the first thoracic vertebra, he will present quadriplegia. When damage occurs below this level, paraplegia results (Fig. 24.1).[1] A SCI injury can be complete or incomplete, with varying degrees of impairment resulting accordingly.

There are approximately 11,000 new cases of SCI in the United States each year that do not result in death. These injuries occur

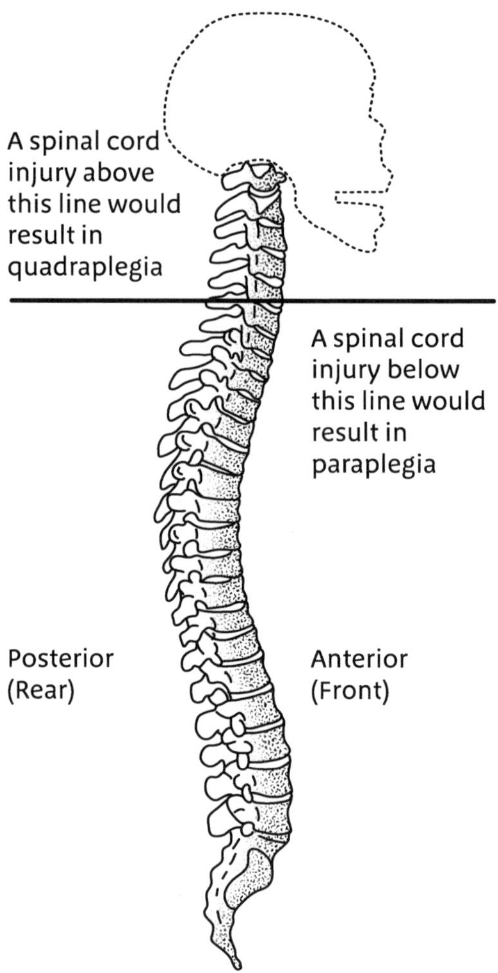



**Figure 24.1.**   Paraplegia and quadriplegia result from damage to different areas of the spinal cord.

most frequently from motor vehicle accidents (46.9% of reported cases since 2000) followed by instances of falling (23.7%) in a way where the spinal cord is damaged.[2] Because individuals with SCI are often confined to a wheel chair and severely physically and psychologically hampered, there is great interest and need for methods to repair the spinal cord.

## 24.2. Present Technology and Limitations

At present, there is no way to repair damage to a spinal cord. Individuals with quadriplegia or paraplegia are often wheelchair-bound and some are ventilator-dependent. Steroids are used in high doses following injury, to minimize secondary injury effects after SCI. The complications of using high doses of steroids, however, may outweigh the benefits.[3] Many SCI patients utilize narcotic analgesics to alleviate chronic pain. These medications have many unwanted side affects and a potential for addiction. They also do not improve severe pain and are thus susceptible to overuse.[4,5] Surgical practices including spinal stabilization and decompression surgeries are occasionally performed but do not conclusively improve neurological outcomes. In several studies, early surgery (taking place within 24 hours of the injury) proved to be more successful than late surgery or no surgery, but results are inconclusive with respect to the general effects of surgical intervention.[3]

Injuries to the spinal cord present a challenge because the injured nerves in the central nervous system (CNS) cannot repair themselves or regenerate, unlike cells in the peripheral nervous system or other areas of the body. The CNS cells in lower-order animals, however, have the capability to repair. Therefore, research is currently taking place to understand the differences in regeneration potential of neural cells.[6] Cells in the CNS are also unique in that they have a high rate of metabolism and therefore are vulnerable to ischemia if the blood flow they require for energy is reduced.[7] Affected individuals are therefore susceptible to hypotension. Inflammation is another secondary effect of SCI, and non-steroid anti-inflammatory methods have proven to be helpful in lessening this complication. Minocycline is one such drug that works in part by altering cytokines and inhibiting apoptosis.[8] Some therapeutic drugs that could prove to be vital in SCI are blocked by the blood–spinal cord barrier that normally protects nerve cells from harmful substances.[1] Other secondary effects of SCI that need to be controlled medically include pressure sores, thrombus formation, incontinence, muscle spasms, and infections of the urinary tract, kidneys, and lungs. Rehabilitation including physical therapy to maximize any mobility that exists after injury is also imperative. This can also include speech and language therapy.

Until recently, a SCI such as paraplegia was considered incurable and permanent. Presently, there exists no way to reverse such an injury. There is now hope, however, that new therapies and approaches to cure a SCI might be successful. Interest and research was directed towards this area when actor Christopher Reeve was paralyzed while competing in an equestrian tournament in 1995. The shattering of his first and second vertebrae left him paralyzed and confined to a wheel chair for the remainder of his life. Because he could not move his hands or any part of his body from his neck down, he operated his wheelchair by blowing air through a straw. Reeve's experiences prompted him to become Chairman of the American Paralysis Association, Vice Chairman of the National Organization on Disability, a cofounder of the Reeve–Irvine Research Center, and creator of the Christopher & Dana Reeve Foundation, which ignited research efforts on SCI through funding.[1,9] His contribution to this field of research has made an impact by bringing attention to an issue that was thought to be an elusive impossibility, and by funding research efforts so that someday a cure may exist. Today, even after Reeve's death, there are myriad aspects of research that are taking place in the area of SCIs.

## 24.3. Proposed Approaches

### 24.3.1. *Axon regeneration*

An axon is the part of a neuron that conducts signals away from the cell body. Axons are encased by a myelin sheath that serves to insulate the axon and increase the speed of impulses along the fiber. In the central nervous system, the sheath is composed of glial cells called ogliodendrocytes. By regenerating the axon, spinal-cord injury repair may be feasible. Attempts to repair remaining axons after SCI have included re-myelination, regeneration, and improving conduction. Many elements of the CNS, however, including the central myelin and glial scars that form during recovery from a SCI, inhibit axon regeneration. For example, inhibitory proteoglycans are found within glial scars and must be neutralized for regeneration of the axon to occur.[7] Chondroitin sulphate is an inhibitory proteoglycan that has been neutralized by chondroitinase in animals, resulting in an improvement of axonal regeneration.[10]

Synthetic materials have also been made for the purpose of axon regeneration including collagen, Neurogel, Matrigel, fibronectin mats, carbon fibers, Millipore, and nitrocellulose. Successful regeneration from brainstem neurons has resulted from the use of a synthetic hydrogel channel made with poly(2-hydroxyethyl methacrylate-co-methylmethacrylate).[11] Growth factors, which are naturally occurring proteins that stimulate cellular differentiation, are also being investigated for the enhancement of axon regeneration. Nerve growth factor (NGF), one of the first identified neurotrophic factors, can aid in the regeneration of sensory and sympathetic axons. It is, however, inadequate for the repair of motor function. Cells can be modified to produce neurotrophic factor, thought to regenerate motor axons.[12–14] Overall, clinical studies using neurotrophic factors for SCI are subject to adverse side effects and lack of favorable results.[15]

### 24.3.2. *Immune-based therapy*

Immune-based therapy, including the injection of "alternatively activated" macrophages, is being investigated for the repair of SCI. A macrophage is the part of the immune system (a type of white blood cell) that phagocytizes pathogens. They also stimulate the response of lymphocytes such as T cells and B cells. Macrophages normally promote healing in the body but are suppressed in the spinal cord, offering the possibility of cell regeneration. By injecting macrophages directly into the site of injury, regeneration of nervous tissue may be achievable.

Other possibilities for repair using the immune system include systemic injection of dendritic cells specific to CNS antigens or T cell-based vaccinations.[16]

### 24.3.3. *Cell and tissue methods*

Because the CNS is not a conducive environment for axon growth, the idea of connecting the part of the spinal cord that contains axons, to an area that permits axon regeneration, has arisen. The cephalad cord in rats and monkeys has been connected to peripheral nerves in which axons can grow. It has been shown that even without the motor neuron of the ventral horns of the spinal cord, the upper motor neuron can still innervate a muscle.[17] Other research in this area is being investigated as well, including peripheral nerve grafts and the olfactory ensheathing glia (OEG) transplants. OEG cells accompany growing olfactory axons into the adult mammalian CNS and may be responsible for aiding in axonal regeneration. Further research is being done on the physiology of these cells to conclude whether they act as passageways for axon regeneration, or are critical factors in the process.[18] These types of therapies attempt to overcome the environment that the CNS provides to favor the regeneration of cells and tissues.

### 24.3.4. *Nanoparticles*

Cerium oxide nanoparticles could eventually offer neuroprotection and retention of function to spinal cord neurons and become part of a therapy for individuals with SCI. Cerium oxide is a rare-earth oxide that has already been shown to have applications in biology because of its unique properties. (See Chapter 4 for a review of nano-ceria nanoparticles and their potential use in repair of neuronal tissues.) In a serum-free cell culture of dissociated adult spinal cord neurons, a single dose of nano-ceria has been shown to promote long-term survival and growth of cells while maintaining normal electrical activity. Single-action potentials were generated by these cells, which corresponded to control adult rat CNS cultures with intact spinal cord neurons.[19] Although the mechanism for neuroprotection is not entirely clear, research is currently being done to understand the behavior of the nanoparticles and how they can eventually be used *in vivo* for the treatment of SCIs. A proposed explanation for the cell's neuroprotective function is that the mixed valence states of cerium allow the nanoparticles to scavenge free radicals from the culture. The cerium ion is then reduced in a series of surface reactions, indicative of an auto-catalytic reaction of the ceria nanoparticles.[19] This application is not a cure for a SCI in its entirety, but instead offers the possibility of a method to alleviate symptoms and loss of function that come with an injury. These nanoparticles may also prove to be beneficial in the alleviation of ischemic events that are hallmarks of SCIs, as well as aid in reducing other secondary symptoms of SCI.

### 24.3.5. *Stem cells*

Neural stem cells (NSC) are multipotent neural progenitor cells, meaning that they have the potential to differentiate into many different types of cells. A potential application for using NSC cells is to engineer them to act as guides for axons into the spinal cord below the site of injury, through the use of chemical and mechanical signals. The cells could also supply a source of new neurons in which damaged communication in the spinal cord could be repaired. Finally, stem cells have the potential to secrete neurotrophic substances that could promote repair and regeneration of axons, damaged tissues, and cells.[20] The improvement of the spinal cord, however, will most likely depend not only on the use of stem cells but on other factors as well, such as the reversal of de-myelination, repair of damaged cellular elements, and reduction of glial scarring.

Another approach in SCI repair research is to use a polymer scaffold seeded with NSCs as an implant for the spinal cord. The inner portion of the implant is a porous polymer that corresponds to the grey matter (unmyelinated neurons) of the spinal cord. The outer portion of the implant includes pores for axonal guidance and corresponds to the white matter (myelinated neurons). This layer allows fluid transport and helps in preventing scar tissue from forming. This type of scaffold has been implanted in adult rats and shown to promote long-term improvement in motor function.[21]

These scaffolds are another component that may be used in the future in a multifaceted approach for treating SCIs.

Marrow stromal cells (MSC) derived from bone marrow are a source of pluripotent stem cells that can differentiate into any of the three germ layers. These cells can be grown in large quantities *in vitro* into multiple mesodermal cell types. They can also be grown into neuron-like cells with the capability to develop into mature neurons. Action potentials, however, cannot be generated in these neurons because the cells lack the ion channels necessary for establishment of an ion gradient. Despite this hindrance, these cells can be injected directly into an injured spinal cord and may have potential for a functional outcome.[22]

Embryonic stem cells (ESC), derived from blastocysts, are pluripotent and very flexible. (See Chapter 2 for a discussion of some technical issues associated with this type of approach.) One use of embryonic stem cells is that they can be used to create oligodendrocytes to remyelinate axons. Early-stage oligodendrocytes derived from embryonic stem cells have been injected into the spinal cords of mice with no myelin tissue. After transplantation, full-grown oligodendrocytes formed, which myelinated the axons.[23] The possibilities of ESC are endless because they can differentiate into any cells of the endoderm, ectoderm, and mesoderm (the three primary germ layers). Research is currently occurring in this area but at a slower pace than would be ideal because of ethical and moral issues, as well as numerous technical barriers, as discussed in Chapter 2.

## 24.4. Ethical Issues

Whenever potential stem cell applications arise, ethical issues arise as well. This conflict is especially evident with respect to embryonic stem cells. An individual who is already living could benefit from the use of embryonic stem cells in terms of quality and duration of life. However, it is argued that this would be at the expense of the initiation of a viable human being. A balance must be found between developing accurate portrayals of foreseen benefits of embryonic research, and clarifying the value of an embryo and its potential.

If adult stem cells are used, rather than embryonic stem cells, most ethical dilemmas are avoided. Also, autografts do not have issues with rejection, whereas stem cells from an embryo are unlikely to be immunologically compatible. Therefore, in most cases the use of embryonic stem cells conceivably requires use of immunosuppressant drugs that have no shortage of negative side effects. With the gain in ethicality and immunological compatibility from using adult stem cells, however, comes a loss in potential for multiplication and differentiation. Adult stem cells do not have the differentiating capabilities of embryonic stem cells, and therefore are not applicable in terms of potential to numerous biomedical advances at this time.

A recent option in the source of stem cells is to harvest stem cells from umbilical cord blood. Umbilical cord blood banking is becoming a popular solution to the issue of ethicality with respect to ESC. This method requires planning, foresight, and expense, because a parent must decide prior to delivery of the placenta whether or not they plan to bank the cord blood from their baby's umbilical cord. Therefore, although this option may be used regularly in the future, individuals cannot take advantage of this stem cell technology at present unless their cord blood has already been banked. This option is also costly and therefore not every individual is in a position to take advantage of this opportunity.

Another ethical dilemma that arises with respect to SCI is the rights of individuals with this type of disability. Should people with severe SCI be allowed to work, and if so, what level of injury should be acceptable to an employer or insurance company? Also, if an individual with a SCI has lower productivity than the average worker but expends comparable effort, should the compensation for the two individuals be equivalent? Legislation for such handicapped individuals and situations needs to be further defined to take into account the changing views of society on the status of the disabled.

An issue that is at the forefront of ethical and moral dilemmas in the realm of SCI is the issue of euthanasia. When individuals are paralyzed from the neck down, their quality of life is substantially decreased. Even with the most positive outlooks, individuals can become discouraged, and find that their new life is not worth living with the injury they possess. When individuals become incontinent, unable to breathe on their own, confined to wheelchairs, and completely reliant on others for daily activity, it is not hard to understand why suicide looks like a potential option. These people, however, do not even possess the ability to commit suicide, and so assisted suicide is often a request. With severe suffering, there are some physicians in select areas of the world who will perform this procedure. Assisted suicide is legal in some countries. This issue is so controversial because, among other reasons, some people feel that euthanasia undermines the importance and value of a human life. Overall, this practice is not widely accepted and it is possible for individuals who are paralyzed and psychologically stable to enjoy a different but fulfilling life.

## 24.5. Economic Issues

When an individual suffers a SCI, they also suffer a loss of function that can range from minor to severe, depending on the site and extent of injury. This detriment can be psychological, physical or both and can affect the individual's work and productivity as well as cause him to lose his job and status in society entirely. Not taking into account these economic losses, an individual could potentially have a substantial monetary dependability on personal care, medical equipment, and prescription drugs, because of his SCI. For example, a 25-year-old paraplegic has an expected lifetime cost of US$977,142. This cost would be directly attributable to yearly health-care and living expenses stemming from the individual's SCI. For a similar individual with an injury to the upper cervical vertebrae, resulting in quadriplegia, these same expenses would be estimated at US$2,924,513.[24] Although the cost of research is high, the benefits to sufferers of SCI would be considerable — economically, psychologically, and physically. Research funding in this area has increased because, in part, of the efforts and media attention Christopher Reeve brought to understanding this debilitating injury. The commercial risk/reward ratio for this type of cure, however, is generally not favorable and is therefore usually avoided by investors and corporations with only financial concerns.

## 24.6. New Developments

The prognosis for patients who suffer spinal cord injuries has generally been poor, although survival rates and long-term clinical outcomes have improved over the past 20 years.[25] However, to date no successful treatment for severe spinal cord injuries that allows the spinal cord to recover to a fully functioning normal status has been discovered.[26] Recently, multi-disciplinary approaches have begun to take shape, with tremendous effort being put towards not only advancing traditional surgical treatment, but also research and translation of cell therapies, tissue engineering, and regenerative medicine techniques. Stem cells, nano-materials, and tissue engineering scaffolds are areas currently under intense investigation and have exhibited promising potential for spinal cord injury repair.

Research into stem cell therapy has been shown to have huge potential for treating SCIs. A viable method in which stem cell transplantation is used to treat SCI should replace cells at the site of injury, offer neuroprotection and trophic support, and facilitate axonal growth.[27] A variety of stem cells from various sources have been studied, with positive results in both animal models and human trials; however a consensus on a premier cell type for treating SCI has not been established.[27] For example, olfactory ensheathing cells (OSCs) found along the olfactory nerve are known to exhibit robust axonal growth *in vitro* and have been shown to promote re-myelination in the spinal cord of animal models of SCI.[28] Lima *et al.* recently conducted a study in which olfactory mucosa autografts, which contain neural stem cells (NSCs), were transplanted into patients with debilitating SCIs.[29] Following transplantation and rehabilitation, patient evaluation demonstrated an improved electrophysiological response in the lower extremities, and improved bladder control. This study is an example of the diverse sources of stem cells studied for SCI repair, including fetal nervous tissue, embryonic stem cells (ESCs), bone marrow mesenchymal stem cells (MSCs), Schwann cells, and peripheral nervous tissue.[26]

SCIs are commonly caused by traumatic injuries, but the type of lesion directs the pattern of scar formation and inflammatory response elicited.[25] The level of the spine at which a SCI occurs may play a role in determining which cell types may be preferential for healing a lesion and therefore better define indications for superior outcomes in SCI treatment. Recently, Sharp *et al.* transplanted human ESC-derived oligodendrocyte progenitor cells (hESC-OPC) into both cervical and thoracic spinal cord contusion rat models.[30] The findings in this study showed that the hESC-OPC-transplanted rats had improved recovery, although interestingly it was observed that the transplanted rats with cervical contusions exhibited increased locomotor improvement compared to the transplanted rats with thoracic contusions. This study shows that the success of a stem cell-based therapy may be heavily dependent upon pathogenesis of a SCI, and further studies are therefore needed to determine the optimal cell therapeutic technique for a particular SCI. Continued advancements in the area of stem cell research will be needed to gain the level of understanding needed to move stem cell therapy from the laboratory to the clinical setting. The complexity of both stem cell biology and the human nervous system make this a challenging task. Stem cell therapy offers a tremendous amount of potential for successfully treating and repairing SCIs, and in combination with advanced biomaterials, gene therapy, and other tissue engineering modalities the future for SCI repair looks increasingly encouraging.

## 24.7. Summary

Although there is currently no cure for SCI, research efforts in the area appear promising. Clinical trials are currently taking place using activated macrophages, peripheral nerve grafts, OEG cells, and other types of stem cells. There is not likely to be one cure or therapeutic approach for a SCI, but instead many techniques working in concert to better and prolong the lives of individuals who undergo such an injury. A multifaceted approach is needed because there are a myriad secondary effects attributed to a SCI, along with physically and psychologically related obstacles. As stem cell research becomes more prevalent and accepted ethically, the efforts to improve this type of injury will increase and a solution to this problem will become attainable.

## References


1. Christopher & Dana Reeve Foundation (2006). *Christopher Reeve Spinal Cord Injury and Paralysis Foundation.* [Online]. Available at: http://www.christopherreeve.org. [Accessed 9 November 2010].

2. Jackson, A.B., Dijkers, M., DeVivo, M.J., *et al.* (2004). A demographic profile of new traumatic spinal cord injuries: change and stability over 30 years, *Arch. Phys. Med. Rehabil.*, **85**, 1740–1748.

3. Lim, P.A. and Tow, A.M. (2007). Recovery and regeneration after spinal cord injury: a review and summary of recent literature, *Ann. Acad. Med. Singapore*, **36**, 49–57.

4. Henwood, P. and Ellis, J.A. (2004). Chronic neuropathic pain in spinal cord injury: the patient's perspective, *Pain Res. Manage.*, **9**, 39–45.

5. Sagen, J. (1998). "Transplantation strategies for the treatment of pain", in Freeman, T.B. and Widner, H. (eds), *Cell Transplantation for Neurological Disorders*, Humana Press, New York, NY, pp. 231–253.

6. Park, K.I., Ourednik, J., Ourednik, V., *et al.* (2002). Global gene and cell replacement strategies via stem cells, *Gene Ther.*, **9**, 613–624.

7. Tsai, E.C. and Tator, C.H. (2005). Neuroprotection and regeneration strategies for spinal cord repair, *Curr. Pharm. Design*, **11**, 1211–1222.

8. Lee, S.M., Yune, T.Y., Kem, S.J., *et al.* (2003). Minocyline reduces cell death and improves functional recovery after traumatic spinal cord injury in the rat, *J. Neurotrauma*, **20**, 1017–1027.

9. Reeve, C. (1998). *Still Me*, Random House, New York, NY.

10. Zuo, J., Neubauer, D., Dyess, K., *et al.* (1998). Degradation of chondroitin sulfate proteoglycan enhances the neurite-promoting potential of spinal cord tissue, *Exp. Neurol.*, **154**, 654–662.

11.  Tsai, E.C., Dalton, P.D., Shoichet, M.S., *et al.* (2004). Synthetic hydrogel channels facilitate regeneration of adult rat brainstem motor axons after complete spinal cord transaction, *J. Neurotrauma*, **21**, 789–804.

12.  Lindsay, R.M. (1988). Nerve growth factors (NGF, BDNF) enhance axonal regeneration but are not required for survival of adult sensory neurons, *J. Neurosci.*, **8**, 2394–2405.

13.  Chun, L.L. and Patterson, P.H. (1977). Role of nerve growth factor in the development of rat sympathetic neurons *in vitro*. I. Survival, growth, and differentiation of catecholamine production, *J. Cell Biol.*, **75**, 694–704.

14.  Namiki, J., Kojima, A. and Tator, C.H. (2000). Effect of brain-derived neurotrophic factor, nerve growth factor, and neurotrophin-3 on functional recovery and regeneration after spinal cord injury in adult rats, *J. Neurotrauma*, **17**, 1219–1231.

15.  Penn, R.D., Kroin, J.S., York, M.M., *et al.* (1997). Intrathecal ciliary neurotrophic factor delivery for treatment of amyotrophic lateral sclerosis (Phase I Trial), *Neurosurgery*, **40**, 94–99.

16.  Schwartz, M. and Yoles, E. (2006). Immune-based therapy for spinal cord repair: autologous macrophages and beyond, *J. Neurotraum.*, **23**, 360–370.

17.  Brunelli, G. (2005). Research on the possibility of overcoming traumatic paraplegia and its first clinical results, *Curr. Pharm. Design*, **11**, 1421–1428.

18.  Moreno-Flores, T.M., Diaz-Nido, J., Wandosell, F., *et al.* (2002). Olfactory ensheathing glia: drivers of axonal regeneration in the central nervous system?, *J. Biomed. Biotechnol.*, **2**, 37–43.

19.  Das, M., Swanand, P., Bhargava, N., *et al.* (2007). Auto-catalytic ceria nanoparticles offer neuroprotection to adult rat spinal cord neurons, *Biomaterials* **28**, 1918–1925.

20.  Barami, K. and Diaz, F.G. (2000) Cellular transplantation and spinal cord injury, *Neurosurgery*, **47**, 691–700.

21.  Teng, Y.D., Lavik, E.B., Qu, X., *et al.* (2002). Functional recovery following traumatic spinal cord injury mediated by a unique polymer scaffold seeded with neural stem cells, *Proc. Nat. Acad. Sci. USA*, **99**, 3024–3029.

22. Hofstetter, C.P., Schwarz, E.J., Hess, D., *et al.* (2002). Marrow stromal cells form guiding strands in the injured spinal cord and promote recovery, *Proc. Nat. Acad. Sci. USA*, **99**, 2199–2204.

23. Liu, S., Qu, Y., Steward, T.J., *et al.* (2000). Embryonic stem cells differentiate into ogliodendrocytes and myelinate in culture and after spinal cord transplantation, *Proc. Nat. Acad. Sci. USA*, **97**, 6126–6131.

24. National Spinal Cord Injury Statistical Center, Birmingham, Alabama (2007). [Online]. Available at: www.nscisc.uab.edu. [Accessed 16 July 2010].

25. Gupta, R., Bathen, M.E, Smith, J.S., *et al.* (2010). Advances in the management of spinal cord injury, *J. Am. Acad. Orthop. Surg.*, **18**, 210–222.

26. Zhang, N., Wimma, J., Qian, S.-J., *et al.* (2010). Stem cells: current approach and future prospects in spinal cord injury repair, *Anat. Rec.*, **293**, 519–530.

27. Sahni, V. and Kessler, J.A. (2010). Stem cell therapies for spinal cord injury, *Nat. Rev. Neurol.*, **6**, 363–372.

28. Jain, K.K. (2009). Cell therapy for CNS trauma, *Mol. Biotechnol.*, **42**, 367–376.

29. Lima, C., Escada, P., Pratas-Vital, J., *et al.* (2010). Olfactory mucosal autografts and rehabilitation for chronic traumatic spinal cord injury, *Neurorehab. Neural Rep.*, **24**, 10–22.

30. Sharp, J., Frame, J., Siegenthaler, M., *et al.* (2010). Human embryonic stem cell-derived oligodendrocyte progenitor cell transplants improve recovery after cervical spinal cord injury, *Stem Cells*, **28**, 152–163.

# Spinal Cord Repair for Paraplegics



## CHAPTER 25

Jeffrey Calhoun

## 25.1. Introduction

It is estimated that 253,000 people in the United States suffer from spinal cord injury (SCI) and there are nearly 11,000 SCI new cases each year. Paraplegics, or those who have suffered a loss of movement or sensation in the lower body, account for approximately 43% of the SCI population, while quadriplegics, or those who have suffered a loss of movement and sensation in the arms and legs, account for the remaining 57%. The group most at risk for spinal cord injuries is males (representing 82% of all SCI cases) between the ages of 16 to 30. The most common causes for spinal cord injuries are vehicular accidents representing 37% of SCI cases; acts of violence 28%; falls 21%; sports-related spinal cord injuries 6%; and other causes approximately 8%. With the advent of safer automobiles and more stringent safety standards in the automotive industry, the impact of vehicular accidents on the spinal cord injury population has shown a slight decline in recent years.[1]

The cost of spinal cord injuries in the United States is astounding, both in the financial cost to the injured person, and the rehabilitation time necessary for even partial recovery in most cases. The average length of stay in a hospital immediately following a spinal cord injury is 15 days. The average stay in a rehabilitation unit following the initial hospital stay is 44 days. Costs for the initial hospital stay are estimated at US$140,000, with average first-year expenses for SCI estimated at US$198,000. Paraplegics average a lifetime cost of treatment for injury of

US$417,000 when injured at the age of 25, while quadriplegics average around US$1.35 million for cost of lifetime treatment when injured at the same age.[1] Spinal cord injuries have an enormous cost to both health-care providers and to injury sufferers. This chapter reviews current treatments for spinal cord injuries below the neck — paraplegia — and summarizes new research directions in the field of spinal cord repair.

## 25.2.  Present Medical Solutions

The spinal cord is a long tubular bundle of nerve cells and nerve fibers that relays all information from the body to the brain and from the brain to the rest of the body, and is enclosed and protected within the vertebrae of the bony spinal column. The spinal column consists of 31 segments, each a part of one of four different regions: the cervical region designated by C1 through C8; the thoracic region designated by T1 through T12; the upper lumbar region designated by L1 through L5; and the sacral segments S1 through S5. The emphasis of this chapter is on injuries to the thoracic, lumbar, and sacral regions of the spinal cord. Each spinal segment within this region is labeled with an abbreviation followed by a number to designate its location within the region, for example T2 (see Fig. 25.1).

When the spinal cord is injured, the transfer of motor and sensory information can be impaired over the region of injury. Injury may be caused by many modes: vascular insult, contusions, and bruising to the cord may cause scar tissue build up and thus prevent motor and sensory impulses from traveling along the scarred region. Injuries such as violence, vehicular injury, and others are listed in the first part of this review. Paraplegia refers specifically to injuries to the thoracic, lumbar, and sacral regions of the spinal cord, which affect the region below the site of injury. Sufferers from paraplegia retain nerve sensory information and motor control of their arms. Injuries to the cervical vertebrae are the most severe and cause impairment of function in the arms, legs, trunk, organs, and pelvic region, resulting in quadriplegia.

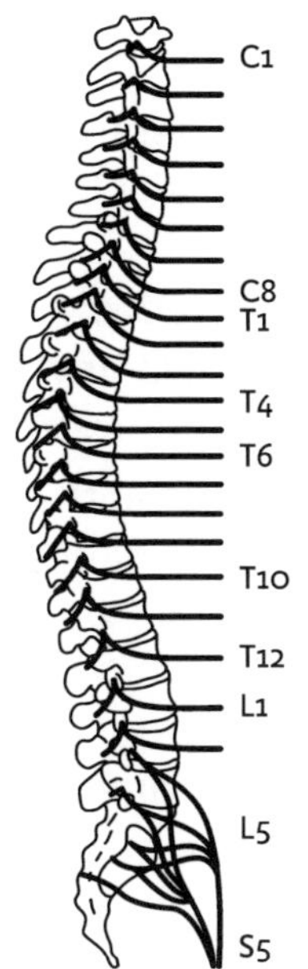


**Figure 25.1.**   Spinal column segments and relative location.

The proper treatment of any disease or illness first requires a definition of what specific factors or milestones make up a successful treatment regime. When applied to spinal cord injuries, this particular statement is somewhat vague at best, and the factors are often hard to define in providing a complete statement of care for the SCI sufferer. The American Spinal Injury Association, or ASIA, has provided the most complete set of standards by which to gauge and assess SCI, and these are used by the National Model System Spinal Cord Injury Database from which the statistics surrounding SCI are published. Neurological examinations recommended by ASIA consist of both sensory and motor examinations, and provide a quantitative representation of neurological deficit. Recovery of a patient is gauged by the difference in motor and sensory scores between successive testings provided by ASIA.

Two main types of spinal cord injuries exist: complete and incomplete injury. Complete spinal cord injury means having no voluntary motor or conscious function below the site of injury. Incomplete injury means that some varying degree of motor and sensory information is retained.[1] The degree of the injury is measured by ASIA standards to determine which method of care is appropriate for the SCI sufferer.[2]

The standard of care in North America following a spinal cord injury is to administer a high dose of methylprednisolone in non-penetrating acute SCI. This preferred procedure is based upon results from the National Acute Cord Injury Studies, or NASCIS, which determined that, if when injected within eight hours of injury, methylprednisolone can significantly improve recovery. These findings are confirmed by animal studies which show that infusions of methylprednisolone can potentially offer protective qualities to neurons within the spinal cord through its anti-inflammatory effects.[3,4] The administration of methylprednisolone injections post injury is not without scrutiny, however, because three NASCIS studies failed to conclude positively that methylprednisolone infusion did in fact improve recovery. The use of such steroids is not without risk either, considering the vulnerable condition of most SCI patients.[5]

Neural prosthetics combine neuroscience and biomedical engineering to create prosthetic devices capable of improving or replacing the function of an impaired nervous system. Lumbar anterior root implants and hand-grasp prostheses are implantable devices that use electrical and mechanical stimulation to aid or assist paralysis victims in regaining or restoring movement to paralyzed limbs. Sacral anterior root stimulators are implantable electrical devices which improve bladder emptying, assist in defecation, and allow male patients to sustain a full erection. Implantable neural stimulating devices help improve the lives of SCI sufferers by allowing them to regain some movement and motor control in limbs that were once incapable of such tasks. These devices are not without their shortcomings, however. Devices intended to help paraplegics may be awkward and somewhat unreliable. Complete motor control is never fully regained in these cases, but at the present time these methods are the best that are available.

Rehabilitation is the most common form of recovery for SCI injuries used by all patients at some point to improve their quality of life. Rehabilitation allows most paraplegics to regain some sensory and motor function back to their limbs, and significantly reduce musculoskeletal or neuropathic pain.[6] Early rehabilitation in spinal cord injury has been found to be more effective in hastening and promoting improvement in function following traumatic spinal cord injury.[7] Incomplete injuries respond better than complete injuries, with the use of a rehabilitation program, and specialized instruments such as the GIGER MD® to regain function in muscles. In any case, a combination of therapies involving rehabilitation have been the most effective means of restoring motor control and sensory information to paralyzed limbs.

## 25.3. New Technologies

Spinal cord injury sufferers face a long road to recovery and in the majority full functionality is never regained in limbs below the site of injury. New therapies being developed point towards drugs that promote regeneration of damaged nerve cells, and transplantation strategies to restore motor function. Perhaps the most promising, and most controversial path for spinal cord repair, is using stem cells to regrow damaged nerves and receptor sites. Recent studies have encouraged belief in the effectiveness of stem cells for nerve-cell regeneration. Animal models, using transplanted human neural stem cells after spinal cord injury resulting in paralysis, have shown a significant improvement in function of the hind legs after transplantation when compared to the control group.[8]

Stem cells differ according to their origin in different locations within the human body (see Chapter 2 for details). Use of early embryonic stem cells is the most controversial because of their means of harvesting, but they can differentiate into any type of cell present in the human body. Umbilical cord and adult stem cells are most often of interest because the method of harvesting these cells raises the fewest ethical issues within the medical community. Stem cells offer a useful tool in the regeneration of tissues for three reasons: they are a precursor to almost any type of cell within the human body; they have yet to play that role and are versatile in nature; and they have the ability to divide and proliferate over long periods of time. The goal of using stem cells as part of a therapeutic process for spinal cord injuries is to regenerate damaged neurons and repair a damaged spinal cord by stimulating axon regrowth to re-establish broken connections. The majority of cells found within the central nervous system are created during the embryonic and postnatal period of life, but in the early 1990s researchers discovered that neural stem cells exist, and that they could be used for regenerative purposes.[9]

In the repair of spinal cord injury animal models have shown promising results in regenerating neural pathways within the spinal cord. In 1999 McDonald and colleagues from the Washington University School of Medicine implanted embryonic stem cells in laboratory rats.[10] The rats were subjected to thoracic spinal cord injury by use of a metal rod which resulted in paralysis. Nine days following the injury, nearly one million embryonic stem cells were transplanted into the syrinx that had formed around the contusion in the spine. The site of injury showed all the normal events following the injury previous to the embryonic stem cell injection. Immediately following the injury some cells died immediately, and within 24 hours a second wave of apoptosis had occurred. Two weeks following the transplantation the implanted stem cells occupied the areas that would normally be subject to glial scarring. Five weeks after transplantation a majority of the stem cells had migrated away from the implantation site and provided neurons and glial cells to the spinal column. The rats regained some limited use of their legs in this study.[10] However, this does not explain the mechanism and factors involved in the observed regeneration of neural cells in the spinal cord. Two possibilities exist with the McDonald study for the regeneration of neural cells: the regaining of partial function could result from the few differentiated neurons; or perhaps the high differentiation of oligodendrocytes re-myelinited enough axons to re-establish communication. The immune system response to such infusions is a concern.

The goal of any spinal cord injury recovery procedure has not changed for some time, but the methods and technologies being used within the field are advancing. Four key principles of spinal cord repair are:

1. protecting surviving nerve cells from further damage;
2. replacing damaged nerve cells;
3. stimulating the regrowth of axons and targeting their connections appropriately; and
4. retraining neural circuits to restore body functions.

Different therapies exist to support the goals outlined above. New developments in preventing additional damage to nerve cells after injury include use of receptor antagonists which selectively block specific types of glutamate receptors responsible for cell self-destruction. Inflammation occurring within the first 12 hours of injury produces a wave of immune cells to protect and clean up nerve cells, but can also lead to further damage. Controlling these immune responses post injury by injecting macrophages directly into the site of injury and selectively boosting T cell response to the site may reduce secondary damage. Apoptosis, or cell self-destruction, can sweep through oligodendrocytes in damaged and nearby tissues, causing further degeneration in cells. Apoptosis-inhibiting drugs are being developed and tested on animal models to control this wave of destruction that can occur post SCI. By controlling apoptosis, scar tissue build up and fluid-filled gaps can be minimized and achieve a more effective means of neural communication throughout the spinal cord. Promoting regeneration of nerve cells within the spinal cord offers the greatest promise for a long-term solution. As an example, cell grafts implanted into the injured spinal cord can act as the media for injured areas to reconnect cut axons.

## 25.4. Ethical Issues

Within the spinal cord injury research community, most testing done is on animals before it is considered for testing on humans.

Caution should be used when using animals to test certain devices or procedures. The first and most basic ethical issues arise in the question of "are we inflicting unnecessary pain and suffering on sentient creatures?" Second, animal studies can only go so far in giving researchers applicable science to use in humans, as such neurological behavior may be much more complex within humans. The most promising stem cells for spinal cord repair have come in the form of embryonic stem cells, which is the most controversial source. An alternative is somatic cell nuclear transfer, or SCNT, which uses genetic material implanted directly into an egg cell, thus bypassing the fertilization of the egg using sperm cells.

The use of animal sources to create stem cells for manipulation in spinal cord injury repair brings up a new set of ethical concerns. Many bioethicists have already called for a ban on species-altering technology, for a multitude of reasons. Creating transgenic combinations of species effectively blurs the line between species and could offer potentially dangerous long-term medical and health risks. Other questions are raised: Will humans adopt physical or behavioral traits from other species? Who will have access to these technologies? What are the long-term socio-economic impacts? What legal controls should be used in such processes? The easiest option ethically is to use the patients' own stem cells for regeneration of their own tissues. See Chapters 2 and 24 for additional discussion of these issues.

## 25.5. Economic Issues

Spinal cord injuries have large economic consequences on the person involved, their family, friends, and society. Statistics taken in 1998 give the following data for spinal cord injury and its cost to the nation:

1. SCI costs the United States more than US$9.73 billion per year.
2. Annual SCI-related medical care costs average US$1.624 billion per year.

3. Medications and supplies cost US$449 million annually for those beyond their first year post-injury.
4. The average cost of personal assistance is estimated to be US$2.068 billion for those who are beyond the first year post-injury.[3]

Huge indirect costs are also an economic factor in spinal cord injury. Social costs in productivity and to the injured are:

1. Losses of productivity, the indirect costs of SCI, are approximately $2.591 billion nationwide.
2. Over the course of the post-injury lifetime, a person with SCI can expect to expend anywhere from $292,800 to $880,700 for injury-related costs, and to lose anywhere from $296,800 to $440,100 in lifetime earnings because of the injury.[3]

Aside from the direct monetary costs, the result of spinal cord injury on subsequent employment is astounding: 63% of spinal cord injury sufferers remained unemployed eight years after their injury.[1]

## 25.6. Recent Developments

Current medical and surgical procedures for treating SCI offer rather poor prognosis for patients and presently there is no intervention which can provide full recovery to a SCI patient. Tissue engineering has become a promising approach for the development of potential novel treatments for SCI. A wide variety of materials has been found to promote and support nerve tissue repair, particularly by allowing for directional axon growth after SCI. The injection of stem cells and growth factors into a spinal lesion has shown promise for the repair of SCI, although recently it has become widely accepted that an ideal treatment for SCI will consist of a multi-disciplinary combinatorial approach to attain full recovery.

The immediate site of the SCI undergoes a process of cyst formation and glial scarring in which no tissue is present that can promote regeneration.[11] As this process creates a gap in the neuronal tissue and inhibits regeneration, bypassing the defect or reconstructing functional tissue within the defect becomes necessary. As recent research has shown, it may be possible to do so with hydrogel-based biomaterials, consisting of natural or synthetic polymers which can be delivered directly to the injury site *via* injection. Hydrogels used for repairing SCIs may be either pre-formed scaffolds implanted into the site or gel, or gel precursors which assemble *in situ* after injection directly into the injury site. Hydrogels may allow axonal penetration, promote angiogenesis, and reduce necrosis and cavitations at the injury site.[12] The hydrogel scaffolds must be highly specialized to act as successful conduits for neuronal regeneration, and need to possess a high degree of biocompatibility. The scaffolds must be porous to allow cell in-growth and angiogenesis, but also be selectively permeable, to maintain consistency of the cerebral spinal fluid.[11] The swelling properties, strength, degradation rate, deformation, and flexibility are also essential characteristics to consider when designing a hydrogel for SCI repair.[11] The pore size, microtexture, and ordering or alignment of the pore architecture is also important to optimize the growth of tissue in a directional manner.[11] Along with the material properties of the scaffold, the incorporation of bioactive agents, biomolecules, and cells must also be taken into consideration when developing a scaffold for SCI repair.

Numerous polymer materials have been studied for the development of an optimal hydrogel biomaterial for neurological repair, including natural polymers such as collagen and fibrin, and synthetic polymers such as poly-D, L-lactic acid, and polyethylene glycol. Fibrin, a fibrous protein involved in blood clotting, has received a great deal of attention as it is biocompatible, supports tissue in-growth, promotes axonal regeneration, and delays accumulation of reactive astrocytes.[13] To demonstrate the potential of fibrin biomaterials for repair of SCI, three studies will be presented in which three distinct competent fibrin constructs were developed.

In a study by King *et al.*, collagen, fibrin, and fibrin-fibronectin (FB/FN) gels were developed for *in situ* formation of a spinal cord tissue regeneration construct.[14] The three gels were evaluated in a rat model SCI for biocompatibility, neuroprotective capability, and the promotion of axonal growth. The collagen gel was found to be least suitable for axon penetration, while the FB/FN mixture exhibited superior integration into host tissue and promoted robust axonal growth and proliferation. This study was novel in that it was the first to examine *in situ* fibrin gelation in a SCI defect. The study also demonstrated the prospect of developing a successful injectable fibrin gel, which eliminates the need for excision of injury-site tissue.[14]

In a study by Johnson *et al.*, the subacute treatment of a rat SCI was performed using a fibrin scaffold with an incorporated heparin-based delivery system (HBDS) for the controlled release of neurotrophin-3 (NT3).[15] Neurotrophins are proteins which promote the growth, function, and survival of neurons by inhibiting apoptosis. It was shown that the SCI treated with the NT3–fibrin scaffold was capable of promoting a significant increase in neural fiber sprouting even after a treatment was delayed for two weeks. Although other studies have demonstrated that injection of NT3 to the injury site promotes neural regeneration, this study showed the potential for developing a tissue scaffold capable of temporary controlled release of NT3 within the SCI site. As most SCI patients are not able to receive all treatments during acute care, this NT3–fibrin scaffold may allow for development of a superior treatment for subacute SCI.

A breadth of recent research has suggested that stem cell therapy has tremendous potential for treating SCIs, although the appropriate delivery method and environment must be developed for tissue regeneration by stem cells. A fibrin scaffold may serve as a suitable construct for transplanting seeded bone marrow stromal cells (BMSCs) into a SCI, as was demonstrated by Itosaka *et al.*[16] It was found that the fibrin matrix seeded with BMSCs transplanted into a rat model SCI were shown to provide enhanced BMSC survival, migration, and differentiation. Also, the rats with transplanted BMSC-fibrin scaffolds showed significant improvement in locomotor function after sustaining SCI. This study is a good example of the multi-faceted approach required to develop treatment for full recovery of a SCI patient. The three studies presented here are only a few examples of the prospective materials and techniques for conceiving an optimal SCI treatment. Other studies have used various natural and polymeric materials, as well as novel processing techniques such as electrospinning and bio-printing.[11,17]

Nanotechnology, or more specifically nanomedicine, has also become a primary area of interest for future diagnosis, treatment, and repair of spinal cord injuries. Chapter 4 discusses the use of nanoparticles of cerium oxide for scavenging free radicals, which offers great promise in the treatment of neural-based ailments. Nano-structured materials are being investigated for use as neural-tissue scaffolds, imaging-contrast agents (MRI, CT, and PET), stem cell-tracking particles, and drug-delivery systems.[18] Biocompatible nanofibers have been used to mimic the fibrous elements of extra-cellular matrix recreating the scale of features, yielding enormous surface area, and guiding the repair of nervous tissue. Zhu *et al.* used an electrospinning method to create a poly(lactide-co-glycolide) dual layer fibrous patch with an inner layer of aligned nanofibers and an outer layer of random oriented nanofibers.[19] The fibers were incorporated with Rolipram, which promotes metabolism in neurons and suppresses inflammation. It was observed that the Rolipram-incorporated nanofiber patches increased axonal growth, supported angiogenesis, and reduced astrocyte proliferation in a rat model SCI. Improved neurological motor function was also observed in the injured rats after several weeks.[19]

Polyethylene glycol (PEG) nanoparticles have been shown to effectively protect cells damaged by spinal cord injury. One study demonstrated that the use of polyethylene glycol-poly(D,L-lactic acid) block copolymer micelles delivered to the SCI site of a rat model could reduce inflammation and restore locomotor function when administered at an early stage of injury.[20] Cho *et al.* showed similar results with the use of dye-containing PEG-coated silica nanoparticles in a guinea pig model of SCI.[21] The dye-encapsulated particles enabled fluorescence microscopy to visualize spinal cord tissue extracted from the animals to show that the particles preferentially targeted the SCI site.

Nanoparticles of various compositions have also been studied for diagnostic imaging and delivery of therapeutic moieties to the site of a spinal cord injury. Super paramagnetic iron oxide nanoparticles and core-shell (i.e. silica core and gold shell) nanoparticles have been shown to improve imaging contrast of magnetic resonance imaging and photo-acoustic tomography, respectively.[22] Drug-loaded nanoparticles have also been investigated for specific cell targeting in spinal cord injuries by attaching cell-specific antibodies to the particle surface.[22] In addition, drug-loaded nanoparticles have been devised which release the payload of drugs upon exposure to a focused light source of specific wavelength or energy. Nanoparticles have been developed which can incorporate imaging agents, cell targeting molecules, and drug payloads known as multi-modal particles. These multi-modal particles have the potential to allow the development of a single combined diagnostic/therapeutic material for the treatment of SCIs. The application of nanomedicine — in tandem with other modalities such as advanced tissue scaffolds, stem cells, and gene therapy — will most likely drive the innovation of a successful regenerative medicine approach to restore full functionality to SCI patients.

## 25.7. Summary

The ease and relaxation of certain stem cell and fetal research laws in the United States has enhanced research into the repair and regeneration of nerve cells through stem cell therapies for spinal cord injury sufferers. Combination therapies such as physical rehabilitation, coupled with the effects of stem cells to regrow damaged neural pathways, represent the future of research in this field. Other such methods of reducing damage post injury also show promise and more immediate effects when compared to stem cell research benefits, which may be years from implementation.

## References



1. *Spinal Cord Injury Facts & Statistics* (2010). [Online]. Available at: http://www.sci-info-pages.com/facts.html. [Accessed 9 November 2010].

2. Young, W. (2007). *Spinal Cord Injury Levels and Classification.* [Online]. Available at: http://www.sci-info-pages.com/levels.html. [Accessed 9 November 2010].

3. Kruse, D., O'Leary, P., Berkowitz, M., *et al.* (1998). *Spinal Cord Injury: An Analysis of Medical and Social Costs*, Demos Medical Publishing, New York, NY.

4. Braughler, J.M. and Hall, E.D. (1983). Lactate and pyruvate metabolism in injured cat spinal cord before and after a single large intravenous dose of methylprednisolone, *J. Neurosurg.*, **59**, 256–61.

5. Hugenholtz, H. (2003). Methylprednisolone for acute spinal cord injury: not a standard of care, *Canadian Med. Assoc. J.*, **168**, 1145–1146.

6. Yap, E.C., Tow, A., Menon, E.B., *et al.* (2003). Pain during in-patient rehabilitation after traumatic spinal cord injury, *Int. J. Rehabil. Res.*, **2**, 137–140.

7. Sumida, M., Fujimoto, M., Tokuhiro, A., *et al.* (2001). Early rehabilitation effect for traumatic spinal cord injury, *Arch. Phys. Med. Rehabil.*, **82**, 391–395.

8. Cummings, B.J., Uchida, N., Tamaki, S.J., *et al.* (2005). Human neural stem cells differentiate and promote locomotor recovery in spinal cord-injured mice, *Proc. Nat. Acad. Sci. USA*, **102**, 14069–14074.

9. Reynolds, B.A. and Weiss, S. (1992) Generation of neurons and astrocytes from isolated cells in the central nervous system, *science*, **255**, 1707–1710.

10. McDonald, J.W., Xiao-Zhong, L., Qu, Y., *et al.* (1999). Transplanted embryonic stem cells survive, differentiate and promote recovery in the injured rat spinal cord, *Nat. Med.*, **5**, 1410–1412.

11. Madigan, N.N., McMahon, S., O'Brien, T., *et al.* (2009). Current tissue engineering and novel therapeutic approaches to axonal regeneration following spinal cord injury using polymer scaffolds, *Respir. Physiol. Neurobiol.*, **169**, 183–199.

12. Goh, E.L.K., Song, H. and Ming, G.-L. (2009). "Tissue Engineering Application in Neurology", in Meyer, U., Handschel, J., Wiesmann,

H.P., *et al.* (eds.), *Fundamental of Tissue Engineering and Regenerative Medicine*, Springer, Berlin and Heidelberg.

13. Johnson, P.J., Parker, S.R. and Sakiyama-Elbert, S.E. (2009). Fibrin-based tissue engineering scaffolds enhance neural fiber sprouting and delay the accumulation of reactive astrocytes at the lesion in a subacute model of spinal cord injury, *J. Biomed. Mater. Res.*, **92A**, 152–163.

14. King, V.R., Alovskaya, A., Wei, D.Y., *et al.* (2010). The use of injectable forms of fibrin and fibronectin to support axonal ingrowth after spinal cord injury, *Biomaterials*, **31**, 4447–4456.

15. Johnson, P.J., Parker, S.R. and Sakiyama-Elbert, S.E. (2009). Controlled release of neurotrophin-3 from fibrin-based tissue engineering scaffolds enhances neural fiber sprouting following subacute spinal cord injury, *Biotechnol. Bioeng.*, **104**, 1207–1214.

16. Itosaka, H., Kuroda, S., Shichinohe, H., *et al.* (2009). Fibrin matrix provides a suitable scaffold for bone marrow stromal cells transplanted into injured spinal cord: a novel material for CNS tissue engineering, *Neuropathology*, **29**, 248–257.

17. Silva, N.A., Salgado, N.J., Sousa, R.A., *et al.* (2010). Development and characterization of a novel hybrid tissue engineering-based scaffold for spinal cord injury repair, *Tissue Eng. A*, **16**, 45–54.

18. Kubinova, S. and Sykova, E. (2010). Nanotechnology for treatment of stroke and spinal cord injury, *Nanomedicine*, **5**, 99–108.

19. Zhu, Y., Wang, A., Shen, W., *et al.* (2010). Nanofibrous patch for spinal cord regeneration, *Adv. Funct. Mater.*, **20**, 1433–1440.

20. Shi, Y., Kim, S., Huff, T.B., *et al.* (2009). Effective repair of traumatically injured spinal cord by nanoscale block copolymer micelles, *Nat. Nanotechnol.*, **8**, 80–87.

21. Cho, Y., Shi, R., Ivanisevic, A., *et al.* (2010). Functional silica nanoparticle-mediated neuronal membrane sealing following traumatic spinal cord injury, *J. Neurosci. Res.*, **88**, 1433–1444.

22. Provenzale, J.M. and Silca, G.A. (2009). Uses of nanoparticles for central nervous system imaging and therapy, *Am. J. Neuroradiol.*, **30**, 1293–1301.

# Brain–Machine Interfaces

CHAPTER **26**

Matthew Goodman

## 26.1. Introduction

Brain–machine interfaces (BMI) offer the potential to change the world. At the experimental level BMIs have demonstrated the ability to treat blindness,[1,2] deafness,[3] Parkinson's disease,[4] and enable those with physical disabilities such as quadriplegia/paraplegia.[5] Despite these incredible promises and an immense clinical need, the use of BMIs remains almost completely experimental. This is due to four primary factors:

1. Uncertainty of glial and central nervous system (CNS) tissue response to implants.
2. Uncertainty of long-term materials safety and efficacy in the brain environment.
3. Lack of standards in hardware, software, and surgical techniques.
4. Risk associated with brain surgery.

Brain–machine interfaces, also called brain–computer interfaces, is a term applied to several devices for the transduction of electrical signals generated between neurons and another electrical device. The intention is to provide either information about the activity of the region, or stimulation to the brain in order to illicit a certain neurological response. BMIs fall into two broad categories: invasive and non-invasive.

Non-invasive BMIs typically utilize electroencephalogram (EEG) like electrodes placed on the scalp, designed to measure the external signals created by large numbers of neurons. After a large amount of biofeedback-based training, these signals can be modulated by the individual under study. When the modulated waveforms are analyzed by a computer, they can serve as simple control devices for the disabled person. This method is suitable for the 2D control of a mouse cursor[6] or operation of a keyboard.[7] Despite the fact that this method has enabled many computer applications including a spelling device,[8] the resolution and speed with which it can be used is extremely limited.[9] This limit is because surface field measurements are very weak[10] and signals are frequently drowned out by stray electromagnetic (EM) radiation that permeates the modern world.[11] Other non-invasive approaches include using PET or MRI scanning to measure brain activity in certain sections, but they are too bulky to be used as any sort of practical control device.[12] Such non-invasive methods hold two huge advantages: they are considerably cheaper, and do not require any invasive surgery to utilize. However, they have three prime limitations that make them unusable for the most promising BMI applications:

1.  They are only usable for output from the brain, eliminating the possibility of their use for sensory restoration or any type of stimulation of the brain.
2.  They can only measure large gradients in neuronal activity.
3.  They are limited to the rate of 5–25 bits per second under optimum conditions, making them unsuitable for use in the control of prosthetics.[13]

For these reasons they are not discussed further in this chapter. For an excellent review of the field see Wolpaw.[14]


## 26.2. Invasive BMIs

Invasive BMIs present a harsh contrast to non-invasive BMIs, because in the former, electrodes come into physical contact with neuronal tissue, either in the brain itself or in the peripheral nervous system. While this seems a severe approach, the benefits of such techniques yield the most potentially promising treatments. With implanted cortical electrodes, the ability to measure or stimulate the firing of even single neurons is a possibility.[15] Second, the stimulation and projection of data onto a small area of brain tissue is also of great promise. This level of data resolution has enabled some treatments and applications that were inconceivable ten years ago, all due to improvement of signal-processing techniques and miniaturization of electronics.

Design of invasive BMIs is currently a large field of research. The earliest and still most widely employed designs are simple microwire electrodes. Such electrodes typically consist of a thin (about 125 μm) platinum–iridium wire, insulated with parylene or polyimide polymer coatings. The electrodes are inserted into the cortex through a small hole in the skull, and mechanically anchored to the skull. However, these types of BMI have several large drawbacks including fixation to the skull and tendency to deviate from the straight trajectories of their insertion.[16] Despite these drawbacks, microwires remain the only way to detect or stimulate neurons buried below the most superficial surfaces of the brain. However, single microwires are unable to measure close proximity groups of neurons. This deficiency has prompted development of more complex electrode arrays.

The advent of high-precision micro-fabrication techniques led to the design and fabrication of the Utah electrical array (UEA). The array consists of a 4.2 mm × 4.2 mm glass/silicon composite base, from which 25–100 μm silicon needle-type electrodes project. Each needle is approximately 1,500 μm long, 80 μm in diameter at the base, and tapers to a sharp point with a platinum coated tip.[17]

The UEA is particularly interesting due to its use of a glass dielectric between the individual sensing tips. This is responsible for the high impedance between the contacts and results in excellent data collection and stimulation capabilities. The process of array production is lengthy, but the product is extremely precise, biocompatible, and appropriately electrically insulated. It is also mechanically durable, and has been used in numerous studies without mechanical failure.

This geometry presents several key advantages. Because the individual electrodes are spaced within several microns, the sampling and resolution of nervous system events is far superior to that provided by traditional microwires. The ability to sample different groups of nearby neurons has proven invaluable for applications such as control interfaces for prosthetic limbs.

Cortical arrays such as the UEA are connected to a percutaneous pedestal by insulated platinum–iridium microwires. This geometry allows the array to float with the brain, and virtually eliminates issues with micro-motion caused by mechanical fixation to the skull. The geometry also lends itself to the separation of neural tissue during insertion, as opposed to the severing thereof. Another approach, though far less widely used, is the Michigan array. It is fabricated by methods used in the semiconductor industry, then several plate-derived electrode combs are sandwiched to create an array. This approach allows use of existing silicon lithography and etching; however it is less widely used due to concerns of brittleness and fragility of the thin silicon sections.

A similar approach using semiconductor manufacturing techniques is to generate a very thin group of electrodes that are flexible.[18] This same design incorporates polyimide-coated sections that could be used to engineer bioactive interfaces. This array, while promising, is not as easily implantable as the UEA. Because it is highly flexible, holes for the three electrodes must first be punctured through the pia with either a #11 scalpel, or a relatively stiff 100 $\mu$m tungsten wire. This is a huge drawback in design; however, electrodes lasting more than a year have been manufactured and tested.[19] This was accomplished by incorporating neural growth factor (NGF) into the implant, likely eliciting a favorable healing response that offset the trauma of insertion.

## 26.3. Surgical Techniques

One of the largest causes of variance of BMI response is the technique of insertion of the electrode arrays. Studies show a wide variety of installation techniques. Rates of array insertion vary from a carefully metered 100 µm/second to pneumatically stapled 8.3 meter/second.[20] This variance is illustrative of two different schools of thought on how to minimize initial trauma and thus glial scarring. Proponents of slow insertion say that it allows the most time for nervous/glial tissue to rearrange under the forces exerted by insertion of the array. The chief drawback is that as the array first comes into contact with the cortex, the soft cerebral tissue dimples in a spongy manner, "pinching" the array, and perhaps leaving stress between the tissue and array after the desired depth is reached.

The higher speeds of insertion are achieved by a specialized pneumatic tool that impacts the back of the array, basically launching it into the cortex. The principle drawback of this is that as the base of the electrode reaches the tissue it will finally rest against, excess kinetic energy is imparted into the brain, causing possible trauma and bleeding. However, the increased rate of insertion does not allow the dimpling and related stress/trauma. Further research is needed in this area to find the optimal technique for insertion.

In addition, standards are lacking for electrical stimulation and reading techniques. Neurons deviate from their typical behavior when in the influence of an electrical field. When a neuron is exposed to weak fields, the probability of reaching the action potential is increased. Similarly when a neuron depolarizes to send a signal, there is a measurable change in potential energy from the creation of an ionic potential by the cell. These two simple concepts are what allow electrical interfaces, conductors, and fields to interact with neuronal tissue.

When working with electrical signals in the brain it is crucial to keep the following facts in mind:

1. Electrical stimuli can damage cells.[21]
2. The response of neuronal and glial tissue varies with region, depth, stimulation frequency, and stimulation intensity.
3. Electrical stimuli can cause preferential axon growth.[22]

Stimulating the electrochemistry of the brain is presently an art that requires great care when attempting measurement or stimulation of nervous tissue. Currently, there is no standard for safe working ranges with such electrodes. Many electrical conventions are observed in relation to EEG machines, regarding the ways in which electrical signals are gathered and processed, and ensuring they do not come into direct contact with neural tissue. With the first commercial devices entering the market, the need for such standards is urgent. Cyberkinetics Inc. has begun clinical trials of its BrainGate™ BMI. The product, which is very similar to the UEA, is designed for use in the severely motor-impaired patient to provide mobility and the ability to interact with a computerized interface. The commercial device is available under an Investigational Device Exemption (IDE) from the US Food and Drug Administration (FDA).

Given the importance of minimizing glial scarring, the investigation of optimal surgical techniques must be emphasized, especially before human testing of new devices.

## 26.4. Current Applications in the CNS

Active electrodes have found use in clinical applications, most prevalently in cochlear implants to provide hearing to profoundly deaf patients (see Chapter 30 for details). Cochlear implants have a grid of wires placed in contact with auditory nerve tissue within the cochlea. The electrodes are stimulated by a microphone/transmitting device, typically worn behind the ear. The implant stimulates the auditory nerve to produce noises that are processed by the brain, so that with sufficient training the patient is aided in the perception of hearing. The small audio cues greatly aid in sound recognition and language comprehension as compared to lip reading techniques.[23] These implants are safe, with the chief risk being an increased chance of contracting viral meningitis.[24] Implants in the peripheral nervous system such as these are notably simpler than those used in the CNS, because the mechanisms for repair of damaged tissue are considerably different.

Recently deep brain stimulation has been shown to be very effective in the treatment of Parkinson's disease.[25] (See Chapter 27 for a discussion of this disease.) Computer-based decision-making models predicted that periodic stimulation of the subthalamic nucleus (STN) would substantially decrease the periodic tremor associated with Parkinson's.[26] This prediction and subsequent implant development has led to a new variety of implants, similar in implementation to a pacemaker. Periodic stimulation can be provided to the STN by employing microwire electrodes and a subcutaneously implanted signal generator that is externally programmable. This treatment is effective at treating the characteristic tremor associated with the disease. A tendency for rapid decision-making is observed,[27] giving it some advantages over dopamine-based treatment.

A similar stimulatory device is used to treat severe epilepsy. In this case, the electrode serves as both a sensor for the oncoming seizure, and a stimulation device to derail the potential before it becomes a fully symptomatic seizure.[28] This device has improved the standard of living for severe epileptics.

## 26.5. BMIs and the Future

One of the potential uses of BMIs in the future is control of external devices by internal thought processes, for example, those required for control of prosthetic limbs. This concept has been demonstrated in monkeys, who were able to perform moderately complex manipulations with brain-controlled mechanical arms in real time.[29] There have also been proof-of-concept experiments for visual prostheses in human subjects, as discussed below.

The chief limiting factor in future applications of BMI technology remains the uncertainty about the long-term behavior of chronically implanted electrodes. When properly implanted, they have been shown to be safe for more than a year (the longest trial to date).[30] This remains the largest question to be addressed before the full prospect of such technology can be realized. Obtaining a viable answer to this question is bounded by knowledge of the CNS wound-healing process, particularly knowledge of the mechanisms of glial scarring.

## 26.6. BMIs and Glial Scarring

Implants outside the blood brain barrier (BBB) experience considerably different wound-healing mechanisms than those inside. Glial scarring is a wound-healing mechanism that is invoked by the CNS whenever chemical or physical trauma is detected.[31] Implantation of almost all materials within the CNS experience this process. Many materials elicit toxic effects when implanted cortically.[32] Even with use of non-toxic, biocompatible electrode materials, the scarring appears instigated by the damage of nervous tissue during implantation rather than a response to an implanted material. This makes material selection limited to those that have been shown to be largely inactive when implanted in the brain. They include silicon and silica,[33] platinum, and polyimide.[34]

One of the major obstacles in BMI research is the difficulty of observing the cell–implant interface. Silicon is a hard material compared to cell cultures. Preparing a suitably thin section for optical microscopy requires the removal of the array from the tissue to be sectioned. When doing this much information is lost, such as the interfacial contact of the array material to neuronal cells. Removal of the array may disrupt any ingrown tissue, effectively preventing analysis of the cell–electrode interface. However, with careful technique, as outlined in Rousche *et al.*'s paper, meaningful information can be obtained.[34]

The wound-healing process that takes place in the CNS behind the BBB is dissimilar to that in the rest of the body. The process by which a glial scar is formed (gliosis) is considerably slower than typical wound-healing processes. At least two weeks are required to form a semi-stable glial scar.[35] However if the BBB is compromised, healing times can extend into months.[36] In CNS-based wound healing, there are at least six different cells involved: oligodendrocytes, microglia, meningeal cells, oligodendrocyte precursors, astrocytes, and multipotent progenitor cells.

This list is greatly expanded when the BBB is compromised in any capacity. This compromise is very common in any sort of physical trauma (such as putting electrodes into brain tissue), as there is very close proximity between capillaries, glial, and nerve cells. Many other cells and complications become involved: red blood cells, platelets, macrophages, microphages, cytokines, and blood proteins and proteoglycans.

There are many different signalling pathways that promote various stages of healing and wound response. Chondroitin sulphate proteoglycans (CS-PG) are of large interest because of the conflicting reports of their role in gliosis.[37] Myelin-associated molecules are also prevalent in the area of trauma for quite some time after the actual injury, and appear to play a key role in gliosis. The CNS has considerably lower phagocytic activity than tissues in the body, so the removal of myelin debris is also correspondingly slow. Such debris may be one of the largest factors that cause chronic response to CNS trauma.[38] While it is clear that all of these molecules play a large role in the process of gliosis, many studies targeted at up- or down-regulating one or more factors have failed both to significantly increase axon regeneration and minimize glial scar formation. This is because, like cytokine-regulated immune response, there are several mechanisms that can independently operate to cause glial scar formation.

Understanding and minimizing scar tissue formation is important to clinical use of BMIs, because studies have linked the decrease in long-term electrical performance to glial scarring.[39] Glial scars serve not only as physical barriers to neuron axon growth, but also appear to chemically deter the growth of healthy nerve axons.[40] Since BMI performance requires having proximity of active healthy neurons to the contact surfaces, reduction of glial scar formation is central to the success of chronically implanted BMIs. The only positive effect of glial scarring is anchoring of the implant by the scar tissue to minimize movement. For this reason, the complete elimination of glial scarring is not desirable, but at the very least the reduction of scarring around the electrode tips is essential.

Conveniently, the BMI itself can be used to measure the amount of glial scarring that each electrode experiences.[41] Glial scar tissue is considerably denser and less conductive than native brain tissue, therefore measuring the resistance or impedance between electrodes provides a qualitative measure for the amount of scarring experienced between the electrode tips. Numerous techniques to minimize or eliminate glial scarring have been investigated, with mixed results. Most were investigated with a goal of regeneration and healing of nervous tissue in the spine and CNS. An excellent summary of those treatments is published by David and Lecroix[35] and Fawcett and Asher.[42] The primary approaches include eliminating myelin compounds from the damaged area; alteration of CS-PGs; increasing cAMP; and neutralizing inhibitors in the scar tissue. Other "extrinsic" methods have also been researched and have proved to be the most successful to date.[43] They include destruction or removal of myelin debris and the complete removal of glial cells from the damaged area. But these two approaches, while effective, have a chief drawback: they are not conducive to a clinical setting. Improvement in understanding of these pathways and techniques constitutes one of the major obstacles for regenerative medicine in the CNS, and also for the use of BMIs.

A large body of literature exists which has studied glial scarring in the CNS in the context of accidental trauma, and investigation and desire for nerve regeneration. To date, research of the relation between glial scarring as it is related to BMIs has generally been a follow-up to other research with the electrode arrays. The matter is complicated by the fact that investigation of the silicon–tissue interface is difficult to prepare for light microscopy. Though the geometry of current BMIs has been optimized to minimize scarring,[44] there is still much room for improvement.

While many studies have connected glial scarring and BMI performance, little work has been done in the area of minimizing glial formation specifically to maximize long-term BMI effectiveness. This research could be performed with a relatively small test group. By performing various treatments to individual needles or rows of needles in an array, the need for test animals could be minimized. Measurement of glial scar formation could be performed by the measurement of potential between electrodes on the array.

## 26.7. Neuroplasticity

One of the fascinating capabilities of the human brain is migration of the function of a particular region in response to damage, or to new stimuli. It is the hope of investigators that this particular facility can be harnessed to improve the performance of BMIs. For example, the tongue has been used as an electrical stimulus feedback interface for joint-positioning sensing. What this means is that, with training, the brain has remapped its joint-sensing recognition to electrical signals delivered to the tongue.[45] Studies have been performed in monkeys with an electrode array providing feedback for a particular problem, exhibiting increased learning rates with corresponding stimulation.[46] This was extended to a point where the feedback was purely by electrical stimulation and witnessed "sensing without touching."[47] The ability to project thoughts and wills upon inanimate objects, some speculate, is an extension of the human tool-using nature.[48] This variety of brain plasticity is fascinating, as it suggests the brain has the potential to rewire itself to adopt new stimuli and apply its patterns to them. The hope is that this plasticity can be harnessed in the future to create visual, auditory, and motor prostheses for humans.[49] It is also foreseeable that this sort of neural training with sensory-motor feedback may someday even extend human perception beyond what we are currently capable of experiencing.

## 26.8. Ethical Issues

The primary ethical issue is the balance between uncertainties of long-term effectiveness of BMIs *versus* the amount of human benefit that could come from these technologies. However, long-term safety and effectiveness is impossible to establish without eventual human trials. Brain surgery is never undertaken lightly; many even say that there is no such thing as routine brain surgery. A decision to perform brain surgery involving implants of unknown lifetime and stability is a difficult ethical issue. Investigators must be careful that patients and families involved in such studies are fully aware of the risks and the uncertainty as to the long-term effects. While this is true of any investigational device, it is a particularly precarious situation with BMIs. A failed operation could leave someone permanently disabled, brain damaged, or dead. Though many could benefit from BMIs, they are typically not used to treat a life-threatening condition. Installation of BMIs with any metallic component makes the subsequent use of MRI dangerous. While silicon is non-magnetic, it is still a good conductor. No study has been conducted on the subject, but the possibility that there could be induced currents on the electrodes is likely to prohibit the use of MRI on the patient. This correspondingly lowers the ability to examine the brains of those who are already suffering.

## 26.9. New Developments

Since 2000, the technologies supporting BMI have developed at a rapid rate due to the broad range of applications for interfaces to computer systems — including communication, biofeedback, and operation of control systems — which all have implications for commercial and medical markets as well as the military, robotics, and even space flight. Perhaps the most exciting applications of this technology remain in providing significant life-changing remediation for victims of neuronal dysfunction, hearing and vision impairment, as well as individuals requiring artificial limbs, or have sustained significant injury to the brain or spinal cord. The promise of these applications has attracted a great deal of research funding which, combined with our understanding of genetics, stem cells, and the complexity the human brain, have led to creative developments in the field of BMI. The latest developments fall principally into one of the following four categories:

1. New and diversified technological applications.
2. Significant advancements in the functionality of these various applications (e.g. signal processing, or precision of the neuronal interfaces).
3. Improvements in the techniques for successfully implementing BMI devices.
4. Development of biomaterials of a greater biocompatibility.

As the rapid expansion of technology suggests, there are many areas of research that warrant further investigation. The development of prosthetics and other medical devices, along with improvements in biocompatibility, have reached a crescendo of activity which will eventually lead to the introduction of products for health-care applications.

The development of an interface for controlling prosthesis is an encouraging area for application of BMIs, as it could greatly improve prosthetic control and coordination, therefore greatly improving the user's quality of life.[50–53] Two main design concepts under investigation are neural signal extraction for motor commands, and signal insertion for sensory perception, as these concepts have been identified as essential principles of a functional interface.[54] Progress is being made in extraction of the appropriate signals from the brain to direct motor control and coordination of the prosthesis. Preliminary studies have been carried out demonstrating a hierarchical brain–computer interface with lower-level functional stimulation of muscle groups for locomotion.[55] The three-tiered system developed by Zhang *et al.* mimics the multiple levels of motor function and control found in the central nervous system.[55] It is important to understand and consider the bi-directional nature of the nervous system when designing BMIs, as the nervous system works on feedback loops involving both the afferent and efferent neural pathways. Processing of sensory stimuli for cognition and feedback *via* motor function has also been an area that has experienced a number of recent advances.[56–58] Gaining a better understanding of the brain–computer interface cycle will allow for the development of a seamless interface between the nervous system and computer. Such a BMI must be able to measure brain activity, classify data, provide feedback, and yield an effect relative to the feedback.[56] Implantable brain computer interfaces for the monitoring and control of physiological functions and disease is another area of interest, with potential applications for patients suffering from Parkinson's disease, epilepsy, and other CNS pathologies.[57]

Advances in signal processing and system controls are of great importance, as these are currently some of the major limiting factors for BMI development. A number of challenges exist in extracting data, correctly processing the data, and developing complex models to calculate appropriate feedback from the processed signals. Current research is focusing on enhancing the understanding of neural representation of causality, workspace, and global-directed navigation.[59] The knowledge of such principles plays an important role in the development of algorithms for modeling neural pathways and neural modulation. Such technologies must also be incorporated to more efficiently detect and amplify neural signalling to allow for better signal averaging, deceased noise, and improved signal deconvolution. Much more research will be needed, with a high level of cohesion between medical professionals and engineers, to reach the full potential offered by BMIs.

## 26.10. Summary

The promise that BMIs hold for the future is exciting, but the obstacles to implementation are great. As more research is done, BMIs will become a realistic treatment option for many conditions. The field of BMIs is broad and rapidly moving. Breakthroughs have been summarized in many different fields: neuroscience,[1] biology,[2] and philosophy,[3] which demonstrate the promise, reward, and corresponding dangers of this new technology.

## References


1. Normann, R. (2007). *Sight Restoration for Individuals with Profound Blindness*, The Center for Neural Interfaces and the John A. Moran Eye Center, University of Utah Press, Salt Lake City, UT.
2. Normann, R.A., Maynard, E.M., Rousche, P.J., *et al.* (1999). A neural interface for a cortical vision prosthesis, *Vision Res.*, **39**, 2577–2587.
3. Wilson, B.S., Finley, C.C., Lawson, D.T., *et al.* (1991). Better speech recognition with cochlear implants, *Nature*, **352**, 236–238.
4. Vingerhoets, F.J., Villemure, J.G., Temperli, P., *et al.* (2002). Subthalamic DBS replaces levodopa in Parkinson's disease: two-year follow-up, *Neurology*, **58**, 396–401.

5. Chapin, J.K. (2000). Neural prosthetic devices for quadriplegia, *Curr. Opin. Neurol.*, **13**, 671–675.

6. Kübler, A., Neumann, N., Kaiser, J., *et al.* (2001). Brain–computer communication: self-regulation of slow cortical potentials for verbal communication, *Arch. Phys. Med. Rehabil.*, **82**, 1533–1539.

7. Obermaier, B., Müller, G. and Pfurtscheller, G. (2003). Virtual keyboard controlled by spontaneous EEG activity, *IEEE T. Neur. Sys. Reh.*, **11**, 422–426.

8. Birbaumer, N., Ghanayim, N., Hinterberger, T., *et al.* (1999). A spelling device for the paralysed, *Nature*, **398**, 297–298.

9. Leuthardt, E.C., Schalk, G., Wolpaw, J.R., *et al.* (2004). A brain–computer interface using electrocorticographic signals in humans, *J. Neural. Eng.*, **1**, 63.

10. Sheikh, H., McFarland, D.J., Sarnacki, W.A., *et al.* (2003). Electroencephalographic (EEG)-based communication: EEG control versus system performance in humans, *Neurosci. Lett.*, **345**, 89–92.

11. Obermaier, B., Neuper, C., Guger, C., *et al.* (2001). Information transfer rate in a five-classes brain–computer interface, *IEEE T. Neur. Sys. Reh.*, **9**, 283–288.

12. Maruishi, M., Tanaka, Y., Muranaka, H., *et al.* (2004). Brain activation during manipulation of the myoelectric prosthetic hand: a functional magnetic resonance imaging study, *Neuroimage*, **21**, 1604–1611.

13. Birbaumer, N. (2006). Brain–computer-interface research: coming of age, *Clin. Neurophysiol.*, **117**, 479–483.

14. Wolpaw, J.R., Birbaumer, N., McFarland, D.J., *et al.* (2001). Brain–computer interfaces for communication and control, *Clin. Neurophysiol.*, **113**, 767–791.

15. Wolpaw, J.R., Birbaumer, N., Heetderks, W.J., *et al.* (2000). Brain–computer interface technology: a review of the first international meeting, *IEEE T. Rahabil. Eng.*, **8**, 164–173.

16. Tian, C.X. and He, J. (2006). Monitoring insertion force and electrode impedance during implantation of microwire electrodes, *Eng. Med. Biol. Soc. Ann.*, **1**, 7333–7336.

17. Jones, K.E., Campbell, P.K. and Normann, R.A. (1992). A glass/silicon composite intracortical electrode array, *Ann. Biomed. Eng.*, **20**, 423–437.

18. Kennedy, J.R. (1989). The cone electrode: a long-term electrode that records from neurites grown onto its recording surface, *Neurosci. Methods*, **29**, 181–193.

19. Rousche, P.J., Pellinen, D.S., Pivin, D.P. *et al.* (2001). Flexible polyimide-based intracortical electrode arrays with bioactive capability, *IEEE T. Bio-Med. Eng.*, **48**, 361–371.

20. Theodore, W.H. and Fisher, R.S. (2004). Brain stimulation for epilepsy, *Lancet Neurol.*, **3**, 111–118.

21. Head, H. and Holmes, G. (1991). Sensory disturbances from cerebral lesion, *Brain*, **34**, 102–254.

22. Asensio-Pinilla, E., Udina, E., Jaramillo, J., *et al.* (2009). Electrical stimulation combined with exercise increase axonal regeneration after peripheral nerve injury, *Exp. Neurol.*, **219**, 258–265.

23. Cohen, N.L., Waltzman, S.B. and Fisher, S.G. (1993). A prospective, randomized study of cochlear implants, *N. Engl. J. Med.*, **328**, 233–237.

24. Teissl, C., Kremser, C., Hochmair, E.S., *et al.* (1999). Magnetic resonance imaging and cochlear implants: compatibility and safety aspects, *J. Magn. Reson. Imaging*, **9**, 26–28.

25. Rodriguez, M.C., Guridi, O.J., Alvarez, L., *et al.* (1998). The subthalamic nucleus and tremor in Parkinson's disease, *Movement Disord.*, **13**, 111–118.

26. Frank, M.J. (2005). Dynamic dopamine modulation in the basal ganglia: a neuro-computational account of cognitive deficits in medicated and non-medicated Parkinsonism, *J. Cognitive Neurosci.*, **17**, 51–72.

27. Jahanshahi, M., Ardouin, C.M.A., Brown, R.G., *et al.* (2000). The impact of deep brain stimulation on executive function in Parkinson's disease, *Brain*, **123**, 1142–1154.

28. Halpern, C.H., Samadani, U., Litt, B., *et al.* (2008). Deep brain stimulation for epilepsy, *Neurotherapeutics*, **5**, 59–67.

29. Carmena, J.M., Lebedev, M.A., Crist, R.E., *et al.* (2003). Learning to control a brain–machine interface for reaching and grasping by primates, *PLoS Biol.*, **1**, E42.

30. Rousche, P.J. and Normann, R.A. (1998). Chronic recording capability of the Utah intracortical electrode array in cat sensory cortex, *J Neurosci. Meth.*, **82**, 1–15.

31. Rudge, J.S., Smith, G.M. and Silver, J. (1989). An *in vitro* model of wound healing in the CNS: analysis of cell reaction and interaction at different ages, *Exp. Neurol.*, **103**, 1–16.

32. Babb, T.L. and Kupfer, W. (1984). Phagocytic and metabolic reactions to chronically implanted metal brain electrodes, *Exp. Neurol.*, **86**, 171–182.

33. Stensaas, S.S. and Stensaas, L.J. (1978). Histopathological evaluation of materials implanted in the cerebral cortex, *Acta Neuropathol.*, **41**, 145–144.

34. Rousche, P.J., Pellinen, D.S., Pivin, D.P. Jr, *et al.* (2001). Flexible polyimide-based intracortical electrode arrays with bioactive capability, *IEEE T. Bio-Med. Eng.*, **48**, 361–371.

35. David, S. and Lacroix, S. (2003). Molecular approaches to spinal cord repair, *Annu. Rev. Neurosci.*, **26**, 411–440.

36. Perry, V.H. and Gordon, S. (1991). Macrophages and the nervous system, *Int. Rev. Cytol.*, **125**, 203–244.

37. Moon, L.D.F., Brecknell, J.E., Franklin, R.J.M., *et al.* (2000). Robust regeneration of CNS axons through a track depleted of CNS glia, *Exp. Neurol.*, **161**, 49–66.

38. Hinterberger, T., Veit, R., Wilhelm, B., *et al.* (2005). Neuronal mechanisms underlying control of a brain–computer interface, *Eur. J. Neurosci.*, **21**, 3169–3181.

39. Polikov, V.S., Tresco, P.A. and Reichart, W.M. (2005). Response of brain tissue to chronically implanted neural electrodes, *J. Neurosci. Meth.*, **148**, 1–18.

40. Silver, J. and Miller, J.H. (2004). Regeneration beyond the glial scar, *Nat. Rev. Neurosci.*, **5**, 146–156.

41. Kübler, A., Kotchouney, B., Kaiser, J., *et al.* (2001). Brain–computer communication: unlocking the locked in, *Psychol. Bull.*, **127**, 358–375.

42. Fawcett, J.W. and Asher, R.A. (1999). The glial scar and central nervous system repair, *Brain Res. Bull.*, **49**, 377–391.

43. Neumann, H., Kotter, M.R. and Franklin, R.J.M. (2009). Debris clearance by microglia: an essential link between degeneration and regeneration, *Brain*, **132**, 288–295.

44. Edell, D.J., Toi, V.V., McNeil, V.M., *et al.* (1992). Factors influencing the biocompatibility of insertable silicon microshafts in cerebral cortex, *IEEE T. Bio-Med. Eng.*, **39**, 635–643.

45. Vuillerme, N., Chenu, O., Demongeot, J., *et al.* (2006). Improving human ankle joint position sense using an artificial tongue-placed tactile biofeedback, *Neurosci. Lett.*, **405**, 19–23.

46. Carmena, J.M., Lebedev, M.A., Crist, R.E., *et al.* (2003). Learning to control a brain–machine interface for reaching and grasping by primates, *PLoS Biol.*, **1**, 193–208.

47. Romo, R., Hernández, A., Zainos, A., *et al.* (2000). Sensing without touching: psychophysical performance based on cortical microstimulation, *Neuron*, **26**, 273–278.

48. Gurfinkel, V.S., Levick, Yu.S. and Lebedev, M.A. (1991). "Body Scheme Concept and Motor Control. Body Scheme in the Postural Automatisms Regulation", in Chernavskii, A.V. (ed.), *Intellectual Processes and Their Modelling*, Nauka, Moscow, pp. 24–53.

49. Schmidt, E.M., Bak, M.J., Hambrecht, F.T., *et al.* (1996). Feasibility of a visual prosthesis for the blind based on intracortical micro stimulation of the visual cortex, *Brain,* **119**, 507–522.

50. Lebedev, M.A. and Nicolelis, M.A.L. (2006). Brain–machine interfaces: past, present and future, *Trends Neurosci.*, **29**, 536–546.

51. Donoghue, J.P. (2002). Connecting cortex to machines: recent advances in brain interfaces, *Nat. Neurosci.*, **5**, 1085–1088.

52. Coyle, S., Ward, T. and Markham, C. (2003). Brain–computer interfaces: a review, *Interdiscipl. Sci. Rev.*, **28**, 112–118.

53. McFarland, D.J. and Wolpaw, J.R. (2010). Brain–computer interfaces for the operation of robotic and prosthetic devices, *Adv. Comput.*, **79**, 196–187.

54. Konrad, P. and Shanks, T. (2010). Implantable brain computer interface: challenges to neurotechnology translation, *Neurobio. Dis.*, **38**, 369–375.

55. Zhang, D., Liu, G., Huan, G., *et al.* (2009). A hybrid FES rehabilitation system based on CPG and BCI technology for coordination: a preliminary study, *Lect. Notes. Artif. Int.*, **5928**, 1073–1084.

56. van Gerven, M., Farquhar, J., Schaefer, R., *et al.* (2009). The Brain–computer interface cycle, *J. Neural Eng.*, **6**, 1–10.

57. Grill, W.M., Norman, S.E., Bellamkonda, R.V., *et al.* (2009). Implanted neural interfaces: biochallenges and engineered solutions, *Ann. Rev. Biomed. Eng.*, **11**, 1–24.

58. O'Doherty, J.E., Lebedev, M.A., Hanson, T.L., *et al.* (2009). A brain–machine interface instructed by direct intracortical microstimulation, *Front. Integr. Neurosci.*, **3**, 1–10.
59. Sanchez, J.C., Principe, J.C., Nishida, T., *et al.* (2008). Technology and signal processing for brain–machine interfaces, *IEEE Signal Proc. Mag.*, **25**, 29–40.

# Stem Cell Research for Treatment of Parkinson's Disease



## CHAPTER 27

Colleen Young and Michael B. Fenn

## 27.1. Introduction

Parkinson's disease (PD) is a chronic, progressive neurodegenerative disorder caused by the degeneration of the mesencephalic dopaminergic neurons found in the substantia nigra region of the brain. These neurons produce the neurotransmitter dopamine which plays a number of important roles in cognition, behavior, and voluntary movement. PD is typically characterized by tremor, rigidity, and hypokinesia.[1] Following the initial onset, later-stage cases of PD present disturbances in gait and balance, and psychological disorders such as depression and dementia. The most extreme cases of PD result in akinesia and failure of autonomic function. To date there is no cure for PD, but stem cell research has been leading the way among many researchers in the scientific and medical communities.

## 27.2. New Developments

The cause of PD is, for the most part, still not understood, and the sporadic nature of the disease adds an increasing degree of difficulty in elucidating the etiology. Several recent studies have, however, indicated that certain genetic factors which cause increased oxidative stress and mitochondrial impairment may play a role in PD.[1] Current treatment involves pharmaceutical-based therapy with dopaminergic agonist, such as levodopa, and inhibitors of dopamine-degrading enzymes such as carbidopa.[2] Surgical techniques currently in use are typically reserved for the most advanced PD cases, with the preferred treatment involving high-frequency deep-brain stimulation of the subthalamic nucleus.[3] Nevertheless, both drug therapy and surgical intervention have numerous limitations and only treat symptoms, as no method has been conceived to prevent or repair damage caused by PD. There is therefore an obvious need for the development of an approach which will offer neuroprotection and regeneration of the damaged nervous tissue.

In the past, cell-based therapies for PD have undergone clinical trials in which human fetal-derived mesencephalic dopamine neurons were grafted onto the striatum of PD patients. Results from such treatments have varied greatly.[1,4] Due to the unpredictable outcomes from fetal tissue transplantation, recent research has focused heavily on stem cell therapy for neuron replacement in PD patients. A great deal of basic scientific research has continued to yield promising potential for stem cell therapy, and the next step is to begin translation of the research to clinical applications.[5] The main goal of stem cell therapy for PD is to establish a treatment which allows for the regeneration and replacement of damaged neurons either by transplanting stem cells of various lineages, or by inducing differentiation of endogenous stem cells by pharmacological means.[6]

Currently, a wide variety of cell lines and cell sources are being investigated, including embryonic stem cells (ESCs); induced pluripotent stem cells (iPSCs); fetal neural stem cells (fNSCs); adult-derived neural stem cells (NSCs); and adult multipotent stem cells.[6] With this in mind it may be apparent that establishing a suitable population of dopaminergic cells, which can easily and efficiently be handled in culture in preparation for transplantation into PD patients, presents a number of challenges. ESCs have demonstrated the capability to generate functional neurons with viable physiological function; however evidence has shown that ESCs present poor cell survival and phenotypic stability, along with the potential for tumor formation.[6] While ESCs do provide a cell source which can undergo unlimited self-renewal, the use of ESCs remains an area of great controversy. Cells reprogrammed to an embryonic-like state, known as induced pluripotent stem cells (iPSCs), provide another promising cell line for neuron replacement as they can be generated from autologous cells, possess potential for unlimited self-renewal, and avoid almost all ethical issues.[6] Despite these positive aspects, iPSCs require complex genetic transfection methods for cellular reprogramming, and therefore have the potential to become tumorigenic.

It has been found that neurogenesis does occur to some extent in certain structures of the adult brain, and therefore this finding provides a basis for further development of treatments involving brain-derived adult NSCs. In a study by Gu *et al.*, rat neural stem cells were genetically modified *in vitro* to express neurotrophin-3 (NT3), a growth factor which promotes neuroregeneration and differentiation in the brain.[7] Upon transplantation of the modified NSCs into a rat model of PD, dopaminergic neuron proliferation increased and yielded recovery of neurological function in the rats. Although it has been observed that adult NSCs can be expanded and modified *in vitro*, human cell sources are limited and non-autologous, and further studies will be needed to evaluate the potential efficacy of adult NSCs for PD treatment.[6]

Adult multipotent stem cells, such as bone marrow-derived mesenchymal stem cells (BMCs) have been recognized as a suitable source of progenitor cells capable of a repairing a variety of tissues. BMCs provide an expandable, easily accessible autologous cell source which can form dopaminergic neurons.[6] BMCs are also non-tumorgenic, and have exhibited neurogenic and neurorescue behavior in a number of animal models.[6,8,9] One group has recently gone as far as developing a standardized process for producing clinical grade BMCs with excellent *in vitro* and *in vivo* neurogenic capacity.[10] In addition, study by Brazzini *et al.* of a human clinical trial found that autologous BMCs derived from PD patients could be safely and effectively implanted, *via* intra-arterial catheterization.[11] This study demonstrated, through radiological evidence, that this innovative BMC delivery method improved neurological function and reduced neurodegeneration in the BMC-implanted PD patients. Adult multipotent stem cells, particularly BMCs, offer a promising source for successful regeneration of damaged nervous tissues in PD patients, although additional studies — including clinical trials — will be needed to fully evaluate such potential.

Due to the vast and varied prospects of stem cell therapy currently under investigation for treating PD, a huge effort will be needed by the scientific and medical community to formulate the optimal therapeutic platform. The requirements which must be met for stem cell therapy to become an approved treatment for PD are great. Not only will it have to be proven safe and effective in PD patients, it will also have to be competitive in terms of associated cost and efficacy when compared to other treatments. Regulatory concerns, economic impact, and ethical issues will need to be taken into consideration once a suitable stem cell therapy regimen for treating PD patients is developed. The pursuit of a cure for PD will continue to present many challenging obstacles; therefore all areas of regenerative medicine, including stem cells, tissue engineering, and gene therapy must be exhuasted in this search. Whatever the final verdict is on a regenerative medicine approach to treating PD, it will almost certainly be the result of a multidisciplinary effort.

## References


1.  Arenas, E. (2010). Towards stem cell replacement therapies for Parkinson's disease, *Biochem. Biophys. Res. Com.*, **396**, 152–156.
2.  Pezzoli, G. and Zini, M. (2010). Levodopa in Parkinson's disease: from the past to the future, *Exp. Opin. Pharmacother.*, **11**, 627–635.
3.  Benabid, A.L., Chabardes, S., Mitrofanis, J., *et al.* (2009). Deep brain stimulation of subthalamic nucleus for treatment of Parkinson's disease, *Lancet Neurol.*, **8**, 67–81.
4.  Martinez-Serrano, A. and Liste, I. (2010). Recent progress and challenges for the use of stem cell derivatives in neuron replacement therapy for Parkinson's disease, *Future Med.*, **5**, 161–165.
5.  Lindvall, O. and Kokaia, Z. (2009). Prospects of stem cell therapy for replacing dopamine neurons in Parkinson's disease, *Trends. Pharmacol. Sci.*, **30**, 260–267.

6.  Meyer, A.K., Maisel, M. and Hermann, A. *et al.* (2010). Restorative approaches in Parkinson's disease: which cell type wins the race?, *J. Neurol. Sci.*, **289**, 93–103.

7.  Gu, S., Huang, H., Bi, J., *et al.* (2009). Combined treatment of neurotrophin-3 gene and neural stem cells is ameliorative to behavior recovery of Parkinson's disease rat model, *Brain Res.*, **1257**, 1–9.

8.  Cova, L., Armentero M.T., Zennaro E., *et al.* (2009). Multiple neurogenic and neurorescue effects of human mesenchymal stem cell after transplantation in an experimental model of Parkinson's disease, *Brain Res.*, **1311**, 12–27.

9.  Glavaski-Joksimovic, A., Virag, T., Mangatu, T.A., *et al.* (2010). Glial cell line-derived neurotrophic factor-screening genetically modified human bone marrow-derived mesenchymal stem cells promote recovery in a rat model of Parkinson's disease, *J. Neurosci. Res.*, **88**, 2669–2681.

10.  Shetty, P., Ravindran, G., Sarang, S., *et al.* (2009). Clinical grade Mesenchymal stem cells transdifferentiated under xenofree conditions alleviates motor deficiencies in a rat model of Parkinson's disease, *Cell Biol. Int.*, **33**, 830–838.

11.  Brazzini, A., Cantella, R., De la Cruz, A., *et al.* (2010). Intra-arterial autologous implantation of adult stem cells for patients with Parkinson's disease, *J. Vasc. Intervent. Radiol.*, **21**, 443–451.

## CHAPTER 28

# Tissue Engineering for Corneal Regeneration

Ayyasamy Aruchamy

## 28.1. Introduction

The cornea is the clear membrane in the front part of the eye's outer wall and it protects the structures inside the eye from dust, germs, and UV radiation.[1] It is a mechanically strong biological tissue with unique structural, optical, and biochemical transport properties. With its curved shape, the cornea is the most powerful element of the eye. Acting as the eye's outermost lens, it contributes two thirds of the eye's focusing power. Healthy corneas have no blood vessels or discolorations. People with healthy corneas can have refractive errors, if its curvature is too much or too small to provide correct focusing. Such refractive errors are common and, in most cases, are easily corrected with lenses or with refractive surgery including laser-based procedures such as PRK, LASIK and LASEK.[2] See Chapter 29 for details. Another approach to correct refractive errors of the cornea is additive refractive keratoplasty. It involves the insertion of a synthetic or biologic material into the cornea to change the refractive power of the eye by altering the curvature of the cornea, or by the refractive index of the material itself.[2,3]

The cornea can suffer various disorders. It can be damaged due to ocular trauma or infection. The corneal tissue repairs and restores itself to its original state when the injuries are minor and infections are treated early. Corneal disorders can be treated with lenses, medications or surgery. In cases of severe injuries or diseases, the cornea may suffer irreversible functional changes from tissue loss, distortion, swelling, deep scarring or opacification (gray patches or cloudiness). In certain cases, corneal corrective surgery can lead to post-operative problems including corneal weakness, corneal haze, epithelial abnormalities, and reduced corneal sensation. When the cornea is highly damaged laterally, blood vessels grow into the cornea (conjunctivalization) in the process of natural healing. All of these above conditions can cause the cornea to scatter or distort light, which results in glare or blurred vision. In severe cases of corneal damage, partial or complete blindness occurs.

When the cornea becomes deeply scarred, distorted, opaque, or swollen, corneal transplantation (penetrating keratoplasty) is required to regain the complete vision. Several other corneal disorders that may require corneal transplants include keratoconus, hereditary corneal failure, corneal dystrophies, scarring due to infections, and corneal failure after other eye surgery and rejection after the first corneal transplant.[4]

In corneal transplantation, a damaged cornea is removed and replaced with a clear donor cornea. Complete or partial corneal tissue replacement with donor tissues is performed worldwide with high success rate (about 90%). Corneal transplantation has been the only long-term solution to recover full function of the cornea. Of all the transplants done today, corneal transplants are by far the most common and successful. However, in some cases, the patient's body rejects the donor cornea and requires immunosuppression to maintain the transplanted cornea. The availability of suitable donor corneas, immunorejection, and the risk of viral infections, are serious limitations.

Of the 45 million blind people worldwide, 10 million are blind due to corneal damage from ocular trauma or disease. Corneal damage is an international crisis in public health. Due to insufficient number of corneal donors, only about 100,000 transplants are performed each year worldwide; of these, 40,000 corneal operations are performed in the United States and 27,500 in Europe.[5] New diseases such as HIV, Creutzfeld–Jacob disease and Hepatitis C as well as increasing use of corrective surgery (LASIK) reduce the number of potential donors, or render corneas unsuitable for transplantation.[6]

There is an urgent need to develop new forms of artificial corneal replacements as alternatives to the donor corneas. The synthetic polymer-based artificial corneas that are currently available have several limitations, including poor integration with the host tissue, and remodelling of the implant. Corneal tissue engineering has the potential to alleviate the problems with allograft transplantation, and to provide an unlimited number of regenerated corneas. Tissue-engineered human corneas can also be used as alternatives to animal models for cosmeto-pharmacotoxicity testing such as Draize irritation testing.[7,8] *In vitro* alternatives to this test are also being developed.[9]

## 28.2. Physiology and Structure of the Human Cornea

The outer layer of the eye comprises three major regions: central cornea, limbus, and sclera. The cornea is the transparent dome-shaped window in the front of the eye (Fig. 28.1). It has no blood

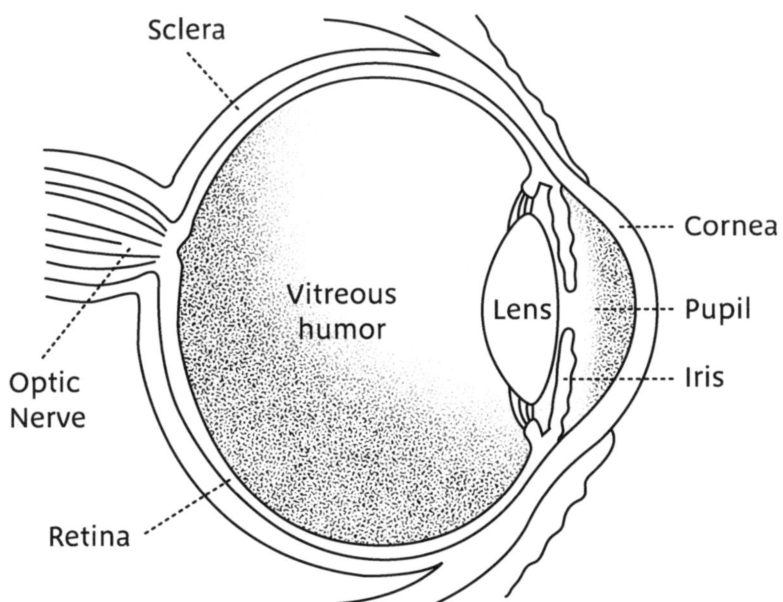


**Figure 28.1.**   Structure of the human eye.

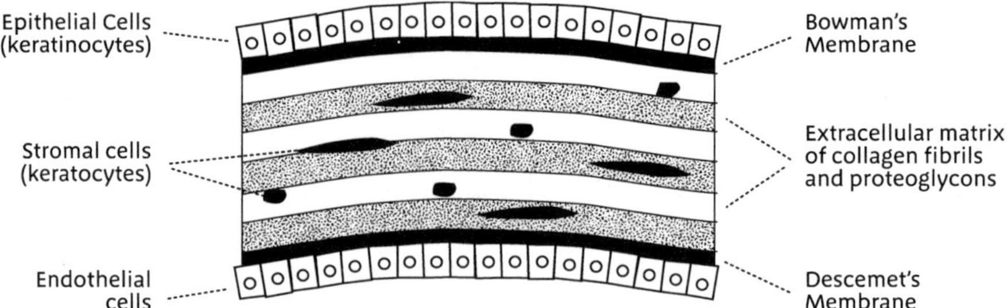



**Figure 28.2.**    Diagrammatic representation of the structure of the human cornea.

vessels (avascular) but tiny vessels at the outermost edge of the cornea provide nourishment, along with the aqueous humor in the back and tear film in the front. It is extremely sensitive and has more nerve endings than anywhere else in the body. Adult human cornea has a diameter of about 11.5 mm and a thickness of 0.5–0.6 mm in the center and 0.6–0.8 mm at the periphery.[1] The limbus is the transition zone of about 1 mm located between central cornea and sclera. The sclera is "the white of the eye," the tough, opaque tissue that serves as the eye's protective outer coat.

The cornea consists of five layers[10] (Fig. 28.2). They are three distinct cell layers (the outer epithelium, the inner endothelium, and the central stroma) separated by specialized extracellular structures known as Bowman's and Descemet's membranes. The epithelium is about 50 μm thick and is composed of keratinocytes of 5–6 cell layers thick. The epithelium makes up about 10% of the thickness of cornea. It is filled with thousands of tiny nerve endings that make the cornea extremely sensitive to pain when rubbed or scratched. The epithelium can regenerate quickly during normal wear and tear, and in cases of minor injuries. The part of the epithelium that serves as the foundation on which the epithelial cells anchor and organize themselves is called the basement membrane. The basement membrane exhibits porosity on the nanoscale. The Bowman's membrane separates the epithelium and the stroma and it is very tough and difficult to penetrate. Its thickness is in the range of 8–14 μm. The stroma is composed of keratinocytes embedded in a dense, highly organized matrix of type I collagen fibrils and proteoglycans. The stroma is 200 layers thick and makes up 90% of the thickness of the cornea, and provides clarity owing to its highly parallel-oriented structure of the collagen fibrils. Descemet's membrane is a thin (10–15 µm) acellular layer separating the stroma and the endothelium. The endothelium is a monolayer of cuboidal cells and it is in contact with the aqueous humor. This layer does not regenerate when damaged.

## 28.3. Corneal Cell Types

Corneal tissue is made up of three major cell types: epithelial cells, corneal fibroblasts (keratinocytes), and endothelial cells. The central cornea is considered to be a differentiated tissue. Cells isolated from the central cornea will not propagate more than two or three passages in *in vitro* culture experiments. Limbal epithelial cells easily go over 12 passages.[11] Stem cells of corneal epithelium exist in the limbal basal regions.[11–16] In the stable state, in which no wound is present in the corneal epithelium, the epithelial cells desquamate from the corneal surface while basal cells in the corneal epithelium actively proliferate to replace them. This results in a natural turnover of epithelial cells. Basal cells of the corneal epithelium are called transient amplifying (TA) cells and are one phase more differentiated than the stem cells. Transient amplifying cells have a determinate life; they can divide only a limited number of times and thus gradually become exhausted by cell division. To compensate for the exhaustion of the population of the basal cells in the corneal epithelium, stem cells in the limbus slowly proliferate, forming daughter cells that migrate inward to repopulate the central part of the cornea. If the corneal epithelium is affected by a large wound, the stem cells actively proliferate to contribute to wound healing. In cases where corneal epithelial stem cells are completely absent because of a limbal disorder, the source of corneal epithelial cells dries up. In these cases, adjacent conjunctival epithelium invariably invades inwardly, and as a consequence, the corneal surface becomes enveloped by vascularized conjunctival scar tissue. The result is severe opacification of the cornea leading to debilitating visual impairment.

Corneal repair may involve using either an artificial prostheses or treatments, and implants involving partial or complete regeneration of the cornea. In several of the treatments, the corneal cell migration and regrowth on and into the implant material provides for sufficient integration of the artificial implant with the corneal epithelium or the stroma.

## 28.4.  Artificial Corneal Implants and Their Limitations

### 28.4.1.  *Corneal onlays and inlays*

Corneal onlays and inlays are essentially contact lenses made of biocompatible materials that resemble the clear surface of the cornea. They are inserted just beneath the eye's surface to provide a permanent correction of the refractive errors of the human cornea. With onlays, some of the epithelial layer can be scraped away and replaced with a layer (contact lens) of the polymer. The polymer lens becomes biointegrated when clear epithelial cells grow over the top of the lens. Inlays, on the other hand, are implanted within the stroma. Corneal implants including onlays and inlays are intended to be permanent, but they can be surgically removed. That is, the procedure is adjustable and reversible with minimal intervention in the central optical zone of the cornea. They differ from surgically implanted lenses because they are not placed behind the cornea. Also, a corneal inlay or onlay is designed to seamlessly "merge" with the eye's surface.[17,18]

Materials used in corneal onlays and inlays should be transparent, non-toxic, flexible, biocompatible, biostable, nutrient permeable, and have surface characteristics that permit the migration and persistent adhesion of corneal epithelial tissue. Various synthetic polymers and collagen–synthetic polymer composites have been investigated for corneal onlay applications. Promising results have been reported for the porous perfluoropolyether (PFPE) and its modified versions.[19]

## 28.4.2. *Artificial corneas*

Artificial corneas are prostheses that are designed to substitute for human donor tissue. They fill the void where people with corneal blindness have rejected donor transplants or where corneal tissue is not readily available. Artificial corneas do not need immunosuppressant therapies usually required with human donor tissue. Materials requirements for artificial corneas are similar in many respects to those for corneal inserts — onlays and inlays. For corneal replacement, the materials should possess the ability to permit corneal cells to migrate and adhere for integration with the tissue. They should be biostable but can be biodegradable (or resorbable) if the corneal regeneration takes place *in vivo* when viable corneal cells are present, and adhere and proliferate in the 3D pores of the implant. They should have sufficient mechanical strength for suturing during implantation.

Several natural and synthetic polymeric materials are in development and in clinical trials to replace or supplement the human cornea. Artificial corneas made from various polymers such as cross-linked collagen, hydrophilic polymers, and hydrogels have been reported. Table 28.1 lists some of the polymer materials being developed for corneal prostheses. For example, a synthetic corneal replacement named AlphaCor[TM] made from the hydrophilic polymer poly(2-hydroxyethyl methacrylate) (PHEMA) has been clinically tested with favorable results.[17]

Stanford University researchers have announced the development of a hydrogel called Duoptix[TM]. The Duoptix[TM] material can swell to a water content of 80% — about the same as biological tissues.[18] It is made of two interwoven networks of hydrogels. One network, made of polyethylene glycol molecules, resists the accumulation of surface proteins and inflammation. The other network is made of molecules of polyacrylic acid, a highly water-absorbing polymer, a strong and stretchy material, able to survive suturing during surgery. This biocompatible hydrogel is transparent and permeable to nutrients. Collagen type I has been successfully employed to produce corneal implants. Collagen has been

**Table 28.1.** Some of the polymer materials under development for corneal prostheses and tissue engineering (TE) of corneas.

| Polymer | Characteristics | Application | Trade name |
| --- | --- | --- | --- |
| Poly(2-hydroxyethyl methacrylate) (PHEMA) | Porous, flexible | Corneal onlay/inlay | — |
| Collagen coated porous perfluoropolyether (PFPE)-based polymer | Porous, flexible | Corneal onlay/inlay | — |
| Polyethylene glycol (PEG)-Polyacrylic acid | Porous, hydrogel | Artificial cornea | Duoptix™ |
| Poly(2-hydroxyethyl methacrylate) (PHEMA) | Porous, flexible | Artificial cornea | AlphaCor™ |
| Polyglycolic acid (PGA) | Porous, flexible | TE scaffold material for cornea | — |
| Collagen type I-composite | — | TE Scaffold material for cornea | — |

cross-linked with other polymer molecules to improve its stability and strength.[20]

While significant improvements have been made in the structure and function of the advanced polymer materials for corneal applications, the polymers still have limitations compared to natural corneal tissue. They are not fully integrated with the human corneal tissue. Tissue overgrowth and surface attachment of proteins and other matter can cause opacity and deterioration with use, although surface modifications can alleviate some of these problems.

## 28.5. Tissue Engineering of the Human Cornea

Tissue engineering of corneas is a paradigm shift in corneal repair and regeneration. Tissue engineering is the design and construction of living, functional constructs in the laboratory so that they can be implanted in the body for the regeneration of malfunctioning or diseased tissues to their original state and function.[21–23] Tissue engineering requires suitable cells that will proliferate and differentiate, and signals to activate the cells and scaffolds to support and co-ordinate the 3D growth of the tissue.

Corneas grown *ex vivo* will substitute for human donor tissue that is used to replace part or full thickness of the damaged or diseased corneas. Tissue engineering of cornea involves the regeneration of the host corneal tissue from corneal epithelial, stromal or endothelial stem cells collected from regions of the cornea, and implanted into a suitable biomimetic scaffold for corneal tissue growth.

Tissue engineering a functional, human corneal implant is particularly demanding because it may require recruitment of several active cell lines, including stromal (in the bulk), epithelial (anterior surface) and endothelial (posterior) cells as well as nerve networks. Furthermore, the tissue-engineered matrix must have low absorption and scattering in the visible region, and acceptable refractive power. Finally, it must have adequate strength to survive both implantation — involving suturing — and the wear and tear of normal use.[20,24,25]

### 28.5.1. *Scaffold materials*

Scaffolds for human cornea must be made of biocompatible materials that are nontoxic, nutrient-permeable, and which have surface characteristics that permit migration and persistent adhesion of corneal cells. Biodegradation or resorbability is not always required. In fact, for the central cornea, the scaffold material should be a biostable or slow degrading polymer implant that will maintain strength, shape, and surface smoothness. Interconnected porosity of larger dimensions than cells is required for cell migration throughout the 3D construct. Nanoporosity is desirable for the diffusion of growth factors into and out of the membrane/scaffold. The scaffold also must have the necessary flexibility to be compatible with other tissues, and have enough strength for suturing during implantation.

Scaffold materials need to be designed with a capability to elicit specific cell response to direct new tissue formation.[23,26] This requires surfaces designed to limit non-specific adsorption of the extra cellular matrix (ECM) proteins while permitting adhesion of the desired cells that proliferate to form new tissues. Surface modification of materials with certain selective ECM proteins or signalling peptides is used for the selective binding of the cells.

Some of the materials listed in Table 28.1 can function as scaffold materials for corneal tissue engineering. In general, standard polymeric substrates need to be modified to suit the particular tissue engineering application. Such modifications include blending with another polymer to form a composite; stabilization of the polymer by cross-linking; and conditioning or derivatizing the surface of the polymer chemically or biochemically to introduce the functionalities needed for cell migration, adhesion and signalling.[27]

Collagen-based composites and polyglycolic acid (PGA) have been investigated as scaffold materials with positive results. Doillon *et al.* have shown that a stabilized type I collagen-glycosaminoglycan scaffold can function well as an *in vitro* corneal stroma. Poly(glycolic acid) (PGA) scaffolds bearing rabbit corneal stroma cells have been investigated to determine if they can be integrated into the ultrastructure of the rabbit corneal stroma without compromising tissue transparency.[28] Results of the study showed it was possible to generate nearly transparent rabbit corneal stroma.

### 28.5.2. *Corneal stem cell sources*

The ocular surface is made up of two distinct types of epithelial cells: conjunctival and corneal epithelial cells. Stem cells for the corneal reside at the corneoscleral limbus.[11–14] Corneal stem cells can come from autologous or homologous sources. An autologous stem cell source will normally be the cells of the intact limbal epithelium of the affected eye. In case of total stem cell deficiency, autologous limbus stem cells of the opposite unaffected eye or homologous limbus stem cells from living related or cadaveric donors can be used.

### 28.5.3. *Tissue culture and transplantation*

A number of investigations have shown that a replica of the human cornea is feasible by tissue engineering. The tissue engineering of the corneal epithelium has been investigated in detail and successful transplantation of the *in vitro* reconstructed epithelia has been reported in recent years.[12,29] Monolayer cultures of corneal epithelial cells can be used to produce the full epithelium consisting of 5–6 cell layers with the basement membrane. The reconstruction of corneal tissues with the stroma, epithelium, and the endothelium has been attempted.[30,31]

- Epithelial layer: This is the easiest corneal layer to be produced by tissue engineering. Human corneal cells and fibroblasts (keratinocytes) have been cultured with growth factors supplemented with serum and antibiotics on nanoporous plastic substrates or on the surface of culture dishes.[11]
- Stroma: Type I collagen gels have been used in the reconstruction of the human cornea. The reconstruction of the stroma has been investigated by culturing a 3D collagen gel containing

human corneal fibroblasts.[32–35] After culturing the stromal construct of collagen–fibroblasts, they were seeded with human corneal epithelial cells and cultured for several days. The formation of an epithelial layer with its basal membrane on the surface of stromal construct has been identified. The formation of a structure with stromal base and epithelial layer with a basement membrane is a great advance towards engineering the complex corneal tissue in the near future.

## 28.6. Tissue Engineering of Carrier-Free Corneal Epithelial Cell Sheets for Transplantation

In order to avoid the post-transplant effects caused by the presence of the substrate material, there are attempts to grow transplantable sheets of corneal epithelium without any carrier or scaffold, and to transfer them directly to the cornea. Successful regeneration and transplantation of multilayered corneal epithelial cell sheets without employing a scaffold has been reported.[12,30–32] The carrier-free cultivation of epithelial sheets involves use of a temperature-responsive culture surface. By lowering the temperature, the researchers were able to detach all the cultured cells from the surfaces as an intact transplantable cell sheet. While the number of processing steps is less, it is also expected that without a substrate, there will be good adhesion between the epithelial sheet and the exposed corneal stroma soon after the operation. In a related work, Amano *et al.* found that epithelial layers produced *in vitro* are similar in morphology to *in vivo* corneas.[33] However, significant differences have been measured in cell density (80% of normal cornea), pump function (55–75% of normal corneas), and some of the HCEC cells are compromised to some extent.

Nishida *et al.* have successfully used autologous oral mucosal epithelial cells as a source of stem cells for the reconstruction of the corneal surface of patients with bilateral total stem cell deficiencies.[31] This indicates that with further understanding of the stem cell sources and better culture techniques, it should be possible to produce corneal tissues from autologous cells for all patients, so that the problem of immunorejection will be practically eliminated.

## 28.7. Challenges and Concerns

Tissue engineering of the human cornea is being actively pursued in various laboratories around the world. Corneal reconstruction is still in its early stage of development but it has been shown to be a viable process. Epithelial layers of human cornea have been produced *in vitro* and successfully transplanted onto patients' corneal surfaces to replace damaged epithelium.[30,31,36] It is a challenge to reconstruct a complete cornea with all of the layers. Much work is needed to understand the process of growing defect-free corneal layers and optimizing the process for large-scale production. Any long-term problems after implantation are not yet known. The major constraints for the corneal reconstruction are the source of corneal stem cells and their culture, and the scaffold materials for the stroma.

## 28.8. Ethical and Socio-Economic Implications

Corneal repair and replacement are one of the most sought-after medical treatments to improve the health and quality of life for a large number of people world-wide. Treatments for the various conditions of the cornea including refractive errors and corneal damage have greatly improved due to advancements in the fields of biomedical engineering, materials science, and tissue engineering. Corneal regeneration by tissue engineering has the potential to revolutionize the treatment of corneal disorders resulting from disease, or from injury or congenital conditions. Tissue engineering can lead to large-scale production of replacement corneas. Serious ethical issues have not arisen so far against this technology because corneal stem cells are relatively easy to harvest from a patient's own tissues or from closely related volunteers. But, as with any emerging technology, it is difficult to predict beforehand if the tissue-engineered cornea will be accepted by all people without concerns relating to health, social or religious sensitivity.

In developed countries, tissue-engineered corneal tissues will provide corneal replacement at affordable cost and without waiting for donor tissues. Health-care costs could be significantly reduced. In poor parts of the developing world where the corneal diseases are serious, especially among laborers, inexpensive corneas could easily be procured and supplied by social service organizations and government agencies. Blindness is a worldwide health crisis and its impact is felt in terms of individual health and income, as well as psycho-socio-economic welfare of families and communities.

Tissue engineering of the human cornea will have important uses in fundamental research as well as in applied fields. Besides the direct application to human corneal regeneration, corneal tissue engineering will have a positive impact with regard to the ethical issues surrounding the use of animals in the pharmacological and toxicological studies. Tissue-engineered human corneas can be used as alternatives to animal models for cosmeto-pharmacotoxicity testing.[7,8] Reconstructed human corneas could be very useful in ophthalmology laboratories to test the toxicity of products associated with contact lenses and artificial corneal implants.[6]

## 28.9. New Developments

Regenerative medicine approaches are currently being used to develop corneal substitutes due to the shortage of suitable human allograft tissue and the limitations of surgical repair.[37] Currently, the most common indication for repair of corneal damage is transplantation of human donor tissue. Transplantation of allogenic corneal tissue is not only limited by the quantity of available viable tissue but also by the complications arising from the transmission of pathogens and possible immune rejection.[37] Corneal regenerative therapies based on the application of stem cells and bioengineered tissue replacements are currently under investigation.

Various diseases can lead to depletion of limbal stem cells and cause corneal opacification and loss of vision, as reviewed by

Kayama *et al.*[38] Conventional corneal transplantation cannot be used for patients with a deficiency in limbal stem cells because conventional allografts lack limbal stem cells. Kayama *et al.* have shown that it is possible to induce ES cells to differentiate towards corneal epithelial cells by culturing them on type IV collagen or introducing the pax6 gene into the ES cells. Their results support the possibility that ES cell-derived corneal epithelial cells hold promise for clinical use.[38]

The functionality of the cornea is directly dependent on its optical clarity, which is maintained by its osmoregulative capability. Groups developing biomimetic constructs for corneal tissue replacement have examined a number of prospective materials with and without the seeding of cells. Much focus has been on trying to replicate the hydrogel-like properties of the cornea with biocompatible polymeric materials. Recently, Mi *et al.* developed a compressed laminin-coated collagen gel seeded with keratocytes during scaffold construction.[39] The collagen gel was shown to exhibit mechanical properties which mimic those of the human cornea. Limbal epithelial cells were then seeded onto the keratocytes-incorporated gel and were found to proliferate, forming stratified layers containing desmosome and hemidesmosome structures. The collagen-gel structures are suggested to be suitable alternatives to human-derived amniotic membrane tissue for the regeneration of corneal tissue.[39] Other studies have used synthetic polymer scaffolds for potentially regenerating corneal tissue. For example, Hong *et al.* found that a polylactic-co-glycolic acid scaffold seeded with autologous adipose-derived stem cells successfully repaired corneal stromal defects in a rabbit model.[40] Such a scaffold seeded with autologous cells offers a promising approach which has a greatly reduced risk of immuno-rejection.

Other approaches include attempting to use human donor corneas which were previously deemed unsuitable for transplantation, by processing them into viable "neo-corneas" fit for corneal replacement.[41] By taking advantage of *in vitro* expansion of human corneal epithelial cells and engineering scaffolds from decellularized discarded donor corneas, Choi *et al.* demonstrated a method that could help increase the supply of donor tissue.[41] Gene therapy is also currently being investigated for corneal repair, including *in vivo* gene delivery methods, using nanoscale polymeric micelles as RNA-delivery vehicles.[42] Current regenerative medicine approaches have a great deal of potential for future clinical applications in corneal regeneration.

## References


1. Records, R.E. (ed.) (1979). *Physiology of the Human Eye and Visual System*, Harper and Row, Hagerstown, MD.
2. Taneri, S., Zieske, J.D. and Azar, D.T. (2004). Evolution, techniques, clinical outcomes, and pathophysiology of lasek: review of the literature, *Survey Ophthalmol.*, **49**, 576–602.
3. Schanzlin, D.J., Asbell, P.A., Burris, T.E., *et al.* (1997). The intrastromal corneal ring segments — phase II results for the correction of myopia, *Ophthalmol.*, **104**, 1067–1078.
4. National Eye Institute (2010). *Facts about the cornea and corneal disease.* [Online]. Available at: http://www.nei.nih.gov/health/cornealdisease. [Accessed 10 November 2010].
5. Cornea Research Foundation (2010). *Cornea Research Foundation of America.* [Online]. Available at: http://www.cornea.org. [Accessed 10 November 2010].
6. Bren, L. (2005). *Keeping human tissue transplants safe. FDA consumer magazine: May–June 2005 issue* [Online]. Available at: http://permanent.access.gpo.gov/lps1609/www.fda.gov/fdac/features/2005/305_tissue.html. [Accessed 10 November 2010].
7. Swanston, D.W. (1983). "Eye Irritancy Testing", in Balls, M., Riddell, R.J. and Warden, A.N. (eds), *Animals and Alternatives in Toxicity Testing*, Academic Press, New York, NY, pp. 337–367.
8. Wilhelmus, K.R. (2001). The draize eye test, *Surv. Ophthalmol.*, **4**, 493–515.
9. Yang, Y., Yang, X., Zhang, W., *et al.* (2010). Combined *in vitro* tests as an alternative to *in vivo* eye irritation tests, *Altern. Lab. Anim.*, **38**, 303–314.

10. Bloom, W. and Fawcett, D. (1975). *A Textbook of Histology*, Chapman and Hall, London.

11. Tseng, S.C.G. (1989). Concept and application of limbal stem cells, *Eye*, **3**: 141–157.

12. Germain, L., Auger, F.A., Grandbois, E., *et al.* (1999). Reconstructed human cornea produced *in vitro* by tissue engineering, *Pathobiology*, **67**, 140–147.

13. Nishida, K. (2003). Tissue engineering of the cornea, *Cornea*, **22**, S28–S34.

14. Daniels, J.T., Dart, J.K.G., Tuft, S.J., *et al.* (2001). Corneal stem cells in review, *Wound Rep. Regen.*, **9**, 483–494.

15. Dua, H.S. and Azuara-Blanco, A. (2000). Limbal stem cells of the corneal epithelium, *Surv. Ophthalmol.*, **44**, 415–425.

16. Lavker, R.M., Tseng, S.C.G. and Sun, T.-T. (2004). Corneal epithelial stem cells at the limbus: looking at some old problems from a new angle, *Experimental Eye Res.*, **78**, 433–446.

17. Lavker, R.M. and Sun, T.-T. (2003). Epithelial stem cells: the eye provides a vision, *Eye*, **17**, 937–942.

18. Hicks, C.R., Crawford, G.J., Lou, X., *et al.* (2003). Corneal replacement using a synthetic hydrogel cornea, Alphacor™: device, preliminary outcomes and complications, *Eye*, **17**, 385–392.

19. Myung, D., Farooqui, N., Zheng, L.L., *et al.* (2009). Bioactive interpenetrating polymer network hydrogels that support corneal epithelial wound healing, *J. Biomed. Mater. Res.*, **90A**, 70–78.

20. Evans, M.D.M., Xie, R.Z., Fabbri, M., *et al.* (2002). Progress in the development of a synthetic corneal onlay, *Invest. Ophthalmol. Vis. Sci.*, **43**, 3196–3201.

21. Li, F., Griffith, M., Li, Z., *et al.* (2005). Recruitment of multiple cell lines by collagen-synthetic copolymer matrices in corneal regeneration, *Biomaterials*, **26**, 3093–3104.

22. Lanza, R.P., Langer, R.S. and Vacanti, J. (2000). *Principles of Tissue Engineering*, 2nd ed., Academic Press, San Diego, CA.

23. Buttery, L.D.K. and Bishop, A.E. (2005). "Introduction to Tissue Engineering", in Hench, L.L. and Jones, J.R. (eds), *Biomaterials, Artificial Organs and Tissue Engineering*, Woodhead Publishing Ltd, Cambridge, UK.

24. Shin, H., Jo, S. and Mikos, A.G. (2003). Biomimetic materials for tissue engineering, *Biomaterials*, **24**, 4353–4364.
25. Allan, B. (1999). Closer to nature: new biomaterials and tissue engineering in ophthalmology, *Br. J. Ophthalmol.*, **83**, 1235–1240.
26. Mooney, D.J. (2002). "Engineering Design Aspects of Tissue Engineering", in McIntire, L.V., Greisler, H., Griffith, L., *et al.* (eds), *WTEC Panel Report on Tissue Engineering Research*, International Technology Research Institute, Layola College, Baltimore, MD.
27. Matsumoto, T. and Mooney, D.J. (2006). Cell instructive polymers, *Adv. Biochem. Eng., Biotechnol.* **102**, 113–137.
28. Doillon, C.J., Watsky, M.A., Hakim, M., *et al.* (2003). A collagen-based scaffold for a tissue engineered human cornea: physical and physiological properties, *Int. J. Artif. Organs* **26**, 764–773.
29. Hu, X., Liu, W., Cui, L., *et al.* (2005). Tissue engineering of nearly transparent corneal stroma, *Tissue Eng.*, **11**, 1710–1717.
30. Germain, L., Carrier, P., Auger, F.A., *et al.* (2000). Can we produce a human corneal equivalent by tissue engineering?, *Prog. Retin. Eye. Res.*, **19**, 497–527.
31. Nishida, K., Yamato, M., Hayashida, Y., *et al.* (2004). Corneal reconstruction with tissue-engineered cell sheets composed of autologous oral mucosal epithelium, *New England J. Med.*, **351**, 1187–1196.
32. Nishida, K., Kohji, Y., Yamato, M., *et al.* (2004). Functional bioengineered corneal epithelial sheet grafts from corneal stem cells expanded *ex vivo* on a temperature-responsive cell culture surface, *Transplantation*, **77**, 379–385.
33. Amano, S., Mimura, T., Yamagami, S., *et al.* (2005). Properties of corneas reconstructed with cultured human corneal endothelial cells and human corneal stroma, *Japanese J. Ophthalmol.*, **49**, 448–452.
34. Minami, Y., Sugihara, H. and Oono, S. (1993). Reconstruction of cornea in three-dimensional collagen gel matrix culture, *Invest. Ophthalmol. Vis. Sci.*, **34**, 2316–2324.
35. Papini, S., Rosellini, A., Nardi, M., *et al.* (2005). Selective growth and expansion of human corneal epithelial basal stem cells in a three-dimensional-organ culture, *Differentiation*, **73**, 61–68.

36. Vasiliev, A.V., Makarov, P.V., Rogovaya, O.S., *et al.* (2005). Repair of corneal defects using tissue engineering, *Biol. Bull.*, **32**, 1–3.
37. Griffith, M., Jackson, W.B., Lagali, N., *et al.* (2009). Artificial corneas: a regenerative medicine approach, *Eye*, **23**, 1985–1989.
38. Kayama, M., Kurokawa, M.S., Ueno, H., *et al.* (2007). Recent advances in corneal regeneration and possible application of embryonic stem cell-derived corneal epithelial cells, *Clin. Opththalmol.*, **1**, 373–382.
39. Mi, S., Chen, B., Wright, B. (2010). *Ex vivo* construction of an artificial ocular surface by combination of corneal limb epithelial cells and a collagen scaffold containing keratocytes, *Tissue Eng.*, **16**, 2091–2100.
40. Hong, J., Xu, J., Sun, X. *et al.* (2009). "Tissue-Engineered Corneal Stroma by Using Autologous Adipose Derived Stem Cell Tissue and Poly(lactic-co-glycolic) Acid", in Dössel, O. and Schlegel, W.C. (eds), *IFMBE Proceedings: Biomedical Engineering for Audiology, Ophthalmology, Emergency & Dental Medicine*, vol. 25/11, Springer, Heidelberg, pp. 1–5.
41. Choi, J.S., Williams, J.K., Greven, M., *et al.* (2010). Bioengineering endothelialized neo-corneas using donor-derived corneal endothelial cells and decellularized corneal stroma, *Biomaterials*, **31**, 6738–6745.
42. Hao, J., Li, S.K., Kao, W.W.Y., *et al.* (2010). Gene delivery to cornea, *Brain Res. Bull.*, **81**, 256–261.

# Development of Lasik: Laser-Assisted *In Situ* Keratomileusis



## CHAPTER 29

Raymond Lee

## 29.1. Introduction

The invention of eyeglasses revolutionized society by allowing the correction of minor abnormalities of the eye using a pair of lenses. As technology and culture evolved over time, people desired a solution that would improve eyesight without having to wear the frames. This led to the development of contact lenses, a much smaller version of eyeglasses that function by being positioned on top of the cornea. Although contacts eliminated the need for frames, they did not liberate a person's need for artificial lenses. However, in 1949, a Colombian professor named José Ignacio Barraquer introduced the idea of refractive cornea surgery to permanently correct a person's vision.[1] Over time, the invention of the laser (Light Amplification by Stimulated Emission of Radiation) allowed Barraquer's theory to become a possibility, because the cornea could now be operated on with fewer complications and with higher precision.[2] This technique would eventually become known as Laser Assisted *In Situ* Keratomeleusis, or LASIK.

Throughout the past 50-plus years, the advancement of technology has assisted in increasing the safety and success rate of LASIK, but because it is a surgical process there will always be risks and these are the main limitations that must be acknowledged. Since the laser was first pioneered by Schawlow and Townes in 1958, it has gone through many evolutions in the surgical field: beginning with the excimer laser, followed by the solid-state laser, and most recently by so-called wavefront technology.[3]

"

The clinical success of LASIK surgery has led to a large increase in its use and an economic drive for the ophthalmology field of medicine. To prevent companies from making misleading claims, the Federal Trade Commission (FTC) monitors the advertisements of every company making sales in the United States, and informs consumers on the product.[4]

## 29.2.  Present Surgical Procedures, Technology, and Limitations

For over 50 years since Professor Barraquer first proposed the idea of keratomileusis (the process of correcting the refraction of the cornea), many scientists contributed to advancing this concept, resulting in refractive eye surgery — LASIK. Currently, the LASIK procedure consists of three stages: preparation and suction ring, microkeratome advancement and flap lifting, and laser treatment and flap realignment.[5] Upon successful completion of all three stages, patients should obtain improvement in their vision.

Stage one of the procedure starts with the preparation of the patient, with the application of a topical anesthetic eye drop and betadine (applied on the eyelid). A sterile drape is then placed over the eyelid, making sure the eyelashes are also covered to prevent interference when using the microkeratome. The lid speculum is then positioned to slowly open the eyelid and maintain its position throughout the surgery. The surgeon then makes pre-place corneal marks on the cornea using specialized LASIK markers to guide the alignment of the flap following surgery. Once the microkeratome settings are calibrated, the suction ring can then be placed close to the limbus. (See Chapter 28 for details of the physiology of the eye.) The intraocular pressure (IOP) measurements are then checked to ensure proper suction before the actual surgical process commences.[6]

There are many different kinds of microkeratomes: non-disposable horizontal (gear driven, automated sliding, and manual); nondisposable vertical (automated and manual); disposable (cable driven, turbine driven, and gear driven); water jet; and laser. A surgeon will usually select the microkeratome based upon specialized training.[7] Stage two starts with the application of proparacaine, a type of lubricant, to wet the corneal surface.[5] After the microkeratome executes the cut, it can be removed along with the suction ring. A dry Murocel® sponge is used to dry the fornices before positioning a modified Chayet ring. Using an irrigating cannula, modified hook, or modified curved tying forceps, the flap is lifted and placed on top of the Chayet ring. A pachymeter is used in obtaining the information needed to calculate and confirm the ablation depth and residual bed, prior to conducting the laser treatment.

Stage three requires the switch to dim oblique illumination to prevent flinching by the patient during the laser treatment procedure. First the laser is centered on the pupil with the head held in place to prevent movement during the operation. Once the laser treatment is completed, a long BSS cannula is used to wet the stomal bed so that the flap can be lifted and then refloated to its original position using a spatula or forceps. Murocel® sponges (or equivalent) are then used to dry the gutter around the flap, and the light source can be switched back to direct illumination mode. After radial and pararadial sweeping, there is a waiting period of two to five minutes, as topical antibiotics, steroids, and nonsteroidal anti-inflammatory eye drops are applied. Once the procedure is completed, the lid speculum and drape can be carefully removed. After several days or weeks, depending on each individual's healing abilities, the patient can determine if the surgical treatment was a success or failure.

## 29.3. Potential Risks

Infections and flap complications are among some of the possible risks that can occur after LASIK surgery. In rare occasions, infections can appear after surgery and lead to pain, redness, a decrease in vision, foreign-body sensations, or discharge.[8] The infections are caused by the introduction of microorganisms in the eye. Depending on the severity of the infection, antibiotic therapy and surgical treatment are used to treat the problem. Flap complications are usually

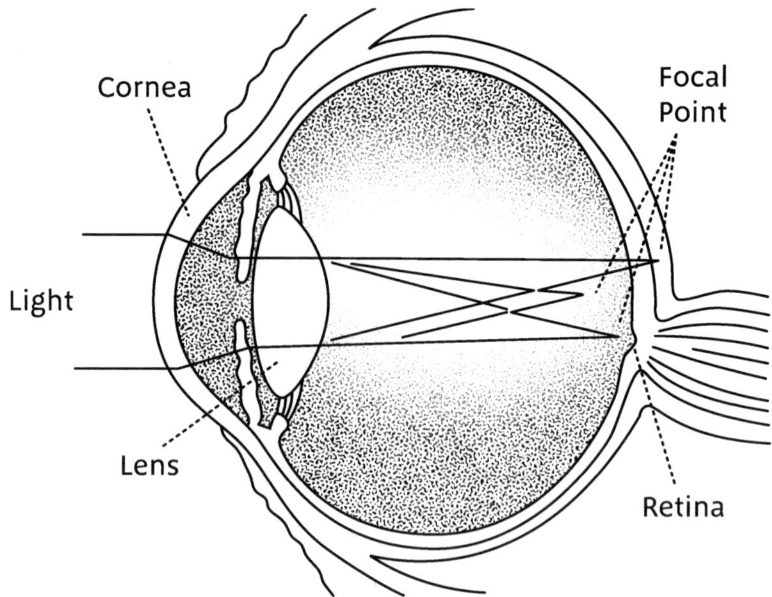


**Figure 29.1.**    Schematic of irregular astigmatism.

caused by improper assembly of the microkeratome, incorrect identification of irregular corneal thickness, keratoconus, and/or previous refractive surgery.[9] Some problems may arise from excessive rubbing of the eyes and blinking. The severity of the complication will determine the course of action to take in correcting it. A regular healing period is most common for allowing any minor complications to subside, although medications and possibly revision surgeries may be needed for more severe complications.

The LASIK procedure cannot correct all vision problems. For example, the current LASIK procedure has difficulty in treating patients with irregular astigmatism (Fig. 29.1) and glaucoma.[10,11]

## 29.4.  New Technology

Technological advances have contributed to the safety and success of LASIK surgery. These include wavefront technology, corneal topography, intrastromal corneal ring segments (ICRS), laser epithelial keratomileusis (LASEK), epikeratoplasty, incisional surgery (radial keratotomy (RK), astigmatic keratotomy (AK), hexagonal keratotomy), orthokeratology, and phakic intraocular lens (IOL). Each of these advanced techniques allows LASIK to undertake different problems that trouble the eyes.

Wavefront technology is an optical technology that allows the evaluation of the optical quality of the eye's wavefront by using the sensors and software to calculate the aberrations of light.[12] In combination with the conventional LASIK procedure, surgeons can modify the laser settings to fit the physical characteristics of each patient, allowing a more precise surgical treatment. The outcome of the procedure yields a sharper image and reduces night-time vision difficulties such as halos and glare. Since wavefront technology is a relatively new technique, the long-term effects must be assessed.[13]

Surgeons have begun to use corneal topography to help select the appropriate vision correction surgical procedure based on patients' needs. Using a projection of illuminated rings that are reflected back and analyzed by a computer, surgeons can determine distortions in the cornea, such as astigmatism, myopia, and hyperopia.[12] With this information, they can advise patients of the most effective procedure to correct their vision. More data is needed to determine if corneal topography can provide an accurate assessment of solutions to surgical problems.

Intrastromal corneal ring segments (ICRS) is an alternative method to treating low levels of myopia (nearsightedness) using an implant that involves corneal rings placed between the layers of the corneal stroma.[14] In order to treat high levels of myopia, it has been proposed to combine both ICRS and LASIK. Using LASIK to create a corneal flap and ICRS to insert the implant allows for a relatively simple reversal procedure, by removal of the implant if the results are less than satisfactory. Thus far, initial studies have shown great promise in treating high myopia, but the long-term effects are still unknown.

Laser epithelial keratomileusis (LASEK) is a new technique created to help correct patients with thin or flat corneas, as the conventional LASIK procedures have surgical limitations in this group. LASEK utilizes a much finer blade known as a trephine, instead of the microkeratome.[15] This allows the surgeon to make a shallower cut, reducing the possibility of infections that could occur following surgery. However, the recovery rate is much slower compared to LASIK, and some discomfort may occur following surgery.

Epikeratoplasty is the surgical process of attaching a lenticule from donor tissue to the de-epithelialized surface of the Bowman's layer.[16] (See Chapter 28 for a discussion of eye physiology.) This procedure is used to reduce myopia and irregular astigmatism of keratoconus. With a low success rate of 33% of patients experiencing vision improvements, and complications such as infections, surgeons have generally abandoned this technique for correcting myopia. However, on rare occasions epikeratoplasty may still be used, due to its ability to treat pediatric and adult aphakia.

Another approach to correcting various problems of the eye is incisional surgery.[16] There are three different kinds of incisional surgery: radial keratotomy, astigmatic keratotomy, and hexagonal keratotomy. Radial keratotomy (RK) is a method that treats myopia through making various incisions into the cornea. This method is relatively simple and has the most long-term follow-up data of any other contemporary refractive procedure. The main drawback of radial keratotomy is its safety and stability. With astigmatic keratotomy (AK), it attempts to correct astigmatic refractive errors by making transverse incisions parallel to the limbus. This method has various outcomes depending on the size of the incision, and will require improved technology and more studies to be able to determine its effectiveness. The third type of incisional surgery is hexagonal keratotomy to treat hyperopia by making circumferential connecting hexagonal peripheral cuts. The result of this surgery showed some success in the beginning, but over time the incisions caused complications with the cornea, and provided various refractive results. This method is rarely used because of the complications.

Orthokeratology is a proposed alternative to avoiding any type of surgery, and instead uses the application of an orthokeratology lens, a flatter, looser, and larger version of the conventional contact lenses[16]. In theory, the lens will alter the cornea over time. Once the vision reaches 20/20, the procedure enters a retainer phase where a special retainer lens is used to maintain the vision status. Due to the lack of studies regarding this method it is uncertain if ortho-keratology is a feasible alternative to obtaining improved vision over refractive surgery, nor have long-term outcomes been evaluated.

The IOL was first studied by Barraquer, but he later abandoned the method due to the lack of technology that could provide enough quality implants.[16] Recent studies have revisited Barraquer's studies utilizing modern advancements. The results reveal the ability to correct a higher refractive error than LASIK or any other type of refractive surgery, along with a high predictability and ability for patients to reverse the outcome. Unfortunately this method has potential risk including the formations of cataracts, *endophthalmitis*, and retinal detachments. The implant has been shown to treat myopia, but researchers are reviewing the possibilities of expanding its range to include the treatment of hyperopia.

Looking into the future of LASIK, there are still areas for improvement in such conditions as chronic dry eyes, decentered ablation, microbial keratitis, and keratecasia.[17]

## 29.5. Economic Issues

As technology continues to improve with advances in materials and surgical techniques, and as the price of lasers decreases, the cost of performing LASIK will decrease. The advances in technology also increase the safety and reliability of eye surgery, escalating the popularity of LASIK as a dependable option to permanent improved vision. With an increasing number of consumers choosing to have permanent vision correction, the number of surgeons skilled in use of LASIK has also increased. Projections predict that the number of vision correction procedures performed will continue to grow by at least 18–26% throughout the next several years. As technology improves, consumers will increasingly have more options for vision correction surgery.

In response to increasing competition, surgeons have cut the costs of performing LASIK surgery. Research showed that the cost of LASIK surgery ranges from US$700 to US$2,300. Although competition may contribute to a reduction in the cost of a typical LASIK procedure, the evolution of technology has had a much greater impact.

Since the first use of lasers in eye surgery, advances in the technology have transformed eye surgery from being a major operation with minimal success rates, to being a routine procedure that effectively corrects vision with minimal risk. Affordability was a major concern when LASIK was first introduced to the public, although current procedures are significantly more affordable. Today patients have a wide range of choices both for elective surgical treatment, and for their surgeon.

## 29.6. New Developments

LASIK surgery has changed the face of refractive eye surgery over the past 20 years, as it has become the single most popular elective surgical procedure. The development and application of excimer lasers provided the basis for the success of LASIK procedures over previous techniques such a radial keratatomy and lamellar surgery.[18] LASIK has been valued highly due to the reduced recovery time and minimal postoperative pain when compared to photorefractive surgery (PRK). Both LASIK and PRK implement excimer lasers, but PRK does not require the creation of a permanent corneal flap as in LASIK, although PRK requires a longer recovery period. A wide spectrum of corneal defects can be repaired using LASIK including myopia, hyperopia, and astigmatism. The success of LASIK is apparent by the high level of positive surgical outcomes providing increased visual acuity, and a 95% patient satisfaction rate.[19] This high degree of consistency and success is due not only to technological advances, but also to the advances in patient selection criteria, which surgeons use to assess the suitability of patients for LASIK procedures. Patients may not be approved for LASIK if they present contra-indications such as glaucoma, immune disorders, certain vascular diseases, and insulin-managed diabetes.[20] As technology continues to advance, patients with such conditions are less likely to be deemed unsuitable LASIK candidates. Innovative technology will also yield an overall improvement in visual performance accompanied by a corresponding reduction in complications. Major advances in refractive surgery have recently been seen in pre-operative planning, intra-operative techniques, and post-operative care.

Understanding the geometry of a patient's corneal topography is critical to the pre-operative selection and planning of an appropriate surgical procedure. As the success of refractive surgery is based on properly altering the relationship between the anterior and posterior cornea, establishing detailed knowledge of the geometric elements of the cornea is of strategic importance to the surgeon.[21] Mapping the cornea and measuring corneal thickness help to determine a patient's eligibility for refractive surgery, and also provide excellent pathological information.[25] In the past, keratometers were used to assess the topography of the cornea. Rarely used today, most surgeons rely on modern corneal topographers, which utilize advanced optical technology, complex data processing algorithms, and 3D image reconstruction.[21]

Recently, a great deal of research and development has spawned a new generation of imaging modalities for mapping corneal topography. Three imaging techniques have recently experienced major technological advancements and are gaining acceptance throughout the field. Scheimpflug imaging devices use optical cross-sectioning, taking series of photographic images, sharing common points and merging them into a single 3D image with a high degree of accuracy.[21] One of the most recent advances in Scheimpflug imaging is the capability to accurately image the posterior corneal surface, as this area is now being used as a responsive indicator for corneal pathologies.[21] Optical coherence tomography (OCT) is another imaging modality which has become popular as its imaging capabilities have progressed. OCT can generate 3D images with micron range resolution from optically scattering media, by employing near-infrared light and an interferometer. Unlike photographic Scheimpflug imaging,

OCT creates an image using interference patterns of scattered light. Anterior-segment OCT allows high resolution imaging of the cornea and accurate evaluation of the iris and pathological depth in the anterior region.[22,23] The development of spectral domain technology will greatly improve OCT imaging capability and make anterior and posterior OCT evaluation routine.[23] Confocal microscopy has also become of interest among many eye surgeons, as it allows high-resolution imaging of thin optical sections of the cornea which can be reconstructed into a 3D image. Confocal microscopy is a commonly used technique in biological and materials sciences, and is becoming more attractive to surgeons due to the multiple imaging modes it offers.[22,23] Confocal is one of the newest applications to imaging technologies for assessing candidates for refractive surgery. Advances in these imaging modalities have allowed for improved surgical outcomes, and when combined with recent developments in wave-front guidance, LASIK procedures may become increasingly patient-specific.[24]

Femtosecond laser technology has been suggested as offering potential advantages over the use of traditional mechanical microkeratomes for creating hinged LASIK flaps. LASIK procedures using femtosecond lasers allow for an "all-laser" approach, in which the flaps are created by ultrafast laser pulses causing photodisruption of the corneal tissue.[25] Following the creation of the corneal flap, the regular LASIK laser remodelling procedure is performed. It has been demonstrated that flaps created by femtosecond lasers are more planar, with a more uniform thickness, and create smoother stromal beds.[25,26] Short-term advantages such as faster visual recovery times have been observed in patients, and a perceived improvement of safety may be provided, compared to that of mechanical microkeratome methods.[26] Although some studies suggest that the use of femotosecond laser enhances visual acuity,[26] other studies have stated that there is no difference in visual outcomes when comparing the two methods.[25,27] A long-term radomized study was conducted by Calvo *et al.*, in which 21 patients underwent LASIK surgery using either a femtosecond laser or a mechanical microkeratotome.[25] Follow-up was completed in regular intervals over a three-year period. The study concluded that femtosecond laser-created flaps did not offer any long-term advantages relating to corneal aberrations or visual acuity. Additional studies will be need to evaluate if advantages exist between the two techniques, although rapidly advancing laser technology and computer-aided surgery will most likely yield an optimized technique, free of mechanical performance.

Refractive surgery is constantly progressing and surgical outcomes are continually improving. Laser technology has allowed for a resurgence in surface ablation procedures such as photorefractive keratectomy, LASEK, and recently epi-LASEK.[28] A better understanding of the biomechanics and wound-healing process of the cornea have also contributed to the progression of refractive surgery and corneal repair. Robotic surgery has recently been under development and may have potential in improving the delicate corneal repair and replacement surgeries.[29] Regenerative medicine may also offer promising solutions to restoring normal vision and replacement of defective corneas (see Chapter 28). Stem cells and tissue-engineered corneal replacements may one day be able to provide improvement in the most severe cases of corneal damage-induced visual impairment.

## 29.7. Summary

The revolutionary idea of refractive eye surgery has provided a permanent alternative to correcting vision plagued by the complications of astigmatism, myopia, and hyperopia. In the past the ability to obtain freedom from wearing eyeglasses or contact lenses was believed to be impossible. With the possible risk of incurring sight-threatening infections, few patients would jeopardize their sight by undertaking surgical intervention. The introduction of LASIK changed everything by utilizing the advanced technology of lasers to reshape the cornea. The procedure currently provides a high success rate with a minimal chance of failure. Depending on an individual's lifestyle or occupation, taking the small risk is often warranted. Athletes, actors, divers, and pilots are ideal candidates because wearing glasses or contact lenses impairs their ability to work. However, since it is still a relatively expensive procedure with a slight chance of complications, LASIK is not the ideal choice for everyone. As research continues to find improvements and further increase safety, reliability, and affordability, eyeglasses and contact lenses may eventually become obsolete.

## References


1. Pallikaris, I.G. and Siganos, D.S. (1997). Laser *in situ* keratomileusis to treat myopia: early experience, *J. Cataract Refract. Surg.*, **23**, 39–49.
2. Pallikaris, I.G., Kymionis, G.D. and Astyrakakis, N.I. (2001). Corneal ectasia induced by laser *in situ* keratomileusis, *J. Cataract Refract. Surg.*, **27**, 1796–1802.
3. Slusher, R.E. (1999) Laser technology, *Revi. Mod. Phys.*, **71**, S471–472.
4. United States Bureau of Consumer Protection (2000). *Office of Consumer and Business Education Facts for Consumers Produced in cooperation with the American Academy of Ophthalmology (AAO)*, GPO, Washington, DC.
5. Farah, S.G., Azar, D.T., Gurdal, C., *et al.* (1998). Laser *in situ* keratomileusis: literature review of a developing technique, *J. Cataract Refract. Surg.*, **24**, 989–1006.
6. Hernández-Verdejo, J.L., Teus, M.A., Roman, J.M., *et al.* (2007). Porcine model to compare real-time intraocularpressure during Lasik with a mechanical microkeratome and femtosecond laser, *Invest. Ophthalmol. Vis. Sci.*, **48**, 68–72.
7. Kakaria, S., Hoang-Xuan, T. and Azar, D.T. (2003). "Microkeratomes", in Azar, D.T. and Koch D.D. (eds), *LASIK: Fundamentals, Surgical Techniques and Complications*, vol. 1, Taylor and Francis, New York, NY, pp. 57–70.
8. Khan, B.F., Chang, M., Jain, S., *et al.* (2003). "Management of infections, inflammation, and lamellar keratitis", in Azar, D.T. and Koch D.D. (eds), *LASIK: Fundamentals, Surgical Techniques and Complications*, vol. 1, Taylor and Francis, New York, NY, pp. 477–490.

9. Roque, M.R., Melki, S.A., Azar, D.T., *et al.* (2003). "Management of Flap Complications in LASIK", in Azar, D.T. and Koch D.D. (eds), *LASIK: Fundamentals, Surgical Techniques and Complications*, vol. 1, Taylor and Francis, New York, NY, pp. 431–462.

10. Braunstein, R.E., Winnick, M. and Greenberg, K.A. (2003). "LASIK Indications, Contraindications, and Preoperative Evaluation", in Azar, D.T. and Koch D.D. (eds), *LASIK: Fundamentals, Surgical Techniques and Complications*, vol. 1, Taylor and Francis, New York, NY, pp. 91–99.

11. O'Keefe, M. and Kirwan, C. (2010). Laser epithelial keratomileusis in 2010 — a review, *Clin. Experimental. Ophthalmol.*, **38**, 183–191.

12. Maeda, N. (2003). "Wavefront Technology and LASIK Applications", in Azar, D.T. and Koch D.D. (eds), *LASIK: Fundamentals, Surgical Techniques and Complications*, vol. 1, Taylor and Francis, New York, NY, pp. 139–151.

13. Maeda, N. (2009). Clinical applications of wavefront aberrometry — a review, *Clin. Experimental. Ophthalmol.*, **37**, 118–129

14. Uceda-Montanes, A., Tomas, J.D. and Alio, J.L. (2006). Correction of severe ectasia after LASIK with intracorneal ring segments, *J. Refract. Surg.*, **24**, 408–411.

15. Lee, S.J., Kim, J.K., Seo, K.Y., *et al.* (2006). Comparison of corneal nerve regeneration and sensitivity between LASIK and laser epithelial keratomileusis (LASEK), *Am. J. Ophthalmol.*, **141**, 1009–1015.

16. Sakimoto, T., Rosenblatt, M.I. and Azar, D.T. (2006). Laser eye surgery for refractive errors, *Lancet*, **367**, 1432–1447.

17. Garg, P., Chaurasia, S., Vaddavalli, P.K., *et al.* (2010). Microbial keratitis after LASIK, *J. Refract. Surg.*, **26**, 209–216.

18. Yeung, K.K., Olson, M.D. and Weissman, B.A. (2002). Complexity of contact lens fitting after refractive surgery, *Am J. Ophthalmol.*, **133**, 607–612.

19. Solomon, K.D., de Castro, L.E.F., Sandoval, H.P., *et al.* (2009). LASIK world literature review: quality of life and patient satisfaction, *Ophthalmology*, **116**, 691–701.

20. Brown, M.C. (2009). An evidence-based approach to patient selection for laser vision correction, *J. Refract. Surg.*, **25**, 661–667.

21. Lam, A.K.C. (2010). New applications in the corneal topography system, *Exp. Rev. Ophthalmol.*, **5**, 115–117.

22. Belin, M.W. and Ambrosio, R. (2010). Corneal ectasia risk score: statistical validity and clinical relevance, *J. Refract. Surg.*, **26**, 238–240.

23. Wylegala, E., Teper, S., Nowinska, A.K., *et al.* (2009). Anterior segment imaging: fourier-domain optical coherence tomography *versus* time-domain optical coherence tomography, *J. Cataract Refract. Surg.*, **35**, 1410–1414.

24. Nubile, M., and Mastropasqua, L. (2009). *In vivo* confocal microscopy of the ocular surface: where are we now?, *Br. J. Ophthalmol.*, **93**, 850–852.

25. Calvo, R., McLaren, J.W., Hodge, D.O., *et al.* (2009). Corneal aberrations and visual acuity after laser *in situ* keratomileusis: femtosecond laser *versus* mechanical microkeratome, *Am. J. Ophthalmol.*, **149**, 785–793.

26. Tanna, M., Schallhorn, S.C., Hettinger, K.A., *et al.* (2009). Femtosecond laser *versus* mechanical microkeratome of visual outcomes at 3 months, *J. Refract. Surg.*, **25**, 668–671.

27. Lee, J.K., Nkyekyer, E.W., Chuck, R.S., *et al.* (2009). Microkeratome complications, *Curr. Opin. Ophthalmol.*, **20**, 260–263.

28. Reynolds, A., Moore, J.E., Naroo, S.A., *et al.* (2010). Excimer laser surface ablation — a review, *Clin. Experiment. Ophthalmol.*, **38**,168–182.

29. Bourges, J.L., Hubschman, J.P., Burt, B., *et al.* (2009). Robotic microsurgery: corneal transplantation, *Br. J. Ophthalmol.*, **93**, 1672–1675.

# Hearing Devices: Experiencing a World of Sound



## CHAPTER 30

Savannah Burnside

## 30.1. Introduction

An estimated 28 million people in the United States have a hearing impairment. In addition there are is increasing number of individuals who will lose their hearing as a result of old age, ear infections or causes such as excessive sound from earphones that damage outer hair cells (OHC). Hearing loss can also be inherited; about 2 to 3 children out of 1,000 will be born deaf or hard-of-hearing in the United States.[1] Genetic disorders such as Waardenburg syndrome or Crouzon syndrome, and infections such as congenital rubella, congenital syphilis, meningitis, toxoplasmosis, rubella, and herpes simplex also lead to loss of hearing.[2,3] Problems that may arise during delivery at birth can result in hearing loss, such as misapplied forceps or prolapsed cord, where the umbilical cord becomes compressed and cuts off the oxygen supply to the undelivered baby.[3]

For hearing to be experienced, all parts of the auditory pathway must work correctly. This pathway includes the external ear, middle ear, inner ear, auditory nerve, and the connection between the auditory nerve and the brain (Fig. 30.1).[2] Damage to any of these components can cause hearing loss, the severity depending on the degree of damage to any of these parts. A healthy young person with normal hearing can perceive a range of frequencies from approximately 20 Hz to 20 kHz. A table comparing the spectrum of hearing frequencies for various mammals and the acoustic spectrum of a piano is shown in Table 30.1.[4]

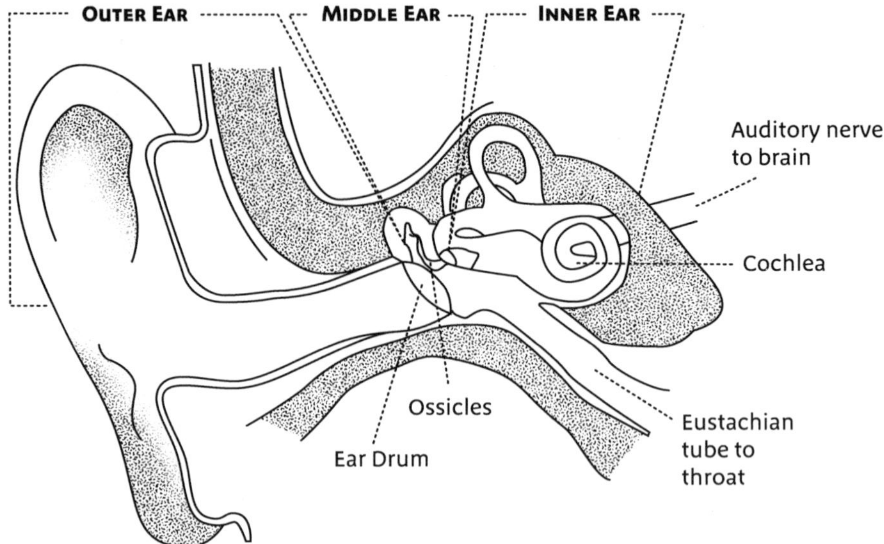


**Figure 30.1.**   A diagram of the ear with pathways including the external ear, middle ear, inner ear, auditory nerve, and the connection between the auditory nerve and the brain.

**Table 30.1.** Frequency spectrum of a human who has regular hearing capabilities compared with other mammals.[4]

| Approximate frequency spectrum | |
| --- | --- |
| Human | 20 Hz–20 kHz |
| Dog | 15 Hz–50 kHz |
| Cat | 60 Hz–65 kHz |
| Dolphin | 150 Hz–150 kHz |
| Bat | 1 kHz–120 kHz |
| Piano | 30 Hz–4 kHz |

The part of speech most affected by damage is the range of higher frequencies. With hearing loss there is usually a problem with the ability to hear weak sounds; however, medium to intense sounds may become excessively loud.[4]

Sensorineural hearing loss results when there is damage to hair cells in the cochlea, the OHC.[5–8] Hearing loss that results from loud noises mainly occurs when the OHC are damaged. It is known that the OHC assist the inner hair cells (IHC) in sensing soft sounds, low-frequencies, and sharpen peaks of the traveling sound wave. The sharpening of the peaks allows one to distinguish among frequencies that lie close together. Because damage to the OHC affects the ability to detect higher frequencies, such damage results in poor speech discrimination. An individual with damage to the OHC experiences a moderate degree of sensorineural hearing loss, 40 decibel hearing loss (db HL) to 60 db HL. However, damage to both the IHC and OHC results in severe hearing loss, ~80 db HL. For comparison, rustling of leaves is about 25 db of sound pressure level (SPL), conversational speech is approximately 50–60 db SPL, and conversational speech exerts a sound pressure of about 10000 μPa.[6] If an individual has moderate or severe hearing loss they cannot understand normal conversation or hear rustling of leaves. The db and sound pressure μPa scale is based on the threshold of hearing of humans, 0 db and 20 μPa of sound pressure.

For a hearing device to help an individual, the device must increase the db SPL over a wide range of frequencies, and accommodate both moderate and severe hearing loss. As a society it is important for all individuals to be able to communicate with their peers; hearing devices make that possible for millions of people world-wide. With current technology, hearing devices are also capable of opening the world of sound to an infant who otherwise would be deaf.

## 30.2.  Current Technology

Past and current assisted-hearing technology lies in six areas based on the location of the device, including: body hearing instruments, eyeglass hearing aids, behind-the-ear (BTE) hearing aids, in-the-ear hearing devices (ITE), completely-in-the-canal (CIC) devices, and implanted hearing devices. All have advantages and disadvantages based on their relative ability to pick up sound through a microphone and ability to emit sound through the receiver to the user. A microphone with a frequency band between 300 Hz and 4000 Hz provides the best range for speech understanding for nearly all hearing-impaired individuals.[7] An optimal receiver should produce the greatest output sound pressure level (SPL) possible. To achieve a higher db sound output, the receiver needs to be large. A more efficient hearing device will have the greatest distance between the microphone and the receiver, and reduction of acoustic feedback. Due to the possible low dexterity of the user, especially the aged or infirm, the size of the device and ability to handle it may affect the choice of hearing device. Location greatly affects the quality of the main function of the device — to pick up and emit sound. All of the devices, and their advantages and disadvantages, are described below.

### 30.2.1. *Body hearing instrument*

A body hearing instrument (BHI) is typically $60 \times 40 \times 15$ mm in size and is worn in the pocket, around the neck or on a belt. The receiver is a custom earmold that fits in the canal and concha.[7] The body instrument offers the greatest possible acoustic amplification due to the maximal distance that lies between the receiver and the microphone. The size of the device includes large controls which make it easy for individuals who have poor dexterity to handle. The size of the body device gives it the advantage of a larger receiver than other devices and in return offers the greatest maximum output sound pressure level.[7] With the wide range of amplification available, body devices are best for individuals with severe hearing impairments.[8] A few drawbacks for the device include amplified noise due to clothes rubbing against the microphone, and since the microphone is not on the head, directional hearing is not possible.

### 30.2.2. *Eyeglass hearing instrument*

The second hearing device described is smaller than the BHI but still fairly large compared to other conventional devices. The eyeglass hearing instrument (EHI), which was first produced in the early 1950s, is integrated into a pair of eyeglasses. All the components of the EHI are built into the frame near the ear (bow). The microphone is placed near the hinge to widen the distance between it and the receiver. Another advantage is that the eyeglass can be combined with a bone conduction receiver, discussed later.[9] Because of the design the hearing device can he hidden. However, if the glasses are misplaced or damaged, then the user will be without both glasses and a hearing device. EHI devices are appropriate for hearing losses greater than 70 db HL12.

### 30.2.3. *Behind the ear device*

The behind the ear (BTE) device is worn behind the ear with a tube connecting to a custom earmold that fits in the ear. Both the microphone and receiver are placed in the casing that sits behind the ear. The BTE is able to utilize a directional microphone which helps the user determine the origin of the sound. A drawback is the effect the tubing has on the quality of natural sounds. However, BTE devices can be used for a range of hearing losses between 25 and 110 db HL.[8]

### 30.2.4. *In the ear hearing instrument*

There are two types of custom hearing devices that sit in the concha or ear canal. The first is the in-the-ear (ITE) hearing instrument — a custom ITE and a semi-modular instrument. For the custom device the shell is initially made with a mold of the user's ear and the shell has the components built directly into it, whereas semi-modular ITE devices are built into a standard housing before attachment to a custom earmold.

The advantage of the custom ITE is that it utilizes all available space in the ear canal. The receiver on the device is placed as deep as possible into the ear canal, while the microphone is placed near the outside by the concha. The merits of ITE devices include enhancement of high frequency input due to the receiver being located near the ear drum (tympanic membrane). In addition, ITE devices are more comfortable to wear. However, with higher levels of amplification the devices can experience acoustic feedback. For the user, acoustic feedback can be annoying "whistling", "howling" or "screeching" sounds. To minimize acoustic feedback ITE devices should be used for hearing losses between ranges of 25–80 db HL.[6]

## 30.2.5. *Completely in the canal hearing instrument (CIC)*

The completely in the canal instrument (CIC) is a fairly new conventional hearing device. The device entered the market in 1993 for individuals for mild or moderate hearing loss. The CIC is tiny enough to fit completely in the canal and is placed near the eardrum. Regardless of the size, CIC devices have an advantage over other hearing devices. The CIC does not experience occlusion, where the user may hear an echo as if they are "talking in a barrel." Occlusions in other devices occur because the ear canal is blocked, so the low frequency vibrations of the user's voice are heard by the user. Since the CIC microphone is positioned so far into the canal it is able to produce more gain and a higher frequency response. The device also is able to take advantage of the natural functions of the outer ear. Similarly, the receiver has an increase in the gain and output in the ear. Because the device fits in the canal and cannot be seen it is the most cosmetically appealing of all the devices. However, the size of the device generates a few disadvantages. With the small distance between the microphone and receiver there is a problem with acoustic feedback. The small size also has additional drawbacks:

1.  the user may not be able to handle the device;
2.  there is no volume control;
3.  batteries are small so need to be changed frequently; and
4.  it is more susceptible to accumulation of ear wax.[6]

## 30.3. New Alternative Devices

### 30.3.1. *Cochlear implants*

Conventional devices are of no use for individuals who suffer from ear malformations or chronic ear drainage. The alternatives are implantable hearing devices. These are designed for individuals who can not benefit from conventional hearing devices such as the BTE, ITE, or CIC, which sit within or on the ear. Cochlear implants are suited for individuals with severe hearing loss while bone-anchored hearing aids (BAHA) are suitable for mild to moderate levels of hearing loss (HL). A multi-channel cochlear implant was developed in the late 1970s.[9] The device contains a platinum electrode array in a silicone cover. The multichannel electrode array is carefully positioned in the cochlea with a retractable platinum guide wire (stylet). As the stylet is removed the electrode begins to curl.

The electrode is then attached to a receiver implanted behind the ear. A transmitting coil is plugged into the receiver/stimulator for use, otherwise the hole can remain capped. The transmitting coil is connected to microphone and speech processor, which can be positioned on the ear or worn around the neck. This device requires an irreversible implantation because the inserted electrode array causes a complete destruction of the inner ear, the cochlea.[7] The electrode array can be removed but must be replaced with another electrode. Candidates for this device can be either adults or children with a severe hearing loss greater than 90 db HL at some frequencies, and greater than 60 db HL at all frequencies. To be considered, adult candidates must have already acquired language and the ability to speak. They must also be willing to take necessary speech classes and training sessions. Although children should be about 18 months old before surgery, nevertheless some as young as six months have been implanted.[7]

There are many advantages in having a cochlear implant over a conventional hearing aid. One of these includes better speech processing, on average, over a hearing aid.[3] This means that people wearing the device were better able to process what was spoken to them. Children who were implanted at adolescence were not as able to achieve understanding of speech with a cochlear implant alone. Studies show that it is better for children to be implanted at a younger age, and this group was better able to achieve speech recognition.[3] The children in this study had congenital deafness, which means they had hearing at birth but had lost it after birth due to some type of trauma. Additionally, individuals with cochlear implants in conjunction with a hearing aid had improved speech perception and were better able to listen with background noise.[10]

### 30.3.2. *Bone anchored hearing aid*

Another implanted device, the bone anchored hearing aid (BAHA) has been in clinical use in Sweden since 1977. For a BAHA implant, a titanium screw is implanted into the temporal bone just behind the ear.[8] The BAHA provides a bone conduction signal which stimulates the skull. The device has two parts: the sound processor which plugs into the screw; and the titanium screw embedded in the mastoid bone of the skull.[11] The BAHA is recommended for patients with hearing losses less than 25 db HL. Patients who use a BAHA device are those who cannot use conventional air conduction devices. Candidates may be individuals with chronic damage to their ears, or children with stenosis of the ear canal.[12] Stenosis occurs when the inner ear does not form correctly during child development; skin or a bony wall may be the cause of the blockage. In a BAHA device, sound is conducted directly from the skull to the cochlea, bypassing the middle and inner ear.[8] Unlike other implants, the BAHA is reversible. In addition there is minimal risk of ear infections. A negative feature of the BAHA is its inability for amplification and limited output, which at high intensities causes distortion.[8] Another disadvantage is the three-month healing time required before the BAHA sound processor can be removed.[6] Cosmetically, protrusion of the titanium screw through the skin may be an issue for users.[5]

### 30.3.3. *Genetic factors*

The current and newer technologies focus on physical devices as means of helping those with hearing loss. However, analysis of the human genome has identified more than 70 gene locations as causes of genetic hearing loss. With this information genetic counselors are able to help couples who may have HL in their family and determine if they are possible carriers of the defective genes. Counseling allows the couples to determine the possibility that their child could inherit HL and make a decision as to have children or prepare them for the possible consequence of a child with HL.[13]

## 30.4. Ethical Issues

Within the deaf community, many do not consider deafness a disability. They feel there is nothing that needs to be "fixed" and that their language is one that is visually given and received. However, individuals of the hearing society have differing, and at times conflicting, views on deafness. Parents with normal hearing may have difficulty accepting a child with a profound hearing loss and many may be shocked, devastated or will grieve at the handicap. In ill-informed societies children may even be isolated from contact with others because the parents believe they are contagious. The parents and family may also treat the individual as if they have a life-threatening illness. Parents may also be in denial and think the child will grow out of the hearing deficit. Because of these stigmas it can be very difficult for deaf individuals to feel accepted.

Individuals who are hard of hearing may have difficulty accepting themselves as deaf. A deaf person must make a personal decision about how to consider life without hearing: is it a disability, or a norm; are they part of the deaf community or part of the hearing community? For children it may be different to an adult deaf person but, regardless, they need their families' support. Other issues arise depending whether the child is born into a hearing family or a deaf family. Hard-of-hearing individuals frequently report rejection or alienation from both groups. Within the deaf community hard-of-hearing individuals are said to be nonmembers of the culture because they still have some residual hearing. A difficult issue within the deaf community is use of hearing devices, especially cochlear implants. The deaf community argues that it is another means to isolate deaf children academically, communicatively, and socially. In the deaf community, calling someone "think-hearing" is saying the person is trying to be someone they are really not, and their membership in the community may be questioned. Issues facing older hard-of-hearing individuals are never clear, but decisions over a child or infant are for the parents to make, and they must decide what is best for the child; professional counseling should be sought.

A deaf child who is born to hearing parents may not be identified as deaf for months or many years, depending on the severity of the hearing loss. Many hearing parents who do not have any other children who are deaf may face issues with denial or shock. Sometimes it may be a teacher who first realizes that a child has problems hearing when the child does not respond. On average, parents begin to suspect that their child has a hearing loss at about 17 months, but many do not confirm their beliefs until a few months later.

Within the deaf community, it is different. In general, individuals who are members of the deaf community prefer to marry someone who is deaf. Nearly 90% of deaf adults prefer to marry a deaf partner.[11,14] In addition to having a deaf spouse, deaf individuals also prefer to have a deaf child. The parents know that the possibility of having a deaf child is high, yet they often oppose genetic counseling. Deaf couples who are carriers of the mutation and choose to have children cause a higher carrier rate of mutations in a gene responsible for hereditary deafness known as Connexin 26. For deaf individuals with deaf parents it is very difficult to determine the chance of having a deaf child.[14]

## 30.5. Economic Issues

For conventional hearing devices, prices ranged from US$600 to US$2,000 in 2000; present costs in 2010 remain the same. Prices vary depending on the manufacturer, type and size of the hearing device[15]. Smaller devices such as the CIC have a higher cost due to the small size and electrical components needed. The larger analog devices cost around US$600. The smaller devices such as the custom ITE require customized fitting, and components must then be fit within the available space by a trained technician. Semi-modular ITE devices may be slightly less costly because of the standard housing, but they are still around the price of the custom ITE at ~US$2,317 (2001).[16] For the last few years hearing aid prices have remained about the same. In 2007 the Mayo Clinic listed hearing devices in the range of US$900–1,200 for analog hearing devices and $1,300–3,000 for digital aids. It is important that individuals considering use of a hearing aid for the first time should consult their physician, or specialist clinic, and discuss the different types of devices and what will be suitable for them. The above figures are the initial costs but parts such as ear molds and batteries need replacing in the future.

The costs of conventional devices are very competitive compared to implantable devices; however, the initial costs may not be the issue, especially if the user benefits more with an implantable device. According to the National Institute on Deafness and Other Communication Disorders (NIDCD) it costs about US$60,000 for the surgery, implant, necessary adjustments, and training for a cochlear implant. In comparison it can cost over a million dollars to support a child that becomes deaf at the age of three. The cost comes from special education and services the child will need over her lifetime.

## 30.6. Recent Developments

With the increasing size of the aging population, hearing loss is set to become a major global health-care issue, and has huge growth potential in terms of the medical device market.[17] Implantable hearing devices are gaining acceptance, with a growing scope of indications, due primarily to advances in technology and encouraging results from clinical studies.[18] The new developments observed in implantable hearing device technology focus on three main goals: increasing efficacy, improving patient quality of life, and the design of more transparent devices.[18] Recently, surgical advancements have been highlighted by the use of bilateral implants, direct electrical stimulation of the brain, and placement of such devices in patients below the age of 12 months.[19] Although progress has been encouraging, not all cases of hearing loss can be sufficiently treated by cochlear implants, such as when severe auditory nerve dysfunction is presented. The efficacy of cochlear implants is also related to the patient's history of auditory pathology. A patient's history of hearing loss must be considered not only when choosing the appropriate surgical intervention, but also in the rehabilitation process, when the brain must be conditioned to recognize speech patterns.[19] Much research has focused on gaining a better understanding of the brain's ability to process speech and recognize sound patterns, as improved knowledge of the neuroscience will allow superior rehabilitation and improved targeting for direct brain stimulation.[19]

Regenerative medicine approaches are also under investigation for treating otological diseases that result in hearing loss. Some research has been focused on the use of tissue engineering methods to repair or replace damaged and diseased auditory structures, such as the development of polymeric hydrogels for tympanic membrane regeneration.[20] Other groups have looked into the use of growth factors such as brain-derived nerve growth factor, which was found to improve prevent hearing loss *in vivo* when applied directly to the cochlea.[21] With the growing incidence of hearing loss in the aging population, new approaches are needed to develop even more functional and cost-effective means to treat patients suffering hearing loss.

## 30.7. Summary

Hearing devices have come a long way during the last century, from ear trumpets to cochlear implants. Now, individuals who are born deaf or who have lost their hearing as a result of a trauma have the opportunity to experience the world of sound again or for the first time. Regardless of the cost of the devices, for most individuals the benefits of restoration of hearing, even if not perfect, are priceless. With new genetic information on the causes of deafness now available, it is hoped that in the near future other treatments may become available to treat or prevent hearing loss. Considering how quickly technology has developed to enhance hearing electronically, the future may lead to a biological solution, such as re-growth of hair cells.

## References


1. National Institute on Deafness and Other Communication Disorders (2007). *Statistics about hearing disorders, ear infections, and deafness.* [Online]. Available at: http://www.nidcd.nih.gov/health/statistics/hearing.asp. [Accessed 10 November 2010].
2. Centers for Disease Control and Prevention (2004). *Developmental disabilities, hearing loss.* [Online]. Available at: http://www.cdc.gov/ncbddd/dd/hi3.htm. [Accessed 10 November 2010].
3. Clark, G. (2003). *Cochlear Implants: Fundamentals and Applications*, Springer-Verlag, New York, NY.
4. McCormick, B. (1991). "Basic Acoustics", in Gregory, S. (ed.), *Constructing Deafness*, Pinter Publishers, London, pp. 84–85.
5. Turner, C.W. (2006). Hearing loss and the limits of amplification, *Audiol. Neurotol.*, **11**, 2–5.
6. Venema, T.H. (1998). *Compression for Clinicians*, Singular Publishing Group, San Diego, CA.
7. Vonlanthen, A. (2000). *Hearing Instrument Technology for the Hearing Healthcare Professional*, 2nd ed., Singular Publishing Group, San Diego, CA.
8. Sataloff, R. and Sataloff, J. (2005). *Hearing Loss*, 4th ed., Taylor & Francis Group, New York, NY.
9. Cohen, N. (2004). Cochlear implant candidacy and surgical considerations, *Audiol. Neurotol.*, **9**, 197–202.
10. Madell, J., Sislian, N. and Hoffman, R. (2004). Speech perception for cochlear implanted patients using hearing aids on the unimplanted ear, *Int. Cong. Ser.*, **1273**, 223–226.

11. Breivik, J.K. (2005). *Deaf Identities in the Making*, Gallaudet University Press, Washington, DC, p. 127.

12. Goldenberg, R.A. (1996). *Hearing Aids: A Manual for Clinicians*, Lippincott-Raven Publishers, Philadelphia, PA, p. 158.

13. Stephens, D. (2002). "Foreword", in Cremers, C. and Smith, R. (eds), *Genetic Hearing Impairment: Its Clinical Presentations*, Karger, Basel, p. ix.

14. Green, G.E. and Cunniff, C. (2002). Genetic evaluation and counseling for congenital deafness, *Adv. Otorhinolaryngol.*, **61**, 230–240.

15. Roland, P., Marple, B. and Meyerhoff, W. (1996). *Hearing Loss*, Thieme Medical Publishers, New York, NY, p. 283.

16. Burkey, J. (2003). *Overcoming Hearing Aid Fears*, Rutgers University Press, New Brunswick, NJ, pp. 135–140.

17. Goldman, D.R. and Holme, R. (2010). Hearing loss and tinnitus — the hidden healthcare time bomb, *Drug Discov. Today*, **15**, 253–255.

18. Haynes, D.S., Young, J.A., Wanna, G.B., *et al.* (2009). Middle ear implantable devices: an overview, *Trends Amplif.*, **13**, 206–214.

19. Moore, D.R. and Shannon, R.V. (2009). Beyond cochlear implants: awakening the deafened brain, *Nat. Neurosci.*, **12**, 686–691.

20. Kim, J.H., Choi, S.J., Park, J.S., *et al.* (2010). Tympanic membrane regeneration using a water-soluble chitosan patch, *Tissue Eng. A.*, **16**, 225–232.

21. Meen, E., Blakley, B. and Quddusi, T. (2009). Brain-derived nerve growth factor in the treatment of sensorineural hearing loss, *Laryngoscope*, **119**, 1590–1593.

# Glass-Ceramic Dental Restorations



## CHAPTER 31

Wolfram Höland and Volker M. Rheinberger

## 31.1. Introduction

This chapter describes the development of special compositions of glass-ceramics that are designed to restore the appearance and function of teeth. This is a significant improvement in enhancing the quality of oral health and quality of life. It is far more cost effective to maintain healthy natural teeth than to replace them with implants or dentures. Millions of glass-ceramic dental restorations have been used in the last decade with extremely high levels of success. The innovative technology presented in this chapter demonstrates the potential of using the infinite variability of glass composition, combined with controlled levels of crystallization, to create unique economical methods to enhance both the quality of life and the length of life.

## 31.2. Concepts and Requirements for Restoring Teeth

Biomaterials used in restorative dentistry are important products for humans who want to regain the full range of chewing movements or improve the esthetic appearance of their teeth. Traumatic dental injuries, the processes of aging, and insufficient oral hygiene may require use of dental inlays, onlays, crowns, veneers bridges, posts or abutments for implant therapy. These materials must fulfill a combination of requirements. They must be able to match or even surpass the esthetic (optical), chemical, and mechanical properties of natural teeth. Chemical durability is essential to prevent the occurrence of caries. Abrasive wear behavior must be similar to that of natural teeth, whereas flexural strength and toughness of dental biomaterials must be higher than that of natural teeth to allow construction of functionally adequate dental bridges, posts or abutments in high load-bearing regions.

Dental biomaterials also have to possess complex optical properties. They must provide an adequate range of light transmission (translucency), which reflects the different levels of translucency occurring in the natural tooth structure. They must exhibit both opalescent properties near the incisal area and fluorescent properties across the entire body of the tooth replacement, similar to natural teeth.

None of the above-mentioned applications require dental biomaterials to be bioactive, for example to directly form a bond with the natural dental tissues. Dental restorative materials are bonded to the natural tooth structure by means of materials that are specifically formulated for this purpose, such as adhesives.

A historical review of the development of biomaterials for restorative dentistry shows that polycrystalline ceramics were used to produce dental restorations as early as the beginning of the 19th century, more than 100 years ago.[1] However, the optical and mechanical properties of these materials did not meet clinical requirements. Esthetic veneering of metal frameworks (crowns and bridges) only became possible after leucite-based ($KAlSi_2O_6$) porcelain-fused-to-metal (PFM) materials were developed.[2]

## 31.3. Glass-Ceramic Dental Restorations

Significant advances in development of materials that match, or even outperform, the optical, mechanical and chemical properties of natural teeth were accomplished with the introduction of glass-ceramics. For the first time, materials capable of meeting the exacting challenges of metal-free tooth replacements became available. Development of these materials followed a complex path. Glass-ceramics for the fabrication of low load-bearing inlays were introduced first, followed by glass-ceramics for crowns and bridges.

This showed the way to gradual replacement of amalgam and metals as framework materials. Furthermore, glass-ceramic dental biomaterials offer an additional advantage over other groups of materials. Because of their advantageous working properties, they allow the application of modern processing techniques, such as viscous flow molding and CAD/CAM machining.

### 31.3.1. *Mica-type glass-ceramics*

DICOR® (Corning Inc/Dentsply Int,) was the first glass-ceramic developed for restorative tooth replacements (Tables 31.1 and 31.2).[3] Crystals of tetrasilicic mica, $KMg_{2.5}Si_4O_{10}F_2$, were precipitated by means of controlled crystallization in the $SiO_2$-$MgO$-$K_2O$-$F$ system. First clinical tests with inlays and crowns were performed as early as 1979. This glass-ceramic had translucent properties and was capable of reproducing the optical characteristics of natural teeth. Flexural strength was around 150 MPa. This material stood out from others because the base glass was processed by means of molding technology. Additional heat treatment was required to convert the base glass into the glass-ceramic. In addition to applications that involved the use of molding technology, machinable types of mica glass-ceramics, such as DICOR® MGC, became available to manufacture dental devices. This was an important development pointing the way towards promising new technologies.

**Table 31.1.** Glass-ceramic type systems — processing technology.

| Crystal phase | Glass-system | Processing |
|---|---|---|
| $KAISi_2O_6$ (leucite) | $SiO_2$-$Al_2O_3$-$K_2O$ | Molding, machining |
| $Li_2Si_2O_5$ (lithium disilicate) | $SiO_2$-$Li_2O$-$P_2O_5$-$Al_2O_3$-$K_2O$ | Molding, machining |
| $Ca_5(PO_4)_3F$ (fluoroapatite) | $SiO_2$-$Al_2O_3$-$Na_2O$-$CaO$-$P_2O_5$-$F$ | Sintering, molding |
| $KAISi_2O_6$ (leucite) and $Ca_5(PO_4)_3F$ (fluoroapatite) | $SiO_2$-$Al_2O_3$-$Na_2O$-$K_2O$-$CaO$-$P_2O_5$-$F$ | Sintering |

**Table 31.2.**   Time line for development of glass-ceramics for dental restorations.

| | |
|---|---|
| 1979 | DICOR® glass-ceramic: first clinical tests as moldable glass-ceramic for dental restoration |
| 1991 | IPS Empress®: moldable leucite based glass-ceramic for inlays, crowns, and veneers |
| 1997 | IPS Empress® Cosmo Post: $ZrO_2$-containing glass-ceramic for fixation of $ZrO_2$ posts |
| 1988 | IPS ProCAD®: machinable leucite based glass-ceramic for inlays, crowns, and veneers |
| 1999 | IPS d.SIGN®: leucite-apatite glass-ceramic to veneer metal frameworks, especially dental crowns and long-span bridges |
| 2000 | IPS Empress® 2/IPS Eris for E2: lithium disilicate glass-ceramic veneered with fluoroapatite glass-ceramic to fabricate dental crowns and small three-unite dental anterior bridges |
| 2004 | IPS Empress® Esthetic: leucite-based glass-ceramic with high optical properties close to the natural tooth |
| 2005 | IPS e.max® Press: lithium disilicate as moldable glass-ceramic for inlays, crowns, and small dental bridges |
| 2005 | IPS e.max® Zir Press: fluoroapatite glass-ceramic to veneer toughness-sintered $ZrO_2$ frameworks (especially crowns and posterior bridges) |
| 2005 | IPS e.max® CAD: machinable lithium disilicate glass-ceramic, especially for dental inlays and crowns |
| 2009 | IPS e.max® Press HT and IPS e.max CAD HT: high translucent (HT) lithium disilicate glass-ceramic as moldable (press) and machinable (CAD) glass-ceramic for dental application |

## 31.3.2. *Leucite-type glass-ceramics*

At the end of the 1980s, special molding technology which eliminated the need for additional heat treatment became available, allowing fabrication of inlays and crowns from leucite, $KAlSi_2O_6$, glass-ceramics.[4,5] Principles of controlled surface nucleation and surface crystallization in base glasses of the $SiO_2$-$Al_2O_3$-$K_2O$-$Na_2O$ system were applied in these glass-ceramics.[6] The resulting product, IPS Empress® (Ivoclar Vivadent AG, Liechtenstein), was introduced worldwide in 1991 (Tables 31.1 and 31.2). From 1991 to 2008, approximately 36.5 million dental restoration units had been

**Table 31.3.**  Clinical applications of special types of glass-ceramics as biomaterials for dental restoration.

| Product | Number of units |
|---|---|
| | (one unit represents one dental crown, inlay, etc.) |
| IPS Empress® | 36.5 million (1991–2008) |
| IPS d.SIGN® | 75.5 million (1999–2008) |
| IPS Empress® CAD (and ProCAD) | 4.5 million (1998–2008) |

produced using this molding technique (Table 31.3). With this material, a lost-wax technique and subsequent viscous flow-molding processes enable fabrication of individualized inlays and crowns, offering a high degree of accuracy of fit. The molding processes are performed in specially designed ceramic press furnaces, for example EP 500, EP 600, EP 3000, and EP 6000 (Ivoclar Vivadent AG), using a temperature range between 1000°C and approx. 1200°C (depending on the specific formulation).

The IPS Empress® glass-ceramic exhibits a coefficient of thermal expansion (CTE) of $15 \cdot 10^{-6}$ $K^{-1}$ m/m to $18.25 \cdot 10^{-6}$ $K^{-1}$ m/m (depending on the leucite content). Flexural strengths are approximately 160 MPa, which can be increased by glazing and additional heat treatment. The material's wear behavior in relation to natural dentition deserves particular mention. Since this glass-ceramic is characterized by similar abrasive properties as natural teeth, the opposing tooth structure of the human dentition is not damaged during mastication.

In 2005, IPS Empress® Esthetics, an even more translucent version of the above leucite glass-ceramic, was developed. Improved optical qualities allow fabrication of dental replacements that are even more similar to natural teeth. Typical applications are dental inlays, veneers, and crowns (Fig. 31.1).

These glass-ceramics are highly machinable, as the leucite crystals incorporated in them exhibit pronounced polysynthetic twinning.[6] Inlays and crowns can be prepared by machining the material with diamond tools in a CEREC unit, such as in Sirona,

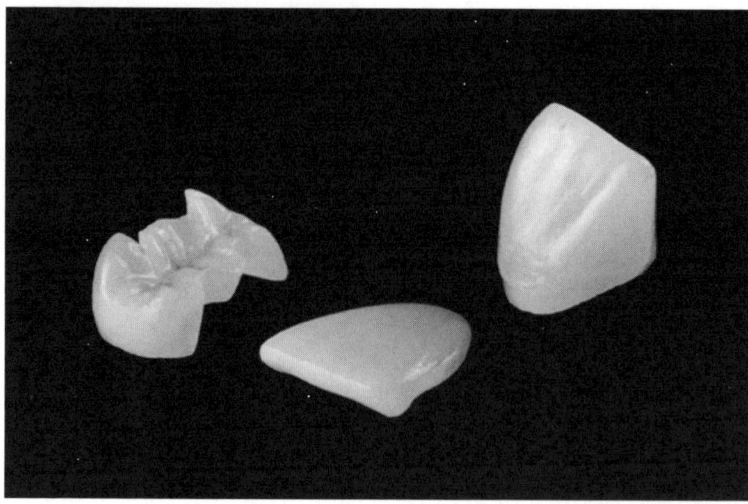

**Figure 31.1.**   Leucite type glass-ceramics: dental inlay, veneer, crown (from left).

Germany. As a result, patients only need one appointment to have their restoration completed.

### 31.3.3. *Lithium disilicate glass-ceramics*

To expand the range of indications of glass-ceramics to include dental bridges, the flexural strength and toughness had to be improved compared to the strength of leucite glass-ceramics. For this purpose, lithium disilicate ($Li_2Si_2O_5$) glass-ceramics came into use. Wide-ranging developments have been undertaken in conjunction with this material system.[7–13] The main crystal phase is produced in the special base glass of the $SiO_2$-$Li_2O$-$K_2O$-$ZnO$-$P_2O_5$-$Al_2O_3$-$La_2O_3$ system by means of heterogeneous nucleation and crystallization. In the process, an interlocking microstructure with a crystal content of more than 60% vol is achieved. The resulting product was called IPS Empress®2 (Ivoclar Vivadent AG) and was launched on the global market in 1999 as a framework material for small three-unit bridges in the anterior region (Table 31.2).

A significant improvement over IPS Empress®2 was achieved in the $SiO_2$-$Li_2O$-$K_2O$-$Al_2O_3$-$ZrO_2$ system and resulted in development of the product IPS e.max® (Ivoclar Vivadent AG) (Tables 31.1 and 31.2)[14–17]. Intensive study of the fracture propagation in this glass-ceramic showed that the high fracture toughness, measured as $K_{IC}$ value, of 2.3 MPa•m$^{0.5}$ (based on the SEVNB method) was caused by crack deviation in the vicinity of the disilicate crystals. When being deviated, the propagating crack loses a considerable amount of energy. This loss of energy results in a high degree of flexural strength (up to 440 MPa) and toughness.

The product group of IPS e.max® encompasses several types of material. The biomaterial IPS e.max® Press presents a moldable glass-ceramic and the biomaterial IPS e.max® CAD, a machinable glass-ceramic. The IPS e.max® Press glass-ceramics are distinguished by the fact that these materials are processed into inlays, crowns, and bridges by the application of molding technology at a temperature of 920°C. The glassy matrix enables the use of viscous flow-molding. However, similar to the initial product, the end product exhibits a dense microstructure, while partial crystal orientation caused by the viscous flow process may be present. The glass-ceramic is veneered with fluoro-apatite glass-ceramic (Fig. 31.2).

Compared to leucite-containing glass-ceramics, it may be more difficult, if possible at all, to machine high-toughness and high-strength lithium disilicate glass-ceramics dental restorations with the CEREC® system. In order to resolve this problem, an intermediate product has been developed which is easier to machine. This lithium metasilicate glass-ceramic has a singular blue color. After the blue glass-ceramic has been machined and tried in the patient's mouth, it undergoes heat treatment at 850°C. During thermal treatment, it is transformed into a white lithium disilicate glass-ceramic.

### 31.3.4. *Fluoro-apatite glass-ceramics*

The needle-like fluoroapatite crystals, contained in glass-ceramics (IPS e.max® Ceram, Ivoclar Vivadent AG) — $Ca_5(PO_4)_3F$ — may

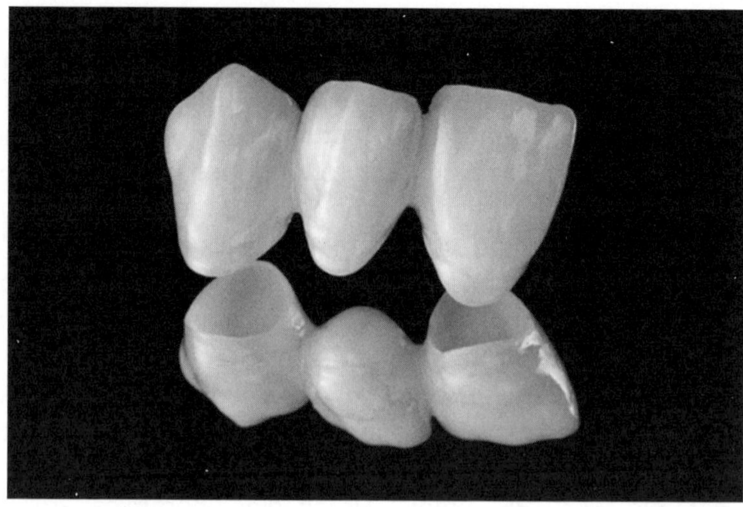

**Figure 31.2.**   Dental bridge reflected on a mirror to show the interior. The bridge consists of a framework of lithium disilicate glass-ceramic veneered with fluoroapatite glass-ceramic.

produce such a favorable light-scattering effect that the resulting translucency corresponds to that of natural teeth.[18,19] The objective was to develop a veneering ceramic that offered tooth-like optical properties. The tooth-like translucency and the CTE of $9.5 \bullet 10^{-6}$ K$^{-1}$ m/m enabled the resulting glass-ceramic to be used on lithium disilicate glass-ceramics as well as on high-toughness $3Y_2O_3$-$ZrO_2$ sintered ceramics. Sintered glass-ceramic devices are achieved by machining open-pore materials and subsequently sintering them to a high density. The high toughness of approx. $4.5$ MPa $\bullet$ m$^{0.5}$ (according to SEVNB) allows the fabrication of metal-free long-span bridges.

### 31.3.5. *ZrO$_2$-containing glass-ceramics*

With the objective of achieving high levels of toughness and a CTE adjusted to that of $ZrO_2$-sintered ceramics, a glass-ceramic containing $ZrO_2$ (IPS Empress® Cosmo Post, Ivoclar Vivadent AG) was developed. Controlled crystallization was used to produce

$Li_2ZrSi_6O_{15}$ as the main crystal phase.[20] This glass-ceramic offers the advantage over other products that it can be pressed onto $ZrO_2$ posts by means of a molding process. The resulting dental posts are suitable for the build-up of devitalized teeth and offer metal-free, highly esthetic solutions for patients.

### 31.3.6. *Leucite–apatite glass-ceramics*

There is still interest from the market in the use of glass-ceramics for veneering metal frameworks. In addition to tooth replacements, glass-ceramic-veneered metal restorations may present an optically esthetic solution for some patients. For instance, long-span bridges and high load-bearing tooth replacements in the posterior region may necessitate a metal–ceramic reconstruction. For this application, a glass-ceramic with coefficient of thermal expansion matched to that of various dental alloys had to be developed. The resulting material is a leucite–apatite glass-ceramic (IPS d.SIGN®, Ivoclar Vivadent AG) derived from the $SiO_2$-$Al_2O_3$-$K_2O$-$Na_2O$-$CaO$-$P_2O_5$-F system (Tables 31.2 and 31.3 ). The leucite content enables this glass-ceramic to attain the required CTE of approx. $12 \bullet 10^{-6}\ K^{-1}$ m/m fluoroapatite needles that provide the product with the required translucent qualities. Nucleation and crystallization in the base glasses of this material are conducted with a twofold reaction. Leucite is formed by surface crystallization, while apatite is precipitated by the mechanisms of internal (volume) processes.

This glass-ceramic can be applied by two different techniques. The sintering technique is the most widely used: the glass-ceramic powder is applied in the form of a high-viscosity, aqueous slip and sintered at temperatures of approximately 830°C. The restoration is built up to resemble the natural tooth by layering a variety of materials in different shades and degrees of translucency.[21,22]

The molding technique is the second method that can be used to process these veneering glass-ceramics. If this technique is used, a glass-ceramic ingot may, for instance, be pressed onto a metal framework according to the principles of the lost wax technique. This method is particularly convenient for large dental laboratories.

## 31.4. Summary

When all the products described above are considered together, glass-ceramic and glass-ceramic composite dental restorative materials cover the entire range of indications for anterior and posterior restorative tooth repair, and therefore, fully fulfill all requirements of patients for functional restoration of teeth. This is a major advance in the health-care of patients world-wide and represents one of the great contributions of glass and glass-ceramic research to the well-being of mankind.

## Acknowledgments

This chapter is an edited version of a section of a review article, "Glass and Medicine", by Larry L. Hench, Delbert E. Day, Wolfram Höland and Volker M. Rheinberger, published in the *International Journal of Applied Glass Science* **1**(1), 104–117 (2010).

## References

1. Gehre, G. (1996). "Keramische Werkstoffe", in Kappert, H. (ed.), *Zahnärztliche Werkstoffe und ihre Verarbeitung. Bd 1: Grundlagen und Verarbeitung*, 6th ed., Hüthig Verlag, Heidelberg, Germany.
2. McLean, J.W. (1972). "Dental Porcelains", in Dickens, D. and Cassels, J.M. (eds), *Dental Materials Research, NBS Publications 354*, National Bureau of Standards, Washington, DC, pp. 77–83.
3. Grossman, D.G. (1972). Machinable glass-ceramic based on tetrasilicic mica, *J. Am. Ceram. Soc.*, **55**, 446–449.
4. Beham, G. (1990). IPS Empress: a new ceramic technology, *Ivoclar Vivadent Report*, **6**, 1–14.
5. Wohlwend, A. and Schärer, P. (1990). Die Empress-Etchnik ein Neues Verfahren zur Herstellung von Vollkeramischen Kronen, Inlays und Facetten, *Quintessenz Zahntechnik*, **16**, 966–978.
6. Höland, W. and Beall, G.H. (2006). *Glass-Ceramic Technology*, John Wiley & Sons, New York, NY.

7. Freiman, S.W. and Hench, L.L. (1972). Effect of crystallization on the mechanical properties of $Li_2O$ — $SiO_2$ glass-ceramics, *J. Am. Ceram. Soc.*, **55**, 86–90.

8. Haedley, T.J. and Loehman, R.E. (1984). Crystallization of a glass-ceramic by epitaxial growth, *J. Am. Ceram. Soc.*, **67**, 620–625.

9. Bengisu, M., Brow, R.K. and White, J.E. (2004). Interfacial reactions between lithium silicate glass-ceramics and Ni-based superalloys and the effect of the heat treatment at elevated temperatures, *J. Mater. Sci.*, **39**, 605–618.

10. Burgener, L.L., Lucas, P., Weinberg, M.C., *et al.* (2000). On the persistence of metastable crystal phases in lithium disilicate glass, *J. Non-Cryst. Sol.*, **274**, 188–194.

11. Deubener, J. (2000). Compositional onset of homogeneous nucleation in (Li, Na) disilicate glasses, *J. Non-Cryst. Sol.*, **274**, 195–201.

12. Beall, G.H. (1989). Design of glass–ceramics, *Solid State Sc.*, **3**, 333–354.

13. Rüssel, C. and Keding, R. (2003). A new explanation of the induction period observed during nucleation of lithium disilicate glass, *J. Non-Cryst. Sol.*, **328**, 174–182.

14. Rheinberger, V.M. (2005). Vom Experiment zur Erfolgsgeschichte-materialforschung eröffnet Heute Breites Indikationsspektrum, *Die Zahnarzt Woche*, **37**, 14–15.

15. Apel, E., Deubener, J., Bernard., A., *et al.* (2008). Phenomena and mechanisms of crack propagation in glass-ceramics, *J. Mech. Behav. Biomed. Mat.*, **1**, 313–325.

16. Kappert, H.F., Schweiger, M. and Rheinberger, V. (2006). Das IPS e.max-System, Werkstoffkundliche Vielfalt, *Dental-Labor.*, **54**, 613–624.

17. Höland, W., Apel, E., van't Hoen, C., *et al.* (2006). Studies of crystal phase formation in high-strength lithium disilicate glass-ceramics, *J. Non-Cryst. Sol.*, **352**, 4041–4050.

18. Schweiger, M. (2006). IPS e.max Ceram, *Ivoclar Vivadent Report*, **17**, 25–36.

19. Höland, W., Ritzberger, C., Apel, E., *et al.* (2008). Formation and crystal growth of needle-like fluoroapatite in functional glass–ceramics, *J. Mater. Chem.*, **18**, 1318–1332.

20. Schweiger, M., Frank, M., Cramer von Clausbruch, S., *et al.* (1998). Microstructure and properties of pressed glass–ceramic core to zirconia post, *Quint. Dent. Technol.*, **21**, 73–79.
21. Höland, W., Rheinberger, V., Apel, E., *et al.* (2007). Principles and phenomena of bioengineering with glass–ceramics for dental restoration, *J. Eur. Ceram. Soc.*, **27**, 1521–1526.
22. Höland, W., Schweiger, M., Watzke, R., *et al.* (2008). Ceramics as biomaterials for dental restoration, *Exp. Rev. Med. Dev.*, **5**, 729–745.

# Pain-Desensitizing Dental Materials



## CHAPTER 32

David C. Greenspan

## 32.1. Introduction

Tooth pain is often due to a clinical condition called dentin hypersensitivity and is a common occurrence in the general adult population. The incidence has been reported to be anywhere from 4% to as high as 57% of adults.[1,2] In periodontal patients, the rate of dentin hypersensitivity can be as high as 98%.[3] The condition has been characterized by a short sharp pain arising from exposed dentin in response to stimuli; typically thermal, evaporative, chemical or tactile.[4,5] The diagnosis of dentin hypersensitivity can only be made when other etiologies or pathologies have been ruled out (differential diagnosis).[6] One objective of this chapter is to review the history of the development of a novel class of biomaterials that have been proven to be highly effective in treatment and prevention of dentin hypersensitivity. The material is 45S5 Bioglass® processed to be a powder that is composed of small particles several micrometers in diameter. The commercial product is trademarked NovaMin®. This novel material contains only calcium, sodium, phosphate, and silica, all as an amorphous matrix, as described in Chapter 3 in this book. The technical description of this material in the dental literature for use in treatment of dentin hypersensitivity is an inorganic, amorphous, calcium, sodium phosphosilicate (CSPS) material. A second objective of this chapter is to describe the mode of action of CSPS particles that results in the superior performance in the reduction of tooth sensitivity by physically occluding dentin tubules.

## 32.2.  Early Development of CSPS (Bioactive Glass) Materials

Until the discovery of this class of materials, all biomaterials were made to be as inert as possible in the human body.[7,8] The initial animal studies with the original CSPS composition (same composition as NovaMin®) demonstrated for the first time that a synthetic biomaterial could actually form a direct chemical bond with bone tissue.[9,10] The seminal discovery that a synthetic biomaterial could form a chemical bond with bone allowed researchers to change their view of how biomaterials could be engineered to interact with the body, rather than finding ways to minimize the interactions between the body and the implanted materials.

Research during the years following this discovery found that the CSPS material went through a series of reactions at the surface of the material that released ions into the surroundings, and that the surface of the CSPS material changed in composition and structure, as summarized in Chapter 3. Studies have shown that sodium is released rapidly from the surface of the material, creating a surface that is depleted in this ion.[11,12] This reaction, which begins immediately upon exposure of the material to the aqueous environment, then allows calcium from the material to diffuse into the surroundings. The net effect of these reactions is a surface that is essentially rich in silica and has a very porous surface. Furthermore, in a closed system the pH of the solution was found to increase from a neutral pH to a basic pH. The extent of the pH rise was found to be related to the buffer strength of the solution, the surface area to volume ratio of the particles, and the composition of the starting CSPS.

These initial reactions form a surface that becomes negatively charged, facilitating the adsorption of proteins, as well as calcium and phosphate ions, from the solution. Over a period of hours a carbonated calcium hydroxyapatite mineral phase will form on the surface of the material, and this becomes the bonding interface. The net effect of these reactions confirmed the mechanism of the bonding interface between the implant surface and bone. More recently, a series of investigations has demonstrated that this specific composition of calcium, sodium phosphosilicate material actually can stimulate and accelerate bone repair and early angiogenesis compared with other bioactive ceramic compounds.[13,14] Chapter 6 discusses recent findings on the role of bioactive glasses in angiogenesis. In addition, studies that simply exposed the ionic reaction products from the 45S5 composition resulted in accelerated bone cell proliferation and maturation, as well as enhanced rates of new blood vessel formation.[15] The mechanisms associated with these properties appear to be directly related to the amount and rate of ionic release from these materials.[16] These findings have resulted in an understanding of the importance of the kinetics and relative critical amounts of ions released from this CSPS composition, and the relation of these reactions to accelerated bone healing.

Following increased commercial use of these materials in the 1990s, especially particulate CSPS materials for treating periodontal and orthopedic bone defects, research led to the characterization of the reactions of small particles of CSPS materials. It became clear that the size distribution of particles and their concentration have a significant effect on the reactivity of the material.[17–19] Because smaller particles have a greater surface area for a given volume of material, it was found that size and concentration of particles were critical in obtaining optimal results in surgical situations. In addition, these studies demonstrated that the effects of particle size would cause different changes in the pH of the reaction solutions, and that the reactions would proceed to varying degrees of completion, depending on the ratio of the surface area of particles to volume of solution.

Interestingly, the reactivity of small particles of the CSPS materials has been shown to result in some anti-microbial properties against oral bacteria[20,21] as well as some transient anti-inflammatory properties.[22] Experiments have shown how the ionic reactions and resultant surface modifications beneficially interact with collagen, allowing for strong bonding between collagen and the surfaces of these materials.[23,24] It was the combination of all of these factors that led to the development of NovaMin® for the treatment of tooth sensitivity.

## 32.3. NovaMin® and Sensitivity — Mode of Action

The currently accepted theory for tooth hypersensitivity is the hydrodynamic theory proposed by Brännström.[25] The premise of the hydrodynamic theory is the concept that open dentinal tubules allow fluid to flow through them, which excites the nerve endings in the dental pulp. Clinical replicas of sensitive teeth viewed under a scanning electron microscope (SEM) reveal varying numbers of open or partially occluded dentinal tubules.[26] These studies also showed that patients with dentin hypersensitivity have a greater number of tubules per area, and the diameter of the tubules is greater than in patients with no sensitivity.[27] In general, tubules are not exposed at the tooth root surface because of the cementum covering the tooth root, or because of a smear layer of dentinal debris that covers the tooth surface and masks the tubules. When the smear layer is present, the fluid flow that can occur through the dentin is only a few percent of that possible following acid removal of the smear layer, which "opens" the tubules.

There are two basic approaches to the treatment of dentinal hypersensitivity. The first approach is to treat the tooth with a chemical agent, such as potassium nitrate or potassium chloride, which penetrates into the dentinal tubules and depolarizes the nerve synapses. This reduces sensitivity by preventing the conduction of pain impulses.[28] There have been numerous clinical studies that tested the efficacy of these agents in reducing dentin hypersen-sitivity. While these materials have consistently shown clinical efficacy in the treatment of sensitivity it may take weeks for the patient to perceive a reduction in pain and sensitivity. The second approach is to treat the tooth with a chemical or physical agent (e.g. potassium oxalate, ferric oxalate, or strontium chloride) to physically occlude dentinal tubules, which reduces sensitivity by prevention of pulpal fluid flow.[29] Although both approaches are effective at reducing or eliminating hypersensitivity, the duration of relief is highly variable. Hypersensitivity usually reappears due to toothbrush abrasion, the presence of acid challenges in the mouth, and/or degradation of the coating material.[30]

The use of calcium, sodium phosphosilicates in periodontal surgery,[31] where tooth sensitivity is routinely found, coupled with the need for improved materials to treat tooth sensitivity, led to the initial investigations of this material for treating tooth sensitivity. Research with CSPS had shown that these materials will actually form a strong attraction with collagen.[23,24,32] Because dentin consists of more than 50% collagen it was believed that the CSPS particles would bind to the exposed dentin surface as well as physically fill the open tubules. It was further hypothesized that the subsequent ionic release and surface reaction would help to form a protective hydroxyl carbonate apatite layer that would impart rapid and continual relief from tooth sensitivity. The earliest studies conducted *in vitro* demonstrated that the material would, in fact rapidly occlude dentin tubules and form a protective layer on the dentin surface.[33] These initial results were encouraging, and led to further research and the commercial development of this technology.

To understand the details of the mechanisms of this action, it is necessary to review the combination of ionic reactions that occur in an aqueous environment, as well as other factors such as pH changes and the surface changes in the particles themselves. All have a critical role in producing the protective effect of this material.

## 32.4. Role of pH and Ionic Release

When particles of the CSPS material are exposed to an aqueous environment, such as water or saliva, there is an immediate release of sodium ions. The release of Na from the particle surfaces increases the pH locally, which can cause a more rapid precipitation of the ions to form the HCA layer. It is well known that calcium phosphates are more soluble in acid environments than in basic or pH neutral solutions,[34] and that precipitation of these ions is accelerated at neutral or basic pH levels. It is the release of Na and increase in pH that allows the conditions for the rapid precipitation of particles and the formation of the calcium phosphate layer (see Chapter 3).

The immediate release of Na from the CSPS particles is necessary but not sufficient for the formation of the calcium phosphate layer. However, the rapid release of sodium allows for the release of calcium and phosphate ions from the particles within minutes of exposure to the aqueous environment, by producing a porous surface layer. An amorphous calcium phosphate layer was found to form on the particle surfaces within an hour of exposure to a simple organic buffer, as discussed in Chapter 3. It is important to realize that this unique attribute of the composition of calcium sodium phosphosilicate sets it apart from all other materials that have been shown to act as physical occluding materials. The slow network dissolution of the particles is critical, because the particles act as reservoirs to continuously release calcium ions and phosphate ions into the local environment, over many days in some cases.

In addition to the release of these ions, the role of soluble silica in the formation of calcium phosphate mineral is important. It is an attribute of the NovaMin® particles that they release silica into the localized oral environment at a concentration between 15 and 40 ppm. This is one of the critical factors in the early stages of the precipitation of an amorphous calcium phosphate phase. In order to be effective at occluding tubules the particles must not only release the proper level of ions over time, but they must be able to remain on the dentin surface over a long period of time. The interactions of the CSPS particles with collagen have been studied by a number of research groups in various *in vitro* models. These studies have all demonstrated that there is a positive interaction between the reacted surface of the particles and collagen, and that this interaction is strongest for the composition used in NovaMin®. These studies have shown that the development of the negative surface charge at the particle surface, due to the initial reactivity, allows for binding to the side groups on type I collagen fibers. Because exposed dentin has a high content of exposed collagen it is reasonable to assume that this is the mechanism that allows the

CSPS particles to attach to and remain on the dentin surface. Once deposited onto the dentin, the particles will continue to provide long-term release of calcium and phosphate into the local environment, leading to long-term protection of the dentinal tubules.

## 32.5. Evidence of Efficacy for CSPS

There are a number of standard *in vitro* models in the area of oral health-care that have been used to demonstrate the mode of action of the various desensitizing agents. The dentin block model has a number of variations to test the ability of materials to occlude tubules and to remain on the dentin surface through various challenges that would normally be found in the oral environment. There has been extensive testing of CSPS using a number of these models and they have repeatedly demonstrated the rapid occlusion of tubules and the persistence of the particles on the dentin surface.[34] These studies have repeatedly shown that a single application of the proper concentration (above 3%) of CSPS either in a daily-use dentifrice or a professionally applied prophylaxis paste is effective at blocking at least 75% of open tubules. In many cases, the single application is sufficient to block over 95% of tubules. Furthermore, these studies have demonstrated that a single application of CSPS in these models will resist repeated acid challenges. When tested in a model where the material is repeatedly applied and alternately subjected to twice-daily acid challenges, the results demonstrate continual blockage of the dentin tubules. Additional testing in these models has shown that the CSPS continues to release calcium ions over a long period compared with other calcium-containing products that release a burst of calcium but then provide little in the way of calcium ions to protect the exposed dentin.

SEM is one of the analytical methods used to evaluate ingredients used for tubule occlusion in treating tooth sensitivity. Figure 32.1(a) is an SEM image that shows a typical prepared dentin block used in the types of *in vitro* studies described above. The piece of dentin has been ground and polished, and then acid

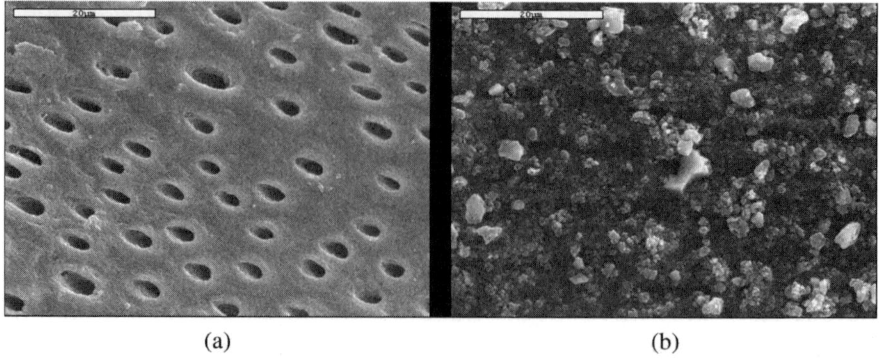

(a)                                                        (b)

**Figure 32.1.**   SEM images from *in vitro* tubule occlusion study showing prepared and treated dentin: (a) prepared dentin slab showing open tubules, (b) dentin after a single application of CSPS for two minutes and 30 second water rinse.

etched to remove the smear layer and to open the dentin tubules. Figure 32.1(b) shows a dentin block that has been treated once with the CSPS material and subjected to a subsequent acid challenge. After the acid challenge the sample was gently rinsed and dried for SEM analysis. Note that the majority of tubules are completely closed and the remainder are at least partially closed. Interestingly, particles are retained on the surface of the dentin block even after rinsing. This evidence substantiates and helps to explain the long-lasting effect of even a single use of the CSPS particles.

While *in vitro* studies are extremely useful in helping to differentiate, rank, and evaluate various materials for their effectiveness in treating tooth sensitivity, the clinical trial is the ultimate test for these materials, and a number have been performed. One study compared the efficacy of a strontium chloride ingredient and a placebo to a daily-use CSPS dentifrice in reducing tooth sensitivity in a six week, prospective study. The results showed that compared to placebo both ingredients reduced sensitivity, but the CSPS material showed statistically greater reductions in sensitivity at the six-week end point compared with the other materials.[35]

## 32.6. Commercialization of CSPS for Treating Tooth Sensitivity

During the past ten years, a number of consumer and professional products containing CSPS have been introduced into the market. The first product developed by NovaMin Technologies was Oravive®, a daily-use, fluoride-free dentifrice that contains 5% of the CSPS ingredient NovaMin®. This product was cleared for use by US Food and Drug Administration (FDA) through a 510K process as a medical device for the rapid and continual reduction of tooth sensitivity through physical tubule occlusion. Other products that have been introduced are listed in Table 32.1, along with a description of the product. Today, CSPS has been formulated into over 15 products and is sold in over 20 countries, including the United States, Canada, India, China, and a number of countries in Europe. These products have proven to be highly effective and the CSPS material has an unparalleled safety profile.

Development of new applications and refinements in the manufacture and use of CSPS continues in the area of oral health-care. Over the past ten years, studies have shown that the surface activity and ionic release from specific compositions of CSPS materials have produced the most effective treatment for tooth sensitivity to date. The interactions of the material with native collagen, along with the reactivity of the particles, ensures that the material will immediately occlude patent dentin tubules and will remain at the

**Table 32.1.**   A list of some of the current CSPS products available.

| Product | Description |
| --- | --- |
| SootheRx (U.S.) | 7.5% CSPS, daily use, professionally supplied |
| X-Pur (Canada) | 5.0% CSPS, daily use, professionally supplied |
| Nanosensitive (Germany) | 7.5% CSPS |
| Sensishield (U.K.) | 5.0% CSPS |
| Nutri-émail (France) | 7.5% CSPS |
| Vantaj (India) | 5.0% CSPS |
| SHY-NM (China) | 5.0% CSPS |
| Odontis Sensiblock (Brazil) | 7.5% CSPS |

dentin surface, allowing this ionic release to build a thin mineral layer that will continue to occlude the tubules and will resist challenges of acidic environments.

The wealth of science behind the development of this class of materials has led to investigations of CSPS for oral health-care applications beyond the treatment of tooth sensitivity. The potential of these materials for re-mineralization of both enamel and dentin has been studied *in vitro* and *in situ* and holds promise.[36,37] In addition, the unique ionic reactions and potential anti-microbial and anti-inflammatory properties might prove useful in treating gingivitis.[38]

## References


1. Rees, J.S. and Addy, M. (2002). A cross-sectional study of dentine hypersensitivity, *J. Clin. Perodontol.*, **29**, 997–1003.
2. Irwin, C.R. and McClusker, P. (1997). Prevalence of dentine hypersensitivity in a general dental population, *K. Ir. Dent. Assoc.*, **43**, 7–9.
3. Chabanski, M.B., Gillam, D.G., Bulman, J.S., *et al.* (1996). Prevalence of cervical dentine sensitivity in a population of patients referred to a specialist periodontology department, *J. Clin. Periodontol.*, **23**, 989–992.
4. Canadian Advisory Board on Dentin Hypersensitivity (2003). Consensus-based recommendations for the diagnosis and management of dentin hypersensitivity, *J. Can. Dent. Assoc.*, **69**, 221–226.
5. Orchardson, R. and Collins, W.J. (1987). Clinical features of hypersensitive teeth, *Br. Dent. J.*, **162**, 253–256.
6. Dowell, P. and Addy, M. (1983). Dentine hypersensitivity — a review. Aetiology, symptoms and theories of pain production, *J. Clin. Periodontol.*, **10**, 341–350.
7. Hench, L.L. and Andersson, Ö.H. (2003). "Bioactive Glasses", in Hench, L.L. and Wilson, J. (eds), *Introduction to Bioceramics*, World Scientific, Singapore, pp. 41–62.
8. Hench, L.L. (1990). Biomaterials, *Science*, **208**, 826–831.
9. Hench, L.L., Splinter, R.J., Greenlee, T.K., *et al.* (1971). Bonding mechanisms at the interface of ceramic prosthetic materials, *J. Biomed. Mater. Res.*, **2**, 117–141.

10. Hench, L.L. and Paschall, H.A. (1973). Direct chemical bonding between bio-active glass–ceramic materials and bone, *J. Biomed. Mater. Res.*, **4**, 25–42.

11. Andersson, Ö.H. and Kangasniemi, I. (1991). Calcium phosphate formation at the surface of bioactive glass *in vitro*, *J. Biomed. Mater. Res.*, **25**, 1019–1030.

12. Pantano, C.G. Jr, Clark, A.E. Jr and Hench, L.L. (1974). Multi-layer corrosion films on Bioglass®, *J. Am. Ceram. Soc.*, **57**, 412–413.

13. Xynos, I.D., Hukkanen, M.J.V., Batten, J.J. *et al.* (2000). Bioglass® 45S5 stimulates osteoblast turnover and enhances bone formation *in vitro*: implications and applications for bone tissue engineering, *Calcif. Tiss. Int.*, **67**, 321–329.

14. Day, R.M. (2005). Bioactive glass stimulates the secretion of angiogenic growth factors and angiogenesis *in vitro*, *Tissue Eng.*, **11**, 768–777.

15. Xynos, I.D., Edgar, A.J., Buttery, L.D., *et al.* (2000). Ionic dissolution products of bioactive glass increase proliferation of human osteoblasts and induced insulin-like growth factor II mRNA expression and protein synthesis, *Biochem. Biophys. Res. Comm.*, **276**, 461–465.

16. Jones, J.R., Sepulveda, P. and Hench, L.L. (2001). Dose-dependent behaviour of bioactive glass dissolution, *J. Biomed. Mater. Res. A.*, **58**, 720–726.

17. Oonishi, H., Hench, L.L., Wilson, J., *et al.* (1999). Comparative bone growth behaviour in granules of bioceramic materials of various sizes, *J. Biomed. Mater. Res.*, **44**, 31–43.

18. Cerruti, M., Greenspan, D. and Powers, K. (2005). An analytical model for the dissolution of different particle size samples of bioglass® in TRIS-buffered solution, *Biomaterials*, **26**, 4903–4911.

19. Arcos, D., Greenspan, D.C. and Vallet-Regi, M. (2003). A new quantitative method to evaluate the *in vitro* bioactivity of melt and sol-gel derived silicate glasses, *J. Biomed. Mater. Res.*, **65**, 344–351.

20. Stoor, P., Soderling, E. and Salonen, J.L. (1998). Antimicrobial effects of a bioactive glass paste on oral microorganisms, *Acta. Odontol. Scand.*, **56**, 161–165.

21. Allan, I., Newman, H. and Wilson, M. (2001). Antibacterial activity of particulate bioglass against supra and subgingival bacteria, *Biomaterials*, **22**, 1683–1687.

22. Rechtenwald, J.E., Minter, R.M., Rosenberg, J.J., *et al.* (2002). Bioglass attenuates a proinflammatory response in mouse peritoneal endotoxicosis, *Shock*, **17**, 135–138.

23. Zhong, J.P., LaTorre, G.P. and Hench, L.L. (1994). "The Kinetics of Bioactive Ceramics Part VII: Binding of Collagen to Hydroxyapatite and Bioactive Glass," in Andersson, Ö.H. and Yli-Urpo, A. (eds), *Bioceramics 7. Proceedings of the 7th International Symposium on Ceramics in Medicine.* Butterworth-Heinemann Ltd, Oxford, pp. 61–66.

24. Oréfice, R., Hench, L.L. and Brennan, A. (2009). Evaluation of the interactions between collagen and the surface of a bioactive glass during *in vitro* testing, *J. Biomed. Mater. Res.*, **90**, 114–120.

25. Brännström, M. and Aström, A. (1972). The hydrodynamics of the dentine; its possible relationship to dentinal pain, *Int. Dent. J.*, **22**, 219–227.

26. Absi, E.G., Addy, M. and Adams, D. (1989). Dentine hypersensitivity: the development and evaluation of a replica technique to study sensitive and nonsensitive cervical dentine, *J. Clin. Periodontol.*, **16**, 190–195.

27. Ishikawa, S. (1969). A clinic-histological study on the hypersensitivity of dentine, *J. Jpn. Stomatol. Soc.*, **36**, 68–88.

28. Schiff, T., Bonta, Y., Proskin, H.M., *et al.* (2000). Desensitizing efficacy of a new dentifrice containing 5.0% potassium nitrate and 0.454% stannous fluoride, *Am. J. Dent.*, **13**, 111–115.

29. Orchardson, R. (2006). Managing dentin hypersensitivity, *JADA*, **137**, 990–998.

30. West, N.X., Hughes, J.A. and Addy, M. (2002). Dentine hypersensitivity: the effects of brushing toothpaste on etched and unetched dentine *in vitro*, *J. Oral. Rehabil.*, **29**, 167–174.

31. Froum, S.J., Cho, S.C., Rosenberg, E., *et al.* (1998). Comparison of Bioglass® synthetic bone graft particles and open debridement in the treatment of human periodontal defects, *J. Perio.*, **69**, 698–709.

32. Efflandt, E.S., Magne, P., Douglas, W.H., *et al.* (2002). Interaction between bioactive glass and human dentin, *J. Mater. Sci. Mater. Med.*, **13**, 557–565.

33. Litkowski, L.J., Hack, G.D., Sheaffer, H.B., *et al.* (1997). "Occlusion of Dentin Tubules by 45S5 Bioglasss®", in Sedel, L. and Rey, C. (eds), *Bioceramics 10. Proceedings of the 10th International Symposium on Ceramics in Medicine*, Elsevier, New York, NY, pp. 411–414.

34. Burwell, A.K., Litkowski, L.J. and Greenspan, D.C. (2009). Calcium sodium phosphosilicate (NovaMin®): remineralization potential, *Adv. Dent. Res.*, **21**, 35–39.

35. Du, M.Q., Bian, Z., Jiang, H., *et al.* (2008). Clinical evaluation of a dentifrice containing calcium sodium phosphosilicate (Novamin) for the treatment of dentin hypersensitivity, *Am. J. Dent.*, **21**, 210–214.

36. Alauddin, S.S., Greenspan, D. and Anusavice, K.J. (2005). *In vitro* human enamel remineralization using bioactive glass containing dentifrice, *J. Dent. Res.*, **84**, 2546.

37. Litkowski, L.J., Quinlan, K.B. and Pandya, A.J. (2007). *In situ* comparison of surfaces by optical profilometry, *J. Dent. Res.*, **86**, 1412.

38. Tai, B.J., Jun, B., Bian, Z., *et al.* (2006). Anti-gingivitis effect of a dentifrice containing bioactive glass (Novamin®) particulate, *J. Clin. Periodontol.*, **33**, 86–91.

# Ethical Considerations in Biomaterials Research and Development



## CHAPTER 33

Mrinal K. Musiband and Subrata Saha

## 33.1. Introduction

Biomedical engineering combines two different fields, medicine and engineering, into a truly interdisciplinary field. It often involves the application of engineering principles for solving human medical problems and has been responsible for many of the dramatic advances in modern medicine. It endeavors to devise novel techniques and devices to alleviate human pain and suffering, and has resulted in improved healthcare and better quality of life for patients.

In spite of the magnificent contributions of the field of biomedical engineering to the betterment of human life, it has also produced new ethical dilemmas and has challenged some of our pre-existing moral standards. The following sections will focus on ethical issues faced by bioengineers in the research and development of some emerging scientific disciplines, including tissue engineering, stem cell research, nanotechnology, animal research and clinical trials. We also point out the need for education on ethics for biomaterials scientists during their training and professional career.

## 33.2. Progress in Biomaterials Research

Development of novel biomaterials that respond to various biological materials such as cells and proteins is likely to raise new ethical questions. The main considerations with any biomaterial intended for human application include the source of the material, its safety, efficacy, and its short and long-term performance.[1]

The 20th century has witnessed significant progress in biomaterials research, which resulted in the development of materials potentially capable of replacing the hard and soft tissues of the human body.[1] Moreover, recent biomedical engineers and healthcare professionals have more in-depth understanding of the molecular/cellular interactions taking place between biomaterials and the human system. The biomaterials arena has progressed in the last century to incorporate technologies and materials that were once strictly reserved for other engineering applications.

## 33.3. Ethical Code of Conduct and Need for Ethical Training for Biomaterials Scientists

Unlike medical "Codes of ethics" that guide clinicians to use treatments individualized to each patient to the best of their knowledge and training, traditional engineering codes of ethics typically govern products and technologies used by large numbers of people such as bridges, buildings, electronic gadgets, and other modern amenities.[1,2] But biomaterials science combines the fields of science and engineering with medicine, and due to this unique interaction it is important to incorporate medical, engineering, and scientific ethical codes and principles into a comprehensive set of guidelines for bioengineers. This need has given rise to the recently developed field of "Bioethics." Bioethics is a quasi-social science that offers guidance in cases of moral conflicts that may develop in medical and biological science, and has helped biomedical engineers to build an ethical foundation. It has also led to the creation of an ethical code that can guide them in this diverse and interdisciplinary field.[3]

According to Pimple, the "Code of ethics" for responsible conduct of research includes scientific integrity, collegiality, protection of human subjects, animal welfare, institutional integrity, and social responsibility.[4] With significant advances in molecular/cell biology and nanotechnology, the need for safe and effective therapies will also create unique ethical situations in the future, and responsible research by biomaterials scientists will necessitate the incorporation of many new rules and regulations into the existing code of ethics. These will be necessary if new-age materials from the emerging areas of science and technology are going to be morally and ethically acceptable to the scientific community and to society.

When engineers apply their expertise to the field of medicine, they often find themselves faced with moral problems in their new venture. Some bioengineers have actually been forced to discontinue their research because they failed to accommodate humane or moral considerations in the conduct of their work. Thus for a well-rounded education, biomedical and clinical engineering students should be introduced to medical ethics as a part of their coursework. As biomedical engineers often work with physicians and other health-care professionals as part of a team with direct responsibility for patient care, this may create ethical issues not encountered in traditional engineering. Hence in order to have well-educated ethical engineers and clinicians, students need repeated and varied exposures to ethics discussions both during their undergraduate and postgraduate studies, and also during their professional lives, for them to be aware of the prevailing ethical standards.

## 33.4. Tissue Engineering

The science of tissue engineering involves the growth and regeneration of malfunctioning/damaged tissues and organs of the human body.[4] In spite of the potential benefits of this emerging field, ethical questions linger, pertaining to the safety of the cells used and the interaction of the cells with the substrate on which the tissue constructs are grown. Most ethical questions relate to the source of the cells: whether they are human cells (autogenous or allogenous), or cells obtained from a different species (xenogenous).[5] For example, usage of embryonic stem cells to create mineralized tooth structures creates a whole new area of debate.[6] The potential to replace lost tissues in the human body can be realized with the combined efforts of scientists involved in stem cell research and tissue engineering.

Novel drug-delivery mechanisms are being developed and will continue to be developed in the future, for successful and controlled tissue regeneration and growth. With such interdisciplinary research being witnessed in tissue engineering, important areas that need to be addressed in this context includes informed consent, safety of the tissue constructs, and their potential effects on patients. Regulations should be reviewed periodically to allow evolution with this emerging field, and national and international organizations need to foster the development of the field of tissue engineering in a way that is responsible and beneficial to all. The general public needs to be informed about relevant advances in the field, as an informed public will be more likely to support, rather than oppose, new developments such as tissue engineering.

## 33.5. Nanotechnology

Nanotechnology is an area of science concerned with very small structures, materials on the nanometer scale (1 nanometer = $1 \times 10^{-9}$ meters), typically less than 100 nm.[4] At this scale, materials behave quite differently from their macroscopic forms. Some of these properties include high surface area, increased or decreased strength, unique optical properties, altered heat or electrical conductivity, increased surface energy, and the property of forming aggregates consisting of numerous individual particles.[7] Nanotechnology aims at designing and/or manipulating materials at their nanoscale that may simulate cellular functions.[8] This may have the potential to specifically tailor nanomachines or "nanorobots" that may be capable of performing specific cellular functions and/or repairing individual, damaged cells.[5] Piehler predicts that in future these "nanorobots" may be used to clear circulatory obstructions, poison cancer cells, and treat infections.[9]

Due to the interdisciplinary nature of nanotechnology, regulation becomes a challenge, and several different regulatory bodies likely linked to the applications of specific nanotechnologies will regulate portions of the field. The special size-related properties of nanoparticles may mean that regulations will have to be modified to take these properties into account. All relevant regulatory bodies should consider whether existing safety regulations are sufficient; they will need to be revised when necessary. Current commercial uses of nanoparticles should be reviewed by appropriate regulatory boards to ensure safety and promote the safe use of nanotechnology under controlled conditions. Regulations should be reviewed periodically to take into account new concerns that may develop.

Equity, privacy, security, environmental consequences, and human–machine interactions are some of the areas that need to be considered. Moore states that the main areas of concern include relations between academic–industry partnerships, the abuse of the technology, social divides, and the very concept of life.[8] According to Weil, intellectual property rights issues related to nanotechnology will be unique because of the complex, interdisciplinary nature of the science.[10] With the introduction of new nanobiomaterials, fresh guidelines and rules to approve these devices need to be developed. Issues such as patient and clinician acceptability of such devices might be a concern in the future, and new regulations to ascertain the safety, efficacy, and potential harmful effects to the human body need to be considered. Further, current graduate training in universities often lacks programs devoted to the philosophy and social science which could support such debate. Emphasis on government support for ethics training related to nanotechnology research needs to be stressed. Protection of intellectual property related to nanotechnology must also be considered as the science progresses. Guidelines to certify and/or approve such futuristic devices may be challenging for future scientists, regulatory agencies, and the industry.

## 33.6. Stem Cell Research, Cloning, and Ethical Concerns

Stem cells refer to cells that can self-renew indefinitely and differentiate into one or more specialized cell types, subsequently being harnessed for cell transplantation, gene therapy, and tissue regeneration applications.[5] Stem cell research has shown potential to provide therapies for many diseases including diabetes mellitus, Parkinson's, and Alzheimer's. However, the source of the therapies, and questions about the potential development of cloned individuals from the stem cells — and the consequences — have raised many ethical questions and moral challenges. Some researchers and ethicists claim that it is unethical to utilize the human embryo as a source of stem cells, as they should be given the status of a human being. Advocates of stem cell research claim that the embryos are a rich source of stem cells, and extracting cells from this source would be a practical solution to create new stem cell lines that may provide many beneficial treatments in the future. The main point of debate is whether life begins at the fertilization of the egg or during the development of the embryo.

Despite the lobbying of religious groups, federal funding for research on some embryonic stem (ES) cells lines was made available during President Bush's tenure in the United States, and the justification provided was that the blastocysts were already destroyed and could therefore be used, but no more new ES cell lines could be generated. Recently, more funding for stem cell research was made available by the new administration under President Obama who, with his advisors, considered that the research was justified by its immense potential.

More ethical issues have been raised related to stem cell research, tissue engineering, and cell-based therapies than any other biomaterial currently in use. With rapid progress in the fields of molecular and cell biology, it may be possible to grow organs from an individual's own cells, or from a different individual or species, and this possibility has led to widespread debate among researchers and academics pursuing such research.

## 33.7. Animal Use in Biomaterials Research and Ethics

In order for any novel biomaterial to be approved for human use, numerous animal tests need to be carried out to ascertain its safety, biocompatibility, and long-term performance. This has been the topic of debate among animal rights activists and proponents of animal research.[11] Animal activists claim that the use of animals for biomedical research is not justified because of the damage or injury to animals that have the ability to feel pain and distress, like humans, and is thus unacceptable and unjustified. Proponents of animal research believe that although this kind of research causes harm to animals, it also helps to improve the lives of humans and animals due to the body of scientific knowledge gained.

Although simulated computer and non-animal models may be used in lieu of live animals, some proponents of animal research claim that the knowledge gained would be limited, and may not always relate to those of live animals.[11] Nevertheless, the long-standing ethical dilemmas related to animal research, and prohibitive rise in costs, may lead to a decline in the use of animals for biomedical research in the future. Most researchers believe that following the rule of the "3 Rs" of animal research — replace, reduce, and refine — will provide a good moral basis for utilizing animals for future biomedical research.[5]

Efforts to reduce the number of animals used for research and to promote the number of non-animal models should be encouraged. Budinger recognizes that there is a dilemma in animal research regarding respect for life and the painless existence of animals versus the justifiable medical need of humans.[12] Furthermore, current US federal regulations minimize unnecessary experiments by demanding scientific justification for the proposed experiment, and have strict requirements on the humane care of animals and cleanliness of the facilities.

## 33.8. Clinical Trials

The ethical and legal issues concerned with human experimentation are even more involved. More than two decades ago, the US Department of Health, Education, and Welfare (DHEW) proposed far-reaching regulations on human research requiring that an ethics committee should review such studies within the organization.[5]

In 1974, the DHEW extended requirements relating to fetuses, pregnant women, prisoners, and the institutionalized mentally disabled. DHEW established the National Commission for the Protection of Human Subjects of Biomedical and Behavioral Research. Currently, biomedical research involving humans is controlled by the Department of Health and human Services (DHHS) and the US Food and Drug Administration (FDA), and both require institutional review boards (IRB) to ensure adequate protection of human subjects.[2]

The IRB determines that the risks to subjects are minimized, the risk–benefit ratio is reasonable, the selection of subjects is equitable, documented informed consent has been obtained from the subject or from a legal representative, the course of the experiment is monitored to protect the safety of the subject, and the confidentiality and privacy of the subject is respected.[7] No data, films, or observational studies of personal habits should be released with personal identification without the informed consent of the subject. Additional safeguards protect the rights of the vulnerable populations. Once a patient has consented to be a subject in a clinical trial, the investigator's responsibility to her is more important than personal academic gains or the financial interest of the sponsor. The subject should be fully informed and supervised throughout the trial, and should have the option to withdraw at any time during the trial. However, no matter how voluminous regulatory forms and procedures become, nothing can substitute for high moral standards among biomedical scientists themselves.

## 33.9. Summary

The future of biomaterials research and development is promising, with many new materials and technologies constantly being developed for human use. It may be highly beneficial to employ nanobiomaterials and tissue-engineered products to fight infections and repair damaged tissues. There have been significant advances in the fields of tissue regeneration and organ transplantation. However, as increasingly sophisticated devices are introduced to medicine, the new technologies may complicate the doctor–patient relationship. Many scientists may face numerous critical ethical and moral questions of whether to emphasize the development of existing biomaterials, or to devote more resources towards the development of growing organs and tissues. Biomedical research is becoming increasingly interdisciplinary in nature, and this changing scenario will probably raise new ethical questions.

While the physician has traditionally been responsible for his patient's well-being, now the bioengineers who develop and design devices, and the clinical engineers who select, maintain, and advise on the use of modern health-care equipment, are also making decisions that affect people's lives. Professionals have a responsibility to serve and protect the public, and engineers in biology and medicine, and other health-care professionals, must be aware of the ethical issues involved, and should prepare themselves to deal with the moral dilemmas. We recommend education in bioethics in undergraduate and postgraduate work; continued education for practicing individuals; and seminars and conferences to keep biomedical engineers and physicians up to date with ethical issues and changing social values, as well as to provide strategies for ethical decision-making.

The health-care industry has become a big business and the bioengineer is not alone in facing new complications. Physicians are finding themselves in uncomfortable roles as they seek a balance between the health needs of their patients, the cost of treatment, the business and legal concerns of the hospitals and their own professional, business, and legal concerns.

Some ethicists believe that the direct interaction between a doctor and patient will be compromised with the advent of these new-age technologies. The key is to strike a balance between the introduction of new technologies and the provision of health-care that is acceptable morally and ethically to all. Due to the constant changes in biomaterials science, and with the introduction of new materials, processing techniques, and technologies, research in this field will be very different compared with existing research practices. The related scientific disciplines of engineering, medicine, and basic sciences will increasingly interact with each other in the future. The introduction of these cutting-edge technologies will change the practice of medicine dramatically, with the possibility for unprecedented consequences and outcomes. Addressing these questions in an ethical framework will be essential for the advancement of the science.

## References


1. Ratner, B.R., Hoffman, A.S., Schoen, F.J., *et al.* (2004). *Biomaterials Science: An Introduction to Materials in Medicine*, 2nd ed., Elsevier, Philadelphia, PA.
2. Saha, S. and Saha, P. (1987). Bioethics and biomaterials, *J. Biomed. Mater. Res.*, **21**, 181–190.
3. Saha, S. and Saha, P. (1997). Biomedical ethics and the biomedical engineer: a review, *Crit. Rev. Biomed. Eng.*, **25**, 163–201.
4. Pimple, K.D. (2002). Six domains of research ethics. A heuristic framework for the responsible conduct of research, *Sci. Eng. Ethics.*, **8**, 191–205.
5. Kashi, A. and Saha, S. (2009). Ethics in biomaterials research, *J. Long-Term Eff. Med.*, **19**, 19–30.
6. Baum, B.J. and Mooney, D.J. (2000). The impact of tissue engineering on dentistry, *J. Am. Dent. Assoc.*, **131**, 309–318.
7. Florczyk, S.J. and Saha, S. (2007). Ethical issues in nanotechnology, *J. Long-Term Eff. Med.*, **17**, 107–113.
8. Moore, F.N. (2002). Implications of nanotechnology applications: using genetics as a lesson, *Health Law Rev.*, **10**, 9–15.
9. Piehler, H.R. (2000). The future of medicine: biomaterials, *MRS Bulletin*, **25**, 67–70.
10. Weil, V. (2003). Zeroing in on ethical issues in nanotechnology, *Proc. IEEE*, **91**, 1976–1979.
11. Saha, P. and Saha, S. (1991). Ethical issues on the use of animals in the testing of medical implants, *J. Long-Term Eff. Med.*, **1**, 127–134.
12. Budinger, T.F. and Budinger, M.D. (2006). *Ethics in Engineering Technologies: Scientific Facts and Moral Challenges*, John Wiley & Sons, New York, NY.

# The Ending

## CHAPTER 34

Larry L. Hench

## 34.1. Introduction

The first 33 chapters of this book describe the marvelous technological advances of the past 40 or so years that have led to a
greatly enhanced quality of life for millions of people. However, life
always has an end. Our remaining challenge is to discuss this end.
It is a difficult subject. It is a subject we generally choose to avoid
or ignore. As pointed out in Chapter 10 of *Science, Faith and
Ethics,*[1] technology in the form of death intervention has become a
revolution in the last few decades. The issue is the moral and legal
rights of the dying.[2-7]

The "right to die" debate is quiet and submerged in social and
personal taboos. Bringing up the subject of death at a dinner party
is even worse than discussing abortion. A person's desire for death
with dignity most often cannot be expressed. The dying often cannot speak. Their wishes cannot be heard; their pain and suffering
cannot be felt. More often than not friends and family are long gone,
or long uncaring. Or, even in the best circumstances, a loving, caring family may be bound by the present medico-legal system to do
nothing, to just stand by and wait. When a patient is admitted to an
intensive care unit (ICU) and put on a life support system numerous legal and moral restraints prevent the removal of that support.
The moral barrier on the slippery slope of decision of when to
remove life support is typically placed to protect the health-care
system rather than the dying patient.

A graphic, moving and thoughtful article by a surgeon, Dr Atul Gawande, in the 2 August 2010 issue of *The New Yorker*[8] says:

> This is a modern tragedy replayed millions of times over. When there is no way of knowing exactly how long our skeins (lives) will run — and when we imagine ourselves to have much more time than we do — every impulse is to fight, to die with chemo in our veins or a tube in our throats or fresh sutures in our flesh.

## 34.2. The Dilemma

The fundamental dilemma that each of us must ultimately face is the balance between quality of life *versus* length of life. For millennia, there was no dilemma. When you became ill or injured, when your lungs stopped breathing and your heart stopped pumping, you died. No breath, no blood, no life — death. There was no decision involved. There were no options. There was no dilemma. Today, modern medicine, with the aid of modern technology, can prolong breathing and heart function indefinitely. A patient living with the assistance of these devices is obviously not well. The person is no longer self-sufficient, may not even be cognitive, but he or she is also not legally dead.

## 34.3. Alternatives

What is an acceptable moral and legal basis for defining death?

As mentioned above, the absence of either a normally beating heart or breathing lungs is no longer a sufficient criterion for defining death. An alternative is to recognize presence or absence of brain function as the critical test of life or death, as discussed by Veatch in Beauchamp, T.L. and Walters, L.[2] among others.[3–7] A brain-oriented definition of death has been adopted as the legal standard of death by many states in America. But does this definition eliminate the dilemma? Not necessarily. There is little moral or legal debate if the patient loses all brain function. A flat-line EEG is seldom debated as the basis for turning off a life support system, even though the decision is still painful for the family. The difficulty is when the patient has brain function but is no longer aware or able to communicate. In such instances brain function is sufficient to control organs and sustain life but is no longer capable of higher-order cognitive processes, such as speech. Evidence of consciousness is missing. There is no indication that such a patient is any longer capable of thinking, feeling, remembering, reasoning, or loving.

The moral, ethical, and legal dilemma is that a person can be alive without thinking. But, can a person *be* a person without thinking?

## 34.4. The Solution

### 34.4.1. *Education*

The first step towards a solution to the dilemma of length of life *versus* quality of life is to recognize and accept that death is a part of life. Each individual must accept that there will never be sufficient medical technology, prayer, or wishful thinking to avoid dying. Teaching and applying the three general ethical principles of respect for autonomy, principle of beneficence, and principle of justice need to be recognized as central to living a modern life. Each person must accept the responsibility of planning for the ending of life, just as we now accept the responsibility of educating ourselves for a productive life. Modern society dictates that individuals have a minimum or ten or more years of education in order to ensure contributions to the welfare of all. Many people elect to have an additional four to eight years of education in order to pursue a high quality of working life. However, it is rare that any person undertakes even short periods of education to ensure that they will have an acceptable departure from life.

A solution to the dilemma of quality *versus* length of life requires education. Learning how and when to die is not easy, but it can be done. The Gawande article[8] is a starting point. He presents the conflict between accepting death *versus* fighting death clearly and with sympathy both for the patient and the family. His surgeon's perspective gives great weight to his reflections on the options that most of us will face at the end of life, for either ourselves or our family members.

The books listed as references to this chapter are a good second step. *Intensive Care: Facing the Critical Choices*[3] is especially recommended. Few patients come out of an ICU alive, and the decision as whether to be admitted or permit a loved one to be admitted to an ICU is likely to be the most difficult decision you or anyone in your family will ever make.

The debates presented in *Contemporary Issues in Bioethics*[2] are sometimes difficult to confront but are useful preparation for making a personal decision regarding the end of life. Veatch's article is especially insightful to understand the distinctions and definitions of death and dying. The book *Ethics in Medicine*[4] is an easier read and a good complement to the Gawande article[8] to obtain a physician's perspective on the issue. Reading the large volume edited by M.M. Uhlmann, *Last Rights? Assisted Suicide and Euthanasia*[5] can be a chore but presents a comprehensive treatment of the pros and cons of intervening to assist the ending of life.

The first two educational steps recommended above are private. They begin the process of understanding your personal perspective towards facing your end. But, seldom is death private. For most people, family and friends are involved in some capacity. For some that is a comfort; for others it is painful. Often it is hard to predict the consequences of your dying, or the loss of a loved one. Grief is all too often mixed with either greed or guilt. Without proper planning, the fiscal consequences of death can be disastrous for a family. The costs of a few final days in an ICU are typically more than the sum of all prior healthcare costs of keeping a person healthy for years.

Guilt is equally unpredictable and potentially disastrous to a family. Either the dying or any one or more of the family members may feel they have not done enough to prevent the illness, or have not provided sufficient care, or resent having to provide so much care, or resent those that have provided the care. The guilt scenarios that can be played out are almost endless.

Thus, the third educational step is to involve family in the discussion even though the ultimate decision is a personal one. These discussions may not be welcomed. They may not prevent onset of greed or guilt entirely but may help decrease the impact of death. Examples of positive effects of such family discussions are a feature of the Gawande article[8] and one of the reasons it is so strongly recommended as a starting point for understanding this subject.

The fourth step is to consult your physician and/or healthcare professionals for guidance on alternatives of home or hospice care. Again, Gawande provides insight on the many options that are now available. Personal and local situations require exploring these options before healthcare needs become an emergency.

An important fifth step if you or your family are religious is to seek counsel from clergy, rabbi, or other religious leaders with whom you feel comfortable. Their professional experience includes years of training and many examples of situations that can guide you to making a final decision that feels right for you. Often it is difficult to accept that is *your* decision. Loved ones may or may not agree with your choice. They may want to influence you to accept life at all cost, regardless of the likelihood that you will not survive an ICU or other interventions for more than a few days. But it is your life, and it is your right to choose how it ends. Prayer and quiet contemplation are often more helpful than confrontation.

### 34.4.2. *Action*

A solution requires action. The educational steps advocated above are important. However, our present day medico-legal system usually requires specific actions to be completed for your decision to be recognized and accepted. At the minimum it is essential to write down your desires and have them witnessed and notarized. However, the complexity of laws in most US states require more than just your written word to protect you from interventions that you do not want, such as resuscitation or admission to an ICU. In the United States, the action that is most reliable is to seek legal counsel and execute advanced directive documents, such as:

1. "Declaration Naming Pre-Need Guardian" to serve legally in the event of your future incapacity. This is the person that you choose and trust to literally put your life in their hands.
2. "Designation of Healthcare Surrogate" to act on your behalf if physicians have determined that you are incapacitated, and to provide informed consent for medical treatment and surgical and diagnostic procedures.
3. "Durable Power of Attorney" for the person you designate to serve as a true and lawful attorney-in-fact to act for you in your name, and on your behalf, to handle financial and other legal, personal, and medical affairs.
4. "Living Will Declaration" that specifically directs all medical personnel on the medical actions or interventions that you do and do not desire to have performed. The document also should designate who you choose to make final decisions to carry out your directive described in the Living Will in the eventuality that you are not able to do so. The Durable Power of Attorney gives that person the authority under law to make those decisions on your behalf as directed in the Living Will.

Typically the same person will be designated as Pre-Need Guardian, Healthcare Surrogate, be specified in your Living Will Declaration, and have Durable Power of Attorney to act on your behalf. Legal counsel will advise if that is best for you depending upon your circumstances and the laws of your state or country.

Each state in the United States has its own laws, and each country has its own variations of law that govern actions that influence life-or-death decisions. Thus, it is important to seek legal counsel and have the documents professionally executed and notarized in order for them to be enforced. The expense incurred is justified to ensure that your desires are legally able to be followed. It is good to have the above documents with you when traveling. A safeguard is to have the originals filed with your legal counsel to protect them from alteration if you are incapacitated.

## 34.5. Summary

Our concern about death with dignity should not be limited to the time of death. Concerns also need to include the quality of life during our final days that lead to the end of life. An outcome on the debate on the ethical issues of death control should be a shift in our attitude towards death and dying. We need to accept, honestly and openly, that dying is the natural, final step of our existence. Protracting life at all costs should become accepted as morally wrong. A report from the Institute of Medicine (IOM) in the United States, titled "Approaching Death: Improving Care at the End of Life,"[9] takes an important step towards bringing discussion of the medical, legal, emotional, and economic issues about ending of life into public awareness. The report concludes:

> In principle, humane care for those approaching death is a social obligation as well as a personal offering from those directly involved. In reality, both society and individuals often fall short of what is reasonably — if not simply — achievable. As a result, people have come both to fear a technologically over-treated and protracted death and to dread the prospect of abandonment and untreated physical and emotional distress.[9]

The IOM report[9] emphasizes several themes that are essential if we are to improve the quality of life at the end of life. They are:

1. "Too many dying people suffer from pain and other distress that clinicians could prevent or relieve with existing knowledge and therapies."
2. "Significant organizational, economic, legal and educational impediments to good care can be identified and, in varying, degrees, remedied."
3. "Important gaps in scientific knowledge about the end of life need serious attention from biomedical, social science and health service researchers."
4. "Strengthening accountability for the quality of care at the end of life will require better data and tools for evaluating the outcomes important to patients and families."

It is more than a decade since the IOM report. Some changes are taking place, as described by Dr Gawande.[8] However, change in attitudes regarding such a significant aspect of life is slow. Many people have come to rely on technology to provide security in their life and do not want to relinquish that dependence at the very end. Religion, and a faith that goes beyond our personal experience and expectations, have been replaced for many people by belief that technology will provide solutions. The reality is that at the end science is a poor substitute for faith. Acceptance of that reality can come with education and action. It is every person's moral responsibility to prepare for their end. That preparation can lead to comfort and acceptance.

The End.

## References


1.   Hench, L.L. (2001). *Science, Faith and Ethics*, Imperial College Press, London.
2.   Beauchamp, T.L. and Walters, L. (eds) (1989). *Contemporary Issues in Bioethics*, 3rd ed. Wadsworth Publishing Co., Belmont, CA.
3.   Raffin, T.A., Shurkin, J.N. and Sinkler, W. III (1989). *Intensive Care: Facing the Critical Choices*, W.H. Freeman and Co., New York, NY.
4.   Heifetz, M.D. (1996). *Ethics in Medicine*, Prometheus Books, New York, NY.
5.   Uhlmann, M.M. (1998). *Last Rights? Assisted Suicide and Euthanasia Debated*, W.B. Eerdmans Publishing Co., Grand Rapids, MI.
6.   Larue, G.A. and Bayly, R. (1992). *Long Term Care in an Aging Society*, Prometheus Books, Amherst, NY.
7.   Kirkwood, T. (1999). *Time of Our Lives*, Weidenfeld and Nicolson, London.
8.   Gawande, A. (2010). Letting go, *The New Yorker*, 2 August 2010, pp. 36–49.
9.   Anon. (1997). *Approaching Death: Improving Care at the End of Life*, The Institute of Medicine, National Academy of Sciences Press, Washington, DC.

# Index